STUDY GUIDE

EIGHTH EDITION

INTRODUCTION TO PROBABILITY AND STATISTICS

STUDY GUIDE

EIGHTH EDITION

INTRODUCTION TO PROBABILITY AND STATISTICS

WILLIAM MENDENHALL
University of Florida, Emeritus

ROBERT J. BEAVER
University of California, Riverside

Prepared by **BARBARA M. BEAVER**
ROBERT J. BEAVER

PWS-KENT PUBLISHING COMPANY   BOSTON

PWS-KENT
Publishing Company

20 Park Plaza
Boston, Massachusetts 02116

Copyright © 1991 by PWS-KENT Publishing Company.

All rights reserved. No part of this book may be reproduced, stored in a retrieval system, or transcribed, in any form or by any means—electronic, mechanical, photocopying, recording, or otherwise—without the prior written permission of PWS-KENT Publishing Company.

PWS-KENT Publishing Company is a division of Wadsworth, Inc.

Printed in the United States of America

91 92 93 94 95 — 10 9 8 7 6 5 4 3 2 1

CONTENTS

1 What Is Statistics?   1

    1.1   The Objective of Statistics   1
    1.2   The Elements of a Statistical Problem   2
           Exercises   3

2 Describing Sets of Data   4

    2.1   Introduction   4
    2.2   A Graphical Method for Describing a Set of Data: Relative Frequency Distributions   4
           Self-Correcting Exercises 2A   8
    2.3   Stem and Leaf Displays (Optional)   9
           Self-Correcting Exercises 2B (Optional)   12
    2.4   Numerical Methods for Describing a Set of Data   13
    2.5   Measures of Central Tendency   14
    2.6   Measures of Variability   16
           Self-Correcting Exercises 2C   20
    2.7   On the Practical Significance of the Standard Deviation   20
    2.8   A Short Method for Calculating the Variance   23
           Self-Correcting Exercises 2D   25
    2.9   A Check on the Calculation of s 25
    2.10   Measures of Relative Standing   26
    2.11   The Box Plot   30
           Self-Correcting Exercises 2E   33
    2.12   The MINITAB Statistical Package   33
           Exercises   35

3 Probability and Probability Distributions   38

    3.1   Introduction   38
    3.2   Probability and the Sample Space   38
    3.3   The Probability of an Event   41
           Self-Correcting Exercises 3A   45
    3.4   Event Composition and Event Relationships   46
    3.5   Conditional Probability and Independence   50
           Self-Correcting Exercises 3B   55
    3.6   Bayes' Rule (Optional)   57
           Self-Correcting Exercises 3C   60
    3.7   Useful Counting Rules (Optional)   61
           Self-Correcting Exercises 3D   64

3.8	Random Variables	65
3.9	Probability Distributions for Discrete Random Variables	67
	Self-Correcting Exercises 3E	71
3.10	Mathematical Expectation for Discrete Random Variables	72
	Self-Correcting Exercises 3F	75
3.11	Random Sampling	76
	Exercises	78

4 Several Useful Discrete Distributions — 83

4.1	Introduction	83
4.2	The Binomial Probability Distribution	83
	Self-Correcting Exercises 4A	97
4.3	The Poisson Probability Distribution	99
4.4	The Hypergeometric Probability Distribution	103
	Self-Correcting Exercises 4B	105
4.5	Other Discrete Distributions	106
	Self-Correcting Exercises 4C	109
	Exercises	110

5 The Normal and Other Continuous Distributions — 114

5.1	Probability Distributions for Continuous Random Variables	114
	Self-Correcting Exercises 5A	119
5.2	The Normal Probability Distribution	120
	Self-Correcting Exercises 5B	124
5.3	Use of the Table of Normal Curve Areas for the Normal Random Variable x	125
	Self-Correcting Exercises 5C	126
5.4	The Normal Approximation to the Binomial Probability Distribution	127
	Self-Correcting Exercises 5D	131
	Exercises	132

6 Sampling Distributions — 135

6.1	Introduction	135
6.2	Sampling Distributions of Statistics	135
6.3	The Central Limit Theorem and the Sampling Distribution of $\bar{x}$	139
	Self-Correcting Exercises 6A	143
6.4	The Sampling Distribution of the Sample Proportion	143
	Self-Correcting Exercises 6B	145
6.5	The Sampling Distribution of the Difference Between Two Sample Means or Proportions	145
	Self-Correcting Exercises 6C	150
6.6	Summary	150
	Exercises	151

7 Large-Sample Statistical Estimation — 153

- 7.1 Introduction — 153
- 7.2 Estimation — 154
- 7.3 Point Estimation — 155
  - Self-Correcting Exercises 7A — 158
- 7.4 Interval Estimation — 159
  - Self-Correcting Exercises 7B — 162
- 7.5 Choosing the Sample Size — 163
  - Self-Correcting Exercises 7C — 165
- 7.6 Summary — 166
  - Exercises — 167

8 Large-Sample Tests of Hypotheses — 169

- 8.1 Introduction — 169
- 8.2 A Statistical Test of an Hypothesis — 169
- 8.3 A Large-Sample Test of an Hypothesis — 174
  - Self-Correcting Exercises 8A — 182
- 8.4 The Level of Significance of a Statistical Test — 182
  - Self-Correcting Exercises 8B — 184
  - Exercises — 185

9 Inference from Small Samples — 188

- 9.1 Introduction — 188
- 9.2 Student's t Distribution — 188
- 9.3 Small-Sample Inferences Concerning a Population Mean — 191
  - Self-Correcting Exercises 9A — 197
- 9.4 Small-Sample Inferences Concerning the Difference Between Two Means $\mu_1 - \mu_2$ — 197
  - Self-Correcting Exercises 9B — 204
- 9.5 A Paired-Difference Test — 204
  - Self-Correcting Exercises 9C — 208
- 9.6 Inferences Concerning a Population Variance — 209
  - Self-Correcting Exercises 9D — 213
- 9.7 Comparing Two Population Variances — 214
  - Self-Correcting Exercises 9E — 218
- 9.8 Assumptions — 219
  - Exercises — 220

10 Linear Regression and Correlation — 224

- 10.1 Introduction — 224
- 10.2 A Simple Linear Probabilistic Model — 225
- 10.3 The Method of Least Squares — 227
  - Self-Correcting Exercises 10A — 229
- 10.4 Calculating s^2, an Estimator of σ^2 — 230
- 10.5 Inferences Concerning the Slope of the Line, β_1 — 231
  - Self-Correcting Exercises 10B — 235

10.6	Estimating the Expected Value of y for a Given Value of x	234
10.7	Predicting a Particular Value of y for a Given Value of x	237
	Self-Correcting Exercises 10C	239
10.8	A Coefficient of Correlation	239
	Self-Correcting Exercises 10D	244
10.9	Regression Analysis Using Packaged Computer Programs	244
10.10	Assumptions	248
	Exercises	249

11 Multiple Regression Analysis — 252

11.1	Introduction	252
11.2	The Multiple Regression Model and Associated Assumptions	252
	Self-Correcting Exercises 11A	254
11.3	A Multiple Regression Analysis	255
	Self-Correcting Exercises 11B	264
11.4	Comparison of Computer Printouts	266
11.5	Problems in Using Multiple Linear Regression Analysis	267
11.6	Some Comments on Model Formulation (Optional)	268
	Self-Correcting Exercises 11C	271
11.7	Summary	272
	Exercises	272

12 Analysis of Enumerative Data — 278

12.1	The Multinomial Experiment	278
12.2	The Chi-Square Test	279
12.3	A Test of an Hypothesis Concerning Specified Cell Probabilities	280
	Self-Correcting Exercises 12A	283
12.4	Contingency Tables	283
	Self-Correcting Exercises 12B	287
12.5	$r \times c$ Tables with Fixed Row or Column Totals: Tests of Homogeneity	287
12.6	Analysis of an $r \times c$ Contingency Table Using Computer Packages	290
	Self-Correcting Exercises 12C	293
12.7	Other Applications	294
	Self-Correcting Exercises 12D	295
12.8	Assumptions	296
	Exercises	296

13 Experimental Design and the Analysis of Variance — 299

13.1	The Design of an Experiment	299
	Self-Correcting Exercises 13A	303
13.2	The Analysis of Variance	303
13.3	The Completely Randomized Design	309
13.4	The Analysis of Variance Table for a Completely Randomized Design	314

13.5	Estimation for the Completely Randomized Design	317
	Self-Correcting Exercises 13B	318
13.6	The Randomized Block Design	319
13.7	The Analysis of Variance for a Randomized Block Design	320
13.8	Estimation for the Randomized Block Design	326
	Self-Correcting Exercises 13C	327
13.9	Some Comments on Blocking	328
13.10	Selecting the Sample Size	329
13.11	Assumptions for the Analysis of Variance	331
	Exercises	332

14 Nonparametric Statistics 335

14.1	Introduction	335
14.2	The Sign Test for Comparing Two Populations	336
14.3	A Comparison of Statistical Tests	338
14.4	The Mann-Whitney U Test for Comparing Two Population Distributions	338
	Self-Correcting Exercises 14A	344
14.5	The Wilcoxon Signed-Rank Test for a Paired Experiment	345
	Self-Correcting Exercises 14B	349
14.6	The Kruskal-Wallis H Test for Completely Randomized Designs	349
	Self-Correcting Exercises 14C	352
14.7	The Friedman F_r Test for Randomized Block Designs	353
	Self-Correcting Exercises 14D	358
14.8	Rank Correlation Coefficient, r_s	357
	Self-Correcting Exercises 14E	358
	Exercises	359

Appendix   Useful Mathematical Results 363

A.1	Introduction	363
A.2	Functions and Functional Notation	363
	Self-Correcting Exercises A	365
A.3	Numerical Sequences	365
A.4	Summation Notation	367
	Self-Correcting Exercises B	371
A.5	Summation Theorems	371
	Self-Correcting Exercises C	375
A.6	The Binomial Theorem	376
	Exercises	377

Solutions to Self-Correcting Exercises   379

Answers to Exercises   406

Tables   411

PREFACE

The study of statistics differs from the study of many other college subjects. One must not only absorb a set of basic concepts and applications but must precede this with the acquisition of a new language.

We think with words. Hence, understanding the meanings of words employed in the study of a subject is an essential prerequisite to the mastery of concepts. In many fields, this poses no difficulty. Often, terms encountered in the physical, social, and biological sciences have been met in the curricula of the public schools, in the news media, in periodicals, and in everyday conversation. In contrast, few students encounter the language of probability and statistical inference before embarking on an introductory college-level study of the subject. Many consider the memorization of definitions, theorems, and the systematic sequence of steps necessary for the solution of problems to be unnecessary. Others are oblivious to the need. The consequences for both types of students are disorganization and disappointing achievement in the course.

This study guide attempts to lead you through the language and concepts necessary for a mastery of the material in *Introduction to Probability and Statistics,* 8th edition, by William Mendenhall and Robert J. Beaver (PWS–KENT, 1991).

A study guide with answers is intended to be an individual student study aid. The subject matter is presented in an organized manner that incorporates continuity with repetition. Most chapters bear the same titles and order as the textbook chapters. Within each chapter, the material both summarizes and reexplains the essential material from the corresponding textbook chapter. This allows you to gain more than one perspective on each topic and, we hope, enhances your understanding of the material.

At appropriate points in each chapter, you will encounter a set of Self-Correcting Exercises in which problems relating to new material are presented. Terse, stepwise solutions to these problems are found at the back of the study guide. You can refer to these at any intermediate point in the solution of each problem or use them as a stepwise check on any final answer. The Self-Correcting Exercises not only provide the answers to specific problems but also reinforce the stepwise logic required to arrive at a correct solution to each problem.

At the end of each chapter, additional sets of exercises can be found. These exercises are provided for the students who feel that further individual practice is needed in solving the kinds of problems found within each chapter. At this point, having been given stepwise solutions to the Self-Correcting Exercises, you are now presented with only final answers to problems. When your answer disagrees with that given in the study guide, you should be able to find your error by recalculating and comparing your solution with the solutions to similar Self-Correcting Exercises. If the answer given disagrees with yours only in decimal accuracy, it can be assumed that this difference is due only to rounding error at various stages in the calculations.

When the study guide is used as a supplement, the textbook chapter should be read first. Then you should study the corresponding chapter within the study guide. Key words, phrases, and numerical computations have been left blank so that you can insert a response. The answers are presented in the page margins. These should be covered until you have supplied a response for each blank. Bear in mind, though, that in some instances more than one answer is appropriate for a given blank. It is left to you to determine whether your answer is synonymous with the answer given within the margin.

Since perfection is something to be desired, we ask that the person who has located an error kindly bring it to our attention.

Barbara M. Beaver
Robert J. Beaver

CHAPTER 1
WHAT IS STATISTICS?

1.1 The Objective of Statistics

Statistics involves sampling from a larger body of data called a
_____ . Consider the following examples of statistical problems: population
1. The gubernatorial preferences of eligible voters in an election are of interest to the politicians of a state. Rather than poll the entire set of voters, questionnaires are sent only to a selected group.
2. A medical experiment was conducted to determine the absorption of a drug in the heart of a rat. A fixed amount of the drug was injected into each of ten rats, the rats were sacrificed, and the amount of drug absorbed by each heart measured.

The following characteristics are common to both of these statistical problems.
    a. The measurements or observations obtained cannot be predicted in advance.
    b. A sample is taken from a larger body of data.
    c. From each element in the sample, one or more measurements or pieces of data are collected.
    d. It is assumed that the conclusions drawn from the study apply to more than those elements within the sample. For example, opinions of all eligible voters would comprise the larger set of measurements of interest to the experimenter. This larger set of measurements is called a population.

A population can exist conceptually or it can exist in fact. In example 2, given above, the measurements constitute a _____ population of   conceptual
measurements made on rats placed in the same experimental situation. On the other hand, consider the problem of estimating the proportion of voters who are in favor of a certain candidate. Here, each voter in favor of the candidate could be counted as a "1," and those opposed or having no opinion as a "0." For this problem, the _____ consists of a set of ones and zeros   population
associated with the eligible voters in the election.

    A _____ is a subset of measurements selected from the   sample
population.
1. One hundred voting residents, chosen as a cross-section of a given city, were polled regarding their opinion about the new city bond issue. These 100 people represent a sample from the (actual, conceptual) population   actual
of voting residents of that city.
2. Examination scores are recorded for 50 students who have been taught

using a certain experimental method. These scores represent a _____ from a(n) (actual, conceptual) population consisting of the large number of measurements that might have been obtained from other students placed under similar conditions.

sample; conceptual

Statistics is concerned with a theory of information and its application in making inferences based on sample information about populations in the sciences and industry.

The *objective of statistics* is to make inferences about a population from information contained in a sample.

1.2 The Elements of a Statistical Problem

We have noted that the objective of statistics is to make inferences about a _____ based on information contained in a _____.

population; sample

Sampling implies the acquisition of data, so statistics is concerned with a theory of information. The attainment of the objective of statistics—inference making—is dependent upon five steps, which we will call the elements of a statistical problem.

The first important task facing the experimenter is one which is many times ignored or performed only cursorily. In order to make inferences about a population, the experimenter must be able to clearly specify the _____ of this population. Further, the experimenter must specify the _____ to be answered about the population using statistical inference. If the population is improperly defined, it may be that the sample will be drawn from the wrong population. Moreover, if the questions to be answered are not accurately stated, inappropriate methods of inference might be used. Before any sampling is done, the problem must be clear in the mind of the experimenter.

nature
questions

The sample contains a quantity of information on which the inference about the population will be based. In fact, information can be quantified as easily as weight, heat, profit, or other quantities of interest. Consequently, the second step in a statistical problem is deciding upon the most economical procedure for buying a specified quantity of information. This is called the _____ procedure or the _____ of the experiment. The cost of the specified amount of information will vary greatly depending upon the method used for collecting the data in a sample.

sampling
design

The third step in a statistical problem involves the extraction of information contained in the _____. By analogy, suppose that information was measured in units of pounds (which it is not). It is not unusual for an experimenter to extract only three pounds from a sample that contains ten pounds of information. Thus extracting information from a sample is equivalent to the problem of extracting juice from an orange. We wish to obtain the _____ amount of information from a given set of data.

sample

maximum

The fourth step in a statistical problem involves the use of the information in a sample to make an _____ about the population from which the sample was drawn. Some inferences, say, estimates of the characteristics of

inference

the population, are very accurate and consequently are good. Others are far from reality and bad. It is therefore necessary to clearly define a measure of goodness for an inference maker. Most people observe the world about them and make inferences daily. Some of these subjective inference makers are very good and accurate; others are very poor. Statistical inference makers are objective rather than subjective, but they vary in their goodness. The statistician wishes to obtain the best inference maker for a given situation.

A measure of the goodness or reliability of an inference is the fifth step in a statistical problem and is always necessary in order to assess its practical value. Thus, inference making is regarded as a two-step procedure. First, we select the best method and use it to make an _____. Second, we always give a measure of the goodness or _____ of the inference.

inference
reliability

The five elements of a statistical problem are (1) the definition of the specific problem and the _____ of interest, (2) the _____ of the experiment, (3) the _____ of the data, (4) the procedure for making inferences about the _____ from information contained in the _____, and (5) the provision of a measure of the _____ of the inference.

population; design
analysis
population
sample
goodness

EXERCISES

1. A market analyst wishes to determine which factors exert the most influence on a purchaser's decision process in selecting his or her new car. Describe the population of interest to the analyst, and indicate what kind of information the analyst might consider collecting.
2. A medical researcher is interested in determining whether a new drug is more effective than those currently available for treating degenerative arthritis. Identify the population or populations of interest to the researcher. Should severity and time since onset of this disease enter into any proposed sampling plan? Why or why not?
3. An agricultural economist is interested in determining the revenue loss to cotton growers in the Imperial Valley of California due to crop infestation by the pink boll worm. Identify the population of interest to the economist. Would his findings apply equally well to cotton growers in Texas?
4. A market research analyst would like to demonstrate that when brand labels on containers of cola softdrinks are removed or hidden, the cola softdrink produced by his company will be preferred over its strongest competitor by people who drink cola softdrinks. Describe the population of interest to this analyst. In taste-testing experiments, the order in which a person samples the items under test may influence his/her ultimate choice. Could this be a potential problem in this research?
5. There is increased emphasis on mainstreaming educable mentally retarded (EMR) students in our schools. In mainstreaming, EMR students are integrated into regular classes when possible, and meet in special classes when necessary. A researcher is interested in assessing the social acceptance of mainstreamed EMR students by their nonretarded peers. Identify the population of interest to this researcher, and indicate any potential problems that you can foresee in sampling this population.

CHAPTER 2
DESCRIBING SETS OF DATA

2.1 Introduction

inferences; population
sample

In Chapter 1, we considered the steps involved in achieving the objective of statistics, which is making _____ about a _____ from information contained in a _____. An obvious but often ignored requirement is that the sample be drawn from the population of interest to the experimenter.

How are inferences made? First, we must be able to describe data in a straightforward pictorial or graphical form. Second, we must be able to reconstruct this visual representation using numerical descriptive measures that describe the salient characteristics of the visual representation. For example, where is the middle of the distribution? Are the measurements tightly grouped or widely scattered? Whether the set of data under consideration comprises an entire population or is merely a sample from a population, we must be able to agree upon numerical measures that describe the data. Inferences about a population can then be made in terms of the population by using the relevant information contained in the sample.

This chapter will deal with two very general methods of describing data:
1. graphical methods
2. numerical methods

2.2 A Graphical Method for Describing a Set of Data: Relative Frequency Distributions

Graphical methods attempt to present the set of measurements in pictorial form so as to give the reader an adequate visual description of the measurements. Let us discuss a graphical method by examining some data.

Example 2.1

The following data are the numbers of correct responses on a recognition test consisting of 30 items, recorded for 25 students:

25	29	23	27	25
23	22	25	22	28
28	24	17	24	30
19	17	23	21	24
15	20	26	19	23

1. First find the highest score, which is _____, and the lowest score, which is _____. These two scores indicate that the measurements have a range of 15.

   30
   15

2. To determine how the scores are distributed between 15 and 30, we divide this interval into subintervals of equal length. The interval from 15 to 30 could be divided into from 5 to 20 subintervals, depending upon the number of measurements available. Wishing to obtain about 7 subintervals, a suitable width is determined by dividing 30 - 15 = 15 by 7. The integer _____ would seem to provide a satisfactory subinterval width for these data.

   2

3. Utilizing the subinterval boundary points 14.5, 16.5, 18.5, 20.5, 22.5, 24.5, 26.5, _____, and _____, we guarantee that none of the given measurements will fall on a boundary point. Thus, each measurement falls into only one of the subintervals or classes.

   28.5; 30.5

4. We now proceed to tally the given measurements and record the class frequencies in a table. Fill in the missing information.

Tabulation of Data for Histogram

Class i	Class Boundaries	Tally	Frequency, f_i	Relative Frequency, f_i/n
1	14.5-16.5	I	1	1/25
2	16.5-18.5	II	2	2/25
3	18.5-20.5	_____	3	3/25
4	20.5-22.5	III	3	_____
5	22.5-24.5	ℍ II	_____	7/25
6	_____	IIII	4	4/25
7	26.5-28.5	_____	3	3/25
8	28.5-30.5	II	2	2/25

III
3/25
7
24.5-26.5
III

5. The number of measurements falling in the ith class is called the ith class frequency and is designated by the symbol f_i. Of the total number of measurements, the fraction falling in the ith class is called the _____ frequency in the ith class. Given n measurements, the relative frequency in the ith class is given as f_i/n. As a check on your tabulation, remember that for k classes,
   a. the sum of the frequencies, f_i, equals _____,
   b. the sum of the relative frequencies, f_i/n, equals _____.

   relative

   n
   1

6. With the data so tabulated, we can now use a frequency histogram (plotting frequency against classes) or a relative frequency histogram (plotting relative frequency against classes) to describe the data. The two histograms are identical except for scale.
   a. Study the following histogram based on our data:

6 / Chap. 2: Describing Sets of Data

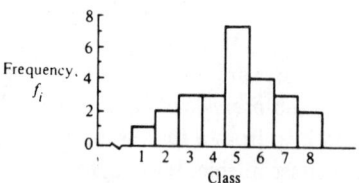

b. Complete the following relative frequency histogram for the same data:

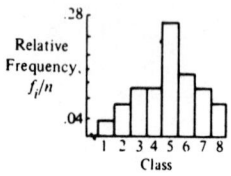

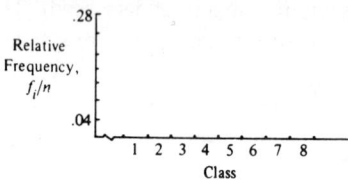

c. When completed, the histograms in parts a and b should appear identical except for scale.
7. By examining the tabulation found in step 4, answer the following questions:

6/25 or 24%

a. What fraction of the students had scores less than 20.5? _____.
b. What fraction of the students had scores greater than 26.5?

5/25 or 20%

_____.
c. What fraction of the students had scores between 20.5 and 26.5?

14/25 or 56%

_____.
8. As the number of measurements in the sample increases, the sample histogram should resemble the population histogram more and more. Thus, to estimate the fraction of students in the entire population that would have scores greater than 26.5, we could use our sample histogram,

5/25; 20%

estimating this fraction to be _____ or _____.
9. A relative frequency histogram is often called a *relative frequency distribution* because it displays the manner in which the data are distributed along the horizontal axis of the graph. The rectangular bars above the class intervals in the relative frequency histogram can be given two interpretations:
a. The height of the bar above the ith class represents the fraction of observations falling in the ith class.
b. The height of the bar above the ith class also represents the probability that a measurement drawn at random from this sample will belong to the ith class.
10. Complete the following statements based on the data tabulation in step 4.
a. The probability that a measurement drawn at random from this data

7/25

will fall in the interval 22.5 to 24.5 is _____.

b. The probability that a measurement drawn at random from this data will be greater than 18.5 is _____ .   22/25
c. The probability that a measurement drawn at random from this data will be less than 24.5 is _____ .   16/25

Example 2.2
The following data represent the burning times for an experimental lot of fuses, measured to the nearest tenth of a second:

5.2	3.8	5.7	3.9	3.7
4.2	4.1	4.3	4.7	4.3
3.1	2.5	3.0	4.4	4.8
3.6	3.9	4.8	5.3	4.2
4.7	3.3	4.2	3.8	5.4

Construct a relative frequency histogram for these data.

Solution
Fill in the missing entries in the table.

Tabulation of Data

Class	Class Boundaries	Tally	Frequency	Relative Frequency	
1	2.45–2.95	I	1	.04	
2	2.95–3.45	III	3	_____	.12
3	3.45–3.95	ℼℼ I	_____	.24	6
4	3.95–4.45	_____	7	.28	ℼℼ II
5	4.45–4.95	IIII	4	_____	.16
6	_____	III	3	.12	4.95–5.45
7	5.45–5.95	I	1	.04	

Relative Frequency Histogram

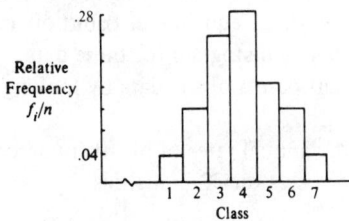

Complete the following statements based on the preceding tabulation.
a. The probability that a measurement drawn at random from this sample is greater than 4.45 is _____ .   8/25
b. The probability that a measurement drawn at random from this sample is less than 3.45 is _____ .   4/25

13/25

5; 20

classes

equal
boundaries
impossible

c. An estimate of the probability that a measurement drawn at random from the sampled population would be in the interval 3.45 to 4.45 is _____ .

In conclusion, the steps necessary to construct a frequency distribution are as follows:
1. Determine the number of classes depending on the amount and uniformity of the data. It is usually best to have from _____ to _____ (give numbers) classes.
2. Determine the class width, by dividing the range by the number of _____ , and adjusting the resulting quotient to obtain a convenient figure. With the possible exception of the lowest and highest classes, all classes should be of _____ width.
3. Locate the class _____ . Class boundaries should be chosen so that it is (impossible, likely) for a measurement to fall on a boundary.

Self-Correcting Exercises 2A

1. The following data are the ages (in months) at which 50 children were first enrolled in a preschool:

38	40	30	35	39
47	35	34	43	41
32	34	41	30	46
55	39	33	32	32
42	50	37	39	33
40	48	36	31	36
36	41	43	48	40
35	40	30	46	37
45	42	41	36	50
45	38	46	36	31

a. Find the range of these data.
b. Using about five intervals of equal width, set up class boundaries used in the construction of a frequency distribution and tabulate the data.
c. Construct a frequency histogram for these data.
d. Describe the salient points of the data by looking at the histogram given in part c.

2. The following are the annual rates of profit (in percentages) on stockholders' equity after taxes for 32 industries:

10.6	10.8	14.8	10.8
12.5	6.0	10.7	11.0
14.6	6.0	12.8	10.1
7.9	5.9	10.0	10.6
10.8	16.2	18.4	10.7
10.6	13.3	8.7	15.4
6.5	10.1	8.7	7.5
11.9	9.0	12.0	9.1

a. Present a relative frequency histogram for these data, utilizing seven intervals of length 2, beginning at 5.55.

Using your histogram (or tabulation), answer the following questions:

b. What is the probability that an industry drawn at random from this distribution has a rate of profit greater than 15.55%?
c. What is the probability that an industry drawn at random has a rate of profit less than 9.55?
d. What is the probability that an industry drawn at random has a rate of profit greater than 9.55 but less than 15.55?

2.3 Stem-and-Leaf Displays

A simple and informative method for displaying a set of data graphically is called the *stem-and-leaf display.* It emerges from a new and quickly developing area of statistics called exploratory data analysis (EDA), whose chief proponent is John Tukey. The objective of EDA is to provide the experimenter with simple techniques which allow him to look more effectively at his data. In particular, the stem-and-leaf display presents a histogram-like picture of the data, while allowing the experimenter to retain the actual observed values of each data point. Since in the process of tabulating the data, we also create a histogram-like picture, the stem-and-leaf display is partly graphical and partly tabular in nature.

In creating a stem-and-leaf display, we must choose part of the original measurement as the *stem* and the remaining part as the *leaf*. Consider for example a set containing four measurements:

$$624, 538, 465, 552$$

We could use the digit in the "hundreds" place as the stem, and the remainder (the digits at or to the right of the "tens" place) as the leaf. In this case, for the observation 624, the digit 6 would be the (stem, leaf) and the digits 24 would be the (stem, leaf). We could also choose to use the digits at or to the left of the "tens" place as the stem and the remainder as the leaf. Then for the observation 465, the digit(s) _____ would be the stem and the digit(s) _____ would be the leaf. The choice of the stem-and-leaf coding depends upon the nature of the observations at hand.

stem
leaf

46
5

The stems and leaves are now used as follows:
1. List all the stem digits vertically, from lowest to highest.
2. Draw a vertical line to the right of the stem digits.
3. For each data point, place the leaf digit of that point in the row corresponding to the correct stem.
4. The stem-and-leaf display may be made more visually appealing by reordering the leaf digits from lowest to highest within each stem row.

Example 2.3
The following data represent the 1989 state gasoline tax in cents per gallon for the 50 United States and the District of Columbia.

10 / Chap. 2: Describing Sets of Data

AL	13.0	IL	13.0	MT	20.0	RI	14.0
AK	8.0	IN	15.0	NE	18.3	SC	15.0
AZ	17.0	IA	18.0	NV	18.0	SD	18.0
AR	13.5	KS	11.0	NH	14.0	TN	17.0
CA	9.0	KY	15.0	NJ	10.5	TX	15.0
CO	18.0	LA	16.0	NM	14.0	UT	19.0
CT	20.0	ME	16.0	NY	8.0	VT	13.0
DE	16.0	MD	18.5	NC	15.7	VA	17.5
DC	15.5	MA	11.0	ND	17.0	WA	18.0
FL	9.7	MI	15.0	OH	14.8	WV	10.5
GA	7.5	MN	20.0	OK	16.0	WI	20.9
HI	11.0	MS	17.0	OR	14.0	WY	8.0
ID	18.0	MO	11.0	PA	12.0		

Source: Federal Highway Administration; National Safety Board; Insurance Institute for Highway Safety

Construct a stem-and-leaf display for the data.

Solution
From an initial survey of the data, the largest observation is 20.9, while the smallest observation is 7.5. The data in this example are recorded in tenths of a cent. Suppose that we were to choose the leading digit, that is, the digit in the "tens" place as the stem. There would be only ____three____ stems for the data, namely "0," "1," and "2". This would not provide a very good visual description of the data. Therefore, we choose to use the digits at or to the left of the "ones" place as the stem. This is also called the "integer part" of the observation. The leaf will be the remaining portion of the observation, that is, those digits falling to the (left, ____right____) of the decimal point.

The stems, from 7 to 20 are listed vertically below, a vertical line is drawn, and the leaves are entered in the correct row. Fill in the missing entries in the stem-and-leaf plot.

LEAF UNIT = 0.1
7 5 REPRESENTS 7.5

Stem	Leaf
7	5
8	0 0 0
9	0 7
10	5 5
11	0 0 0 0
12	0
13	0 5 0 0
14	0 0 8 0 0
15	5 0 0 0 7 0 0
16	0 0 0 0
17	0 0 0 0 5
18	0 0 0 5 3 0 0 0
19	0
20	0 0 0 9

The display can now be redone if necessary, with the leaves reordered from lowest to highest. Fill in the missing entries.

Stem	Leaf
7	_____
8	0 0 0
9	0 7
10	5 5
11	0 0 0 0
12	0
13	0 0 0 5
14	0 0 0 0 8
15	0 0 0 0 0 5 7
16	0 0 0 0
17	0 0 0 0 5
18	0 0 0 0 0 3 5
19	0
20	_____

5

0 0 0 9

From the stem-and-leaf display, we may make the following observations:
1. The lowest state gasoline tax is _____ cents per gallon. The highest gasoline tax is _____ cents per gallon.   7.5   20.9
2. There are (no, some) extreme values (values which are much higher or much lower than all the others).   no
3. The data (are, are not) approximately mound-shaped.   are not
4. Most of the gasoline taxes fall between _____ cents per gallon and _____ cents per gallon.   8   19

The stem-and-leaf procedure can be extended to accommodate data which does not fit easily into the structure described above. If the data consists of a large number of digits, as many as six or seven, the resulting stem-and-leaf display will have either an extremely large number of stems, or will have an adequate number of stems with leaves consisting of four or five digit numbers. This will tend to destroy the clarity and hence the informativeness of the display. In this case, it may be more convenient to use the first one or two digits as the stem, the second one or two as the leaf, and to drop the remaining digits. Hence, we are sacrificing the ability to *exactly* recreate the data from the stem-and-leaf display in order to clarify the presentation.

Example 2.4
The following data represent the planned rated capacity in megawatts (millions of watts) for the world's 27 largest hydroelectric plants.

20,000	5,225	3,409
13,320	5,020	3,300
10,830	4,678	3,200
10,300	4,500	3,200
7,260	4,500	3,000
6,400	4,150	2,715
6,000	4,000	2,700
6,000	3,600	2,700
5,328	3,600	2,700

Source: United States Commission on Large Dams, of the International Commission on Large Dams, Sept., 1989.

12 / Chap. 2: Describing Sets of Data

Construct a stem-and-leaf display for the data.

Solution
We will take the leading digit, (the digits at or to the left of the "thousands" place) to be the stem. Then to simplify the presentation, take the two digits in the "hundreds" and "tens" places as the leaf. For example, the observation 13,320 will have stem 13 and leaf 32. The observation 5225 will have stem _____ and leaf _____ . The stem-and-leaf display is constructed below. As an aid to decoding the stem-and-leaf display, we write

5; 22

LEAF UNIT = 10.0
2 70 REPRESENTS 2700

Stem	Leaf
2	70 70 70 71
3	00 20 20 30 40 60 60
4	00 15 50 50 67
5	02 22 32
6	00 00 40
7	26
8	
9	
10	30 83
11	
12	
13	32
.	.
.	.
.	.
20	00

The stem-and-leaf display indicates a high concentration of capacities between 2000 and 5000 Mwatts, with almost all of the data falling between _____ and _____ Mwatts. There are four unusually large values.

2000; 8000

Sometimes the available stem choices result in a display containing too many stems (and very few leaves within a stem) or in too *few* stems (and many leaves within a stem). In this situation, we may divide the too few stems by stretching them into two or more lines depending on the leaf values with which they will be associated. However, this option tends to complicate the stem-and-leaf procedure and hence minimizes the major advantage of simplicity in the stem-and-leaf display. The interested student should consult the references at the end of Chapter 2 in the text.

Self-Correcting Exercises 2B

1. The following data are the ages in years of 42 students enrolled in an adult education class.

51	32	31	33	23	52
23	21	55	34	38	32
49	35	26	29	50	34
30	19	41	39	41	27
25	21	18	36	35	28
44	44	59	28	23	46
27	37	42	32	43	30

Construct a stem-and-leaf display of the data using the digit in the "tens" place as the stem and the digit in the "ones" place as the leaf.

2. The following data represent the number of United States farms in June, 1989, per state, measured in thousands of farms.

AL	47.0	LA	35.0	OH	87.0
AK	.6	ME	7.3	OK	69.0
AZ	8.1	MD	15.6	OR	37.0
AR	49.0	MA	6.9	PA	54.0
CA	84.0	MI	55.0	RI	.8
CO	27.3	MN	90.0	SC	25.5
CT	4.0	MS	41.0	SD	35.0
DE	3.0	MO	108.0	TN	91.0
FL	41.0	MT	24.7	TX	186.0
GA	48.0	NE	57.0	UT	13.0
HI	4.7	NV	2.5	VT	7.1
ID	22.3	NH	3.2	VA	47.0
IL	86.0	NJ	8.3	WA	38.0
IN	71.0	NM	14.0	WV	21.0
IA	105.0	NY	39.0	WI	81.0
KS	69.0	NC	65.0	WY	8.9
KY	96.0	ND	33.5		

a. Construct a stem-and-leaf display of these data in which the digits in the "tens" place are the stems and the digits in the "ones" place are leaves.
b. Are any outliers apparent? What can be said about the limits within which most of the number of farms per state lie?
c. Would you suggest using other configurations for the stems and leaves for this data set?

2.4 Numerical Methods for Describing a Set of Data

The chief advantage to using a graphical method is its visual representation of the data. Many times, however, we are restricted to reporting our data verbally. In this case a graphical method of description cannot be used. The greatest disadvantage to a graphical method of describing data is its unsuitability for making inferences, since it is difficult to give a measure of goodness for a graphical inference. Therefore, we turn to *numerical descriptive measures*. We seek a set of numbers that characterizes the frequency distribution of the measurements and at the same time will be useful in making inferences.

We will distinguish between numerical descriptive measures for a population and those associated with a set of sample measurements. A numerical descriptive measure calculated from all the measurements in a population is called a

14 / Chap. 2: Describing Sets of Data

parameter
statistics

(statistic, parameter). Those numerical descriptive measures calculated from sample measurements are called _____ .

Numerical descriptive measures are classified into two important types.
1. Measures of *central tendency* locate in some way the "center" of the data or frequency distribution.
2. Measures of *variability* measure the "spread" or dispersion of the data or frequency distribution.

Using measures of both types, the experimenter is able to create a concise numerical summary of the data.

2.5 Measures of Central Tendency

We first consider two of the more important measures of central tendency that attempt to locate the center of the frequency distribution.

The *mean* of a set of n measurements $x_1, x_2, \ldots, x_n$ is defined to be the sum of the measurements divided by n. The symbol $\bar{x}$ is used to designate the sample mean while the Greek letter μ is used to designate the population mean.

The sample mean can be shown to have very desirable properties as an inference maker. In fact, we will use $\bar{x}$ to estimate the population mean, μ. To indicate the sum of the measurements, we will use the Greek letter Σ (sigma). Then Σx_i will indicate the sum of all the measurements that have been denoted by the symbol x. Using this summation notation, we can define the sample mean by formula as

$$\bar{x} = \frac{\sum_{i=1}^{n} x_i}{n}$$

The numbers above and below the Greek letter Σ denote the values of i for which the summation is performed. That is,

$$\sum_{i=1}^{n} x_i = x_1 + x_2 + \ldots + x_n$$

A more detailed discussion of summation notation can be found in the Appendix.

Example 2.5
Find the mean of the following measurements:

2, 5, 7, 10, 11, 13.

Solution

$$\sum_{i=1}^{6} x_i = \underline{\qquad},\qquad\qquad 48$$

$$\bar{x} = \Sigma x_i/n = (\underline{\qquad})/6,\qquad\qquad 48$$

$$\bar{x} = \underline{\qquad}.\qquad\qquad 8$$

In addition to being an easily calculated measure of central tendency, the mean is also easily understood by all users. The calculation of the mean utilizes all the measurements and can always be found exactly.

One disadvantage of using the mean to measure central tendency is well known to any student who has had to pull up one low test score: the mean (is, is not) greatly affected by extreme values. For example, you might be unwilling to accept, say, an average property value of $95,000 for a given area as an acceptable measure of the middle property value if you knew that (a) the property value of a residence owned by a millionaire was included in the calculation and (b) excluding this residence, the property values ranged from $45,000 to $60,000. A more realistic measure of central tendency in this situation might be the property value such that 50% of the property values are less than this value and 50% are greater.

is

The *median* of a set of n measurements $x_1, x_2, \ldots, x_n$ is the value of x that falls in the middle when the measurements are arranged in order of magnitude. When n is *odd*, the median is the measurement with rank $(n + 1)/2$. When n is *even*, the median is the simple average of the measurements with ranks $n/2$ and $(n/2) + 1$; that is, the average of the two middle measurements.

Example 2.6
Find the median of the following set of measurements:

5, 3, 2, 7, 4.

Solution
1. Arranging the measurements in order of magnitude, we have

    2, 3, 4, 5, 7.

2. The median will be the _____ ordered value, since $(n + 1)/2 =$ 6/2 = 3. Hence the median is _____ .

third
4

Example 2.7
Find the median of the following set of measurements:

10, 8, 13, 14, 9, 8.

16 / Chap. 2: Describing Sets of Data

Solution
1. Arranging the measurements in order of magnitude, we have

$$8, \ 8, \ 9, \ 10, \ 13, \ 14.$$

2. Since $n = 6$ is even, the median will be the average of the _____ (third) and _____ (fourth) ordered values. Hence

$$\text{median} = \frac{\underline{\quad(9)\quad} + \underline{\quad(10)\quad}}{2}$$

$$= \underline{\quad(9.5)\quad}.$$

Example 2.8
Find the mean and median of the following data:

$$5, \ 7, \ 8, \ 10, \ 10, \ 11, \ 13, \ 14.$$

1. $\displaystyle\sum_{i=1}^{8} x_i = \underline{\quad(78)\quad}$; $\bar{x} = \Sigma x_i/n = (\underline{\quad 78\quad})/8 = \underline{\quad 9.75\quad}$.

2. To find the median, we note that the measurements are already arranged in order of magnitude and that $n = 8$ is even. Therefore, the median will be the average of the fourth and fifth ordered values.

$$\text{median} = (10 + 10)/2 = \underline{\quad(10)\quad}.$$

In the last example, the mean and median gave reasonably close numerical values as measures of central tendency. However, if the measurement $x_9 = 30$ were added to the eight measurements given, the recalculated mean would be $\bar{x} = \underline{\quad(12)\quad}$, but the median would remain at 10, reflecting the fact that the median is a positional average unaffected by extreme values.

2.6 Measures of Variability

Having found measures of central tendency, we next consider measures of the variability or dispersion of the data. A measure of variability is necessary since a measure of central tendency alone does not adequately describe the data. Consider these two sets of data:

Set I. $x_1 = 9$ Set II. $y_1 = 1$
$x_2 = 10$ $\bar{x} = \underline{\quad(10)\quad}$ $y_2 = 10$ $\bar{y} = \underline{\quad(10)\quad}$
$x_3 = 11$ $y_3 = 19$

Both sets of data have a mean equal to _____. However, the second set of measurements displays much more variability about the mean than does the first set.

In addition to a measure of central tendency, a measure of variability is indispensable as a descriptive measure for a set of data. A manufacturer of machine parts would want very (little, much) variability in her product in order to control oversized or undersized parts, while an educational testing service would be satisfied only if the test scores showed a (large, small) amount of variability in order to discriminate among people taking the examination.

We have already used the simplest measure of variability, the range.

10

little

large

The *range* of a set of measurements is the difference between the largest and smallest measurements.

Example 2.9
Find the range for each of the following sets of data:

Set I.	23	73	34	74
	28	29	26	17
	88	8	52	49
	37	96	32	45
	81	62	23	62

Range = 96 − _____ = _____ .

8; 88

Set II.	8.8	6.7	7.1	2.9
	9.0	0.2	1.2	8.6
	6.3	6.4	2.1	8.8

Range = 9.0 − _____ = _____ .

0.2; 8.8

By examining the following distributions, it is apparent that although the range is a simply calculated measure of variation, it alone is not adequate. Both distributions have the same range, but display different variability.

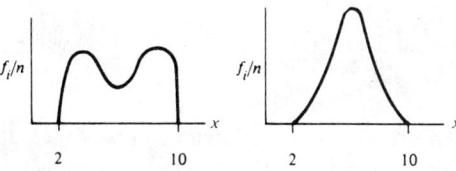

We base the next important measure of variability on the dispersion of the data about their mean. Consider two sets of measurements. The first set consists of N measurements and represents the entire population of interest to the experimenter. The second set is a sample of n measurements taken from this population. Define μ to be the population mean. For a finite population consisting of N measurements,

$$\mu = \sum_{i=1}^{N} x_i/N$$

Recall that the indices above and below the summation sign indicate the values that the subscript i may take on as the summation is performed. That is,

$$\sum_{i=1}^{N} x_i = x_1 + x_2 + \ldots + x_N$$

Similarly, define

$$\bar{x} = \sum_{i=1}^{n} x_i/n = \frac{x_1 + x_2 + \ldots + x_n}{n}$$

to be the sample mean. Finally, we define the quantity $(x_i - \mu)$, or $(x_i - \bar{x})$ (depending upon whether the set of measurements is a population or a sample), as the ith deviation from the mean. Large deviations indicate (more, less) variability of the data than do small deviations. We could utilize these deviations in different ways.

1. If we attempt to use the average of the N (or n) deviations, we find that the sum of the deviations is _____ . To avoid a zero sum, we could use the average of the absolute values of the deviations. This measure, called the *mean deviation*, is difficult to calculate, and we cannot easily give a measure of its goodness as an inference maker.
2. A more efficient use of data is achieved by averaging the sum of squares of the deviations. For a finite population of N measurements, this measure, called the *population variance*, is given by

$$\sigma^2 = \sum_{i=1}^{N} (x_i - \mu)^2/N.$$

Large values of σ^2 indicate (large, small) variability, while (large, small) values indicate small variability.

Since the units of σ^2 are not in the original units of measurement, we can return to these units by defining the standard deviation.

The *standard deviation* σ is the positive square root of the variance. That is,

$$\sigma = \sqrt{\sigma^2} = \sqrt{\Sigma(x_i - \mu)^2/N}.$$

Margin answers: more; zero; large, small

Sect. 2.6: Measures of Variability / 19

For a sample of size n, it would seem reasonable to define the sample variance in a similar way, using $\bar{x}$ in place of μ and n in place of N. However, we choose to modify this definition slightly.

Since our objective is to make inferences about the population based on sample data, it is appropriate to ask if the sample mean and variance are good estimators of their population counterparts, μ and σ^2. The fact is that $\bar{x}$ is a good estimator of μ, but the quantity

$$\Sigma(x_i - \bar{x})^2/n$$

appears to (underestimate, overestimate) the population variance σ^2 when the sample size is small.

| | underestimate |

The problem of underestimating σ^2 can be solved by dividing the sum of squares of deviations by $n - 1$ rather than n. We then define

$$s^2 = \Sigma(x_i - \bar{x})^2/(n - 1)$$

as the sample variance. The sample standard deviation is then

$$s = \sqrt{s^2} = \sqrt{\Sigma(x_i - \bar{x})^2/(n - 1)}.$$

Example 2.10
Calculate the sample mean, variance, and standard deviation for the following data:

4, 2, 3, 5, 6.

Solution
Arrange the measurements in the following way, first finding the mean, $\bar{x} = $ _____ .

| | 4 |

x_i	$x_i - \bar{x}$	$(x_i - \bar{x})^2$
4	0	0
2	-2	4
3	-1	1
5	1	1
6	2	4
$\Sigma x_i = 20$	$\Sigma(x_i - \bar{x}) =$ _____	$\Sigma(x_i - \bar{x})^2 =$ _____

| | 0; 10 |

After finding the mean, complete the second column and note that its sum is zero. The variance is

$$s^2 = \Sigma(x_i - \bar{x})^2/(n - 1) = \underline{\qquad}/4 = \underline{\qquad},$$

| | 10; 2.5 |

while the standard deviation is

1.581

$$s = \sqrt{2.5} = \underline{\hspace{2cm}}.$$

Note: We will introduce a shortcut formula for calculating

$$\sum_{i=1}^{n} (x_i - \bar{x})^2;$$

more examples will be given then.

Self-Correcting Exercises 2C

1. Fifteen brands of breakfast cereal were judged by nutritionists according to four criteria: taste, texture, nutritional value, and popularity with the buying public. Each brand was rated on a 0-5 scale for each criterion and the sum of the four ratings reported. (A high score with respect to the maximum of 20 points indicates a good evaluation of the brand.)

9	8	16	17	10
15	12	6	12	13
10	13	19	11	9

Find the mean and the median scores for these data. Compare their values.

2. The number of daily arrivals of cargo vessels at a West Coast port during an 11-day period are given below:

3	2	0
5	4	4
2	3	2
7	1	

 a. Compare the mean and median for these data.
 b. Calculate the range of the data.
 c. Calculate the standard deviation of the number of arrivals per day during this 11-day period. (As an intermediate check on calculations, remember that the sum of deviations must be zero.)

2.7 On the Practical Significance of the Standard Deviation

Having defined the mean and standard deviation, we now introduce two theorems that will use both these quantities in more fully describing a set of data.

> *Tchebysheff's Theorem:* Given a number k greater than or equal to one and a set of n measurements $x_1, x_2, \ldots, x_n$, at least $(1 - 1/k^2)$ of the measurements will lie within k standard deviations of their mean.

The importance of this theorem is due to the fact that it applies to any set of measurements. It applies to a population using the population mean μ and the population standard deviation σ, and it applies to a sample from a given population using $\bar{x}$ and s, the sample mean and sample standard deviation. Since this theorem applies to _____ set of measurements, it is of necessity a conservative theorem. It is therefore very important to stipulate that _____ _____ $(1 - 1/k^2)$ of the measurements will lie within k standard deviations of their mean.

any

at least

Complete the following chart for the values of k given.

k	Interval $\bar{x} \pm ks$	Interval Contains at Least the Fraction $(1 - 1/k^2)$
1	$\bar{x} \pm s$	_____ 0
2	$\bar{x} \pm 2s$	_____ 3/4
3	$\bar{x} \pm 3s$	_____ 8/9
10	$\bar{x} \pm 10s$	_____ 99/100

Example 2.11
The mean and variance of a set of $n = 20$ measurements are 35 and 25, respectively. Use Tchebysheff's Theorem to describe the distribution of these measurements.

Solution
Collecting pertinent information we have

$$\bar{x} = 35, \quad s^2 = 25, \quad s = \sqrt{25} = 5.$$

1. At least 3/4 of the measurements lie in the interval $35 \pm 2(5)$ or from _____ to _____.

   25; 45

2. At least 8/9 of the measurements lie in the interval $35 \pm 3(5)$ or from _____ to _____.

   20; 50

3. At least 15/16 of the measurements lie in the interval $35 \pm 4(5)$ or from _____ to _____.

   15; 55

Example 2.12
If the mean and variance of a set of $n = 50$ measurements are 42 and 36, respectively, describe these measurements using Tchebysheff's Theorem.

Solution
Pertinent information: $\bar{x} = 42$, $s^2 = 36$, and $s = 6$.

1. At least 3/4 of the measurements lie in the interval $42 \pm 2(6)$ or from _____ to _____.

   30; 54

22 / Chap. 2: Describing Sets of Data

2. At least 8/9 of the measurements lie in the interval 42 ± 3(6) or from

24; 60   _____ to _____ .

3. At least 15/16 of the measurements lie in the interval 42 ± 4(6) or from

18; 66   _____ to _____ .

A second theorem can be used if the distribution of measurements is known to be of a particular form.

Empirical Rule: Given a distribution of measurements that is approximately bell-shaped, the interval

a. $\mu \pm \sigma$ contains approximately 68% of the measurements.
b. $\mu \pm 2\sigma$ contains approximately 95% of the measurements.
c. $\mu \pm 3\sigma$ contains almost all (approximately 99.7%) of the measurements.

This rule holds reasonably well for any set of measurements that possesses a distribution that is mound-shaped. Bell-shaped or mound-shaped is taken to mean that the distribution has the properties associated with the normal distribution, whose graph is given in your text and elsewhere in this study guide.

Example 2.13
A random sample of 100 oranges was taken from a grove and individual weights measured. The mean and variance of these measurements were 7.8 ounces and 0.36 (ounces)^2, respectively. Assuming the measurements produced a mound-shaped distribution, describe these measurements using the Empirical Rule.

Solution
First find the intervals needed.

k	$\bar{x} \pm ks$	$\bar{x} - ks$	to	$\bar{x} + ks$
1	$\bar{x} \pm s$	_____	to	_____
2	$\bar{x} \pm 2s$	_____	to	_____
3	$\bar{x} \pm 3s$	_____	to	_____

7.2; 8.4
6.6; 9.0
6.0; 9.6

Then approximately

a. _____ % of the measurements lie in the interval from _____ to _____ ,

68; 7.2
8.4

b. _____ % of the measurements lie in the interval from _____ to _____ ,

95; 6.6
9.0

c. _____ % of the measurements lie in the interval from _____ to _____ .

100(or 99.7); 6.0
9.6

would not
would not

When n is small, the distribution of measurements (would, would not) be mound-shaped and as such the Empirical Rule (would, would not) be appropriate in describing these data. Since Tchebysheff's Theorem applies to any set of measurements, it can be used regardless of the size of n.

2.8 A Short Method for Calculating the Variance

The calculation of $s^2 = \Sigma(x_i - \bar{x})^2/(n-1)$ requires the calculation of the quantity $\sum_{i=1}^{n}(x_i - \bar{x})^2$. To facilitate this calculation, we introduce the identity

$$\sum_{i=1}^{n}(x_i - \bar{x})^2 = \sum_{i=1}^{n} x_i^2 - \frac{\left(\sum_{i=1}^{n} x_i\right)^2}{n}$$

the proof of which is omitted. This computation requires the following:

1. The ordinary arithmetic sum of the measurements, $\sum_{i=1}^{n} x_i$.

2. The sum of the squares of the measurements, $\sum_{i=1}^{n} x_i^2$.

Note the distinction between $\sum_{i=1}^{n} x_i^2$ and $\left(\sum_{i=1}^{n} x_i\right)^2$ used in the identity given above.

1. To calculate $\sum_{i=1}^{n} x_i^2$, we *first square* each measurement and *then sum* these squares.

2. To calculate $\left(\sum_{i=1}^{n} x_i\right)^2$, we *first sum* the measurements and *then square* this sum.

Example 2.14
Calculate s^2 for Example 2.10.

24 / Chap. 2: Describing Sets of Data

Solution
Display the data in the following way, finding Σx_i and Σx_i^2:

x_i	x_i^2
4	16
2	4
3	9
5	25
6	36
$\Sigma x_i =$ _____	$\Sigma x_i^2 =$ _____

20; 90

1. We first calculate

$$\Sigma(x_i - \bar{x})^2 = \Sigma x_i^2 - (\Sigma x_i)^2/n$$

$$= 90 - (20)^2/5$$

80

$$= 90 - \underline{\qquad}$$

10

$$= \underline{\qquad}.$$

2. Then

10; 2.5

$$s^2 = \Sigma(x_i - \bar{x})^2/(n-1) = \underline{\qquad}/(5-1) = \underline{\qquad}.$$

Example 2.15
Calculate the mean and variance of the following data: 5, 6, 7, 5, 2, 3.

Solution
Display the data in a table.

x_i	x_i^2
5	25
6	36
7	49
5	25
2	4
3	9

28; 148

$$\Sigma x_i = \underline{\qquad} \qquad \Sigma x_i^2 = \underline{\qquad}$$

4.67

$$\bar{x} = \Sigma x_i/n = 28/6 = \underline{\qquad},$$

$$\Sigma(x_i - \bar{x})^2 = \Sigma x_i^2 - (\Sigma x_i)^2/n$$

$$= 148 - (28)^2/6$$

$$= 148 - 130.67$$

17.33

$$= \underline{\qquad},$$

17.33; 3.467

$$s^2 = \Sigma(x_i - \bar{x})^2/(n-1) = \underline{\qquad}/(6-1) = \underline{\qquad}.$$

Self-Correcting Exercises 2D

1. The number of accidental tanker oil spills for the years 1984–1990 are given below:

    36, 48, 45, 29, 49, 35, 65.

    Calculate the variance and the standard deviation of these data.
2. Using the data given in Exercise 2, Self-Correcting Exercises 2C, calculate the variance utilizing the shortcut formula to calculate the required sum of the squared deviations. Verify that the values of the variance (and hence the standard deviation) found using both calculational forms are identical.
3. If a person were concerned about accuracy due to rounding of numbers at various stages in computation, which formula for calculating $\Sigma(x_i - \bar{x})^2$ would be preferred:

    a. $\Sigma(x_i - \bar{x})^2$ or b. $\Sigma x_i^2 - (\Sigma x_i)^2/n$?

    Defend your choice of either a or b.
4. The following measurements represent the times required for rats to run a maze correctly: 5.2, 4.2, 3.1, 3.6, 4.7, 4.8, 4.1. Calculate the sample variance and standard deviation.
5. The heights in inches of five men consecutively entering a doctor's office were 70, 74, 69, 71, 72. Calculate the mean and variance of these heights.

2.9 A Check on the Calculation of s

For mound-shaped or approximately normal data, we can use the range to check the calculation of s, the standard deviation. According to Tchebysheff's Theorem and the Empirical Rule, at least 3/4 and more likely 95% of a set of measurements will be in the interval $\bar{x} \pm 2s$. Hence, the sample range R should approximately equal $4s$, so that

$$s \approx R/4.$$

This approximation requires only that the computed value be of the same order as the approximation.

Example 2.16
Check the calculated value of s for the first set of data given in Example 2.9.

Solution
For these data, the range is 96 - 8 = _____, and   88

$$s \approx 88/4 = \underline{\qquad}.$$ 22

Comparing 22 with the calculated value, 25.46, we (would, would not) have   would not
reason to doubt the accuracy of the calculated value.

26 / Chap. 2: Describing Sets of Data

In referring to the second set of data in Example 2.9, which consists of 12 measurements, we find that the range is 9.0 - 0.2 = 8.8. Hence an approximation to s using $R \approx 4s$ yields

2.2
$$s \approx (8.8)/4 = \underline{\hspace{1cm}}.$$

When compared with the calculated value 3.21, this approximation is not as close as the approximation for the first set of data.

large
Since extreme measurements are more likely to be observed in (large, small) samples, we can adjust the approximation to s by dividing the range by a divisor that depends upon the sample size n. A rule of thumb to use in approximating s by using the range is given in the following table.

n	Divide Range by
5	2.5
10	3
25	4
100	5

Example 2.17
Use the range approximation to check the calculated value of s, 3.21, for the second set of data in Example 2.9.

Solution
We know that $R = 8.8$; hence for $n = 12$ measurements, we use the approximation

2.93
$$s \approx R/3 = (8.8)/3 = \underline{\hspace{1cm}},$$

which more closely approximates the calculated value of s, 3.21, than did the earlier approximation, 2.2.

Example 2.18
Use the range approximation to check the calculation of s for the data given in Example 2.10.

Solution
4
For the five measurements, 4, 2, 3, 5, 6, the range is 6 - 2 = \underline{\hspace{1cm}}. Therefore, an approximation to s would be

1.6
$$s \approx R/(2.5) = 4/(2.5) = \underline{\hspace{1cm}},$$

which closely agrees with the calculated value of 1.581.

2.10 Measures of Relative Standing

Occasionally, we wish to know the relative position of a measurement relative to others in the set. The mean and standard deviation of the set of measurements can be used to calculate a sample z-score used to measure the relative standing of a measurement in a data set. The z-score is defined as follows:

The sample z-score corresponding to a measurement x is

$$z\text{-score} = \frac{x - \bar{x}}{s}$$

A z-score measures the distance between a measurement and the sample _____ measured in units of standard deviation.
The magnitude of a z-score takes on meaning when used in conjunction with Tchebysheff's Theorem and the Empirical Rule. At least 3/4 and more likely 95% of the measurements in a set lie within _____ standard deviations of the mean; therefore, z-scores between -2 and $+2$ are highly likely. At least 8/9, and more likely all of the measurements, lie within _____ standard deviations of the mean; therefore, z-scores larger than 2 but less than 3 in absolute value are not very likely, and z-scores larger than 3 in absolute value are very unlikely.

A set of measurements may contain values that for some reason lie far from the middle of the distribution in either direction. These values lie in the tails of the distribution and are called *outliers*. Sometimes outliers are generated when a mistake is made in recording the data; e.g., a number may be misread or mistyped. Perhaps the environmental conditions under which an experiment is performed changed drastically for a short period of time, and gave rise to one or more _____. *Hence, if a z-score exceeds 2 in absolute value, we would suspect that the measurement is an outlier.* In this case, the experimenter should check to see whether a faulty measurement has been obtained.

The median is defined as the measurement that falls in the middle when all measurements are arranged in increasing magnitude. The median is also defined to be the value of x, such that at most 50% of the measurements are _____ than x and _____ are greater. Therefore, in addition to providing a measure of the center of a set of measurements, the median also allows the user to assess the position of one particular measurement in relation to the others in the set. For example, if the median family income in a particular state was $26,500, then half the families in the state had incomes less than $26,500, and half had incomes _____ than $26,500.

A *percentile* is another measure of relative standing that is most often used for large data sets.

mean

two

three

outliers

less; half

greater

Let $x_1, x_2, \ldots, x_n$ be a set of n measurements arranged in order of increasing magnitude. The *p*th percentile is that value of x such that at most $p\%$ of the measurements are less than x and at most $(100 - p)\%$ are greater than x.

Example 2.19
An elementary school child has scored in the 91st percentile on the SAT tests. What does this mean?

28 / Chap. 2: Describing Sets of Data

Solution
Scoring in the 91st percentile implies that at most 91% of the children taking the examination scored lower than this child, and at most 9% scored higher.

Example 2.20
A set of test scores are approximately mound-shaped with mean 73 and standard deviation 10. If a student scored 93 on the test, into what percentile would his score fall?

Solution
The value $x = 93$ lies two standard deviations above the mean. Hence, using the Empirical Rule and the fact that a mound-shaped distribution is symmetric about its mean we can draw the following conclusions:

1. The percentage of scores falling below 73 is _____ %. [50]

2. The percentage of scores falling between 53 and 93 is approximately 95%. Hence, the percentage falling between 73 and 93 is _____ /2 = _____ % and the percentage of scores below 93 is then [95; 47.5]

$$50 + \underline{} = \underline{} \%.$$ [47.5; 97.5]

3. Then, 97.5% of the scores lie below $x = 93$ and _____ % lie above $x = 93$. [2.5]

Rounding to the nearest percent, the score $x = 93$ lies in the _____ th percentile. [98]

Percentiles provide an excellent way to measure relative standing and are used frequently in the presentation of test scores, sociological data, and medical data. However, the user must be careful when using percentiles with small sets of data. In this situation, the exact percentile may fall between two data points and any value falling between these two points can be taken as the percentile value. Most often, however, percentiles are encountered in the description of large sets of data.

By definition of a percentile, the 50th percentile is also called the _____. The 25th percentile is called the (lower, upper) quartile, while the 75th percentile is called the (lower, upper) quartile. [median; lower; upper]

The upper and lower quartiles, when located along with the median on the horizontal axis of a frequency distribution representing the measurements of interest, divide that frequency distribution into four parts, each containing an equal number of measurements. The upper and lower quartiles are formally defined and calculated as follows:

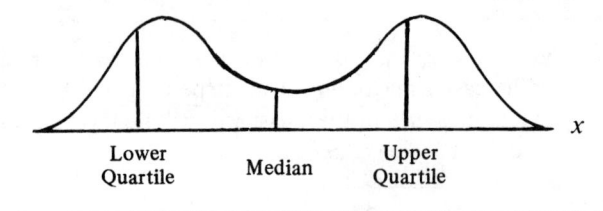

Lower Quartile   Median   Upper Quartile x

Let $x_1, x_2, \ldots, x_n$ be a set of n measurements, arranged in order of increasing magnitude. The lower quartile, Q_L, is a value of x such that at most 1/4 of the measurements are less than x and at most 3/4 are greater than x. The upper quartile, Q_U, is a value of x such that at most 3/4 of the measurements are less than x and at most 1/4 are greater.

When working with a small set of data, the above definition sometimes admits many numbers which would satisfy the criteria necessary for Q_L and Q_U. For this reason, we will avoid this inconsistency by calculating sample quartiles in the following way.

Let $x_1, x_2, \ldots, x_n$ be a set of n measurements arranged in order of increasing magnitude. The lower quartile, Q_L, is the value of x in the $.25(n + 1)$ position and the upper quartile, Q_U, is the value of x in the $.75(n + 1)$ position.

When $.25(n + 1)$ and $.75(n + 1)$ are not integers, the quartiles are found by interpolation, using the values in the two adjacent positions. (This definition of sample quartiles is consistent with that used in the MINITAB package.)

Example 2.21
Find the upper and lower quartiles for the following set of measurements:

$$3, 8, 7, 1, 1, 12, 13, 9, 3, 2, 10$$

Solution
The measurements are first arranged in order of increasing magnitude.

$$1, 1, 2, 3, 3, 7, 8, 9, 10, 12, 13$$

Since $n = 11$, the lower quartile, Q_L, will be in position

$$.25(n + 1) = .25(11 + 1) = 3$$

and the upper quartile, Q_U, will be in position

$$.75(n + 1) = .75(12) = 9$$

Since the quartile positions (are, are not) both integer valued, the quartiles can be read directly as   are

$$Q_L = 2 \text{ and } Q_U = 10$$

Example 2.22
Find the upper and lower quartiles for the following set of prices:

$$\$2.15, 3.50, 6.80, 4.29, 1.67, 2.20, 1.59, 2.98$$

Solution
The measurements are first arranged in order of increasing magnitude.

$$\$1.59, 1.67, 2.15, 2.20, 2.98, 3.50, 4.29, 6.80$$

Since $n = 8$, the lower quartile, Q_L, is found in position

$$.25(n + 1) = .25(9) = 2.25$$

and the upper quartile, Q_U, is found in position

$$.75(n + 1) = .75(9) = 6.75$$

Therefore,

$$Q_L = 1.67 + .25(2.15 - 1.67) = 1.67 + .12 = 1.79,$$

and

$$Q_U = 3.50 + .75(4.29 - 3.50) = 3.50 + .59 = 4.09.$$

The student should note that there are many possible values which will satisfy the definition of a quartile. The author chooses to use the above convention in order to avoid complicated arithmetic calculation. The student should not be surprised to find that other references may choose to calculate quartiles in a different way.

2.11 The Box Plot

The box plot is a technique introduced in exploratory data analysis (EDA) to describe the behavior of the measurements both in the middle and in the _____ of the distribution. The box plot is constructed using the median and two other measures known as *hinges*. The _____ divides the ordered data set into halves; the hinges are the values in the middle of each half of the data. The hinges are similar to the _____, and serve essentially the same purpose. The procedure for calculating the hinges follows.

tails
median

quartiles

---

1. Calculate the position of the median, $(n + 1)/2$, and drop the remainder fraction $1/2$ if there is one. This quantity is called $d(M)$, and measures the depth of the median as measured from each end of the ordered measurements.
2. Calculate the position of the hinges using

$$\frac{d(M) + 1}{2}.$$

The hinges are the values in position $[d(M) + 1]/2$, as measured from each end of the ordered data set.

For a data set with $n = 20$ measurements, the position of the median is $(20 + 1)/2 = 10.5$, and the depth of the median is $d(M) =$ _____. The position of the hinges is found to be

$$\frac{d(M) + 1}{2} = \frac{10 + 1}{2} = \underline{\qquad}.$$

10

5.5

Since the position of the hinges as measured from each end of the ordered data set ends in the fraction 1/2, the lower hinge will be the average of the 5th and _____ measurement and the upper hinge will be the average of the _____ and 16th measurement in the ordered data set.

6th
15th

The dispersion of the measurements is measured in terms of the difference between the hinges, called the *H-spread*, which is approximately equal to $Q_U - Q_L$, the *interquartile range*. The H-spread can now be used as a yardstick to detect outliers.

The *inner fences* are defined as:

> lower inner fence = lower hinge − 1.5(H-spread)
> upper inner fence = upper hinge + 1.5(H-spread).

The *outer fences* are defined as:

> lower outer fence = lower hinge − 3(H-spread)
> upper outer fence = upper hinge + 3(H-spread).

If a measurement lies between the inner and outer fences, it is called a suspected _____. If a measurement lies outside the outer fences, it is (a suspected, _____ _____) outlier.

outlier
an unquestionable

Constructing a Box Plot

1. Calculate the median, the upper and lower hinges, and locate them on a horizontal line representing the scale of measurement.
2. Draw a box whose ends are the upper and lower hinges. Draw a vertical line through the box at the position of the median.
3. Locate the inner and outer fences on the horizontal line. Determine any suspect and extreme outliers and locate them on the horizontal line.
4. The data values closest to, but still inside the inner fences are called *adjacent values*. Locate the adjacent values on the box plot, and draw a dashed line from the box to the corresponding adjacent values.

Example 2.23
Construct a box plot for the following $n = 27$ measurements, the planned rated capacity in megawatts (millions of watts) for the world's 27 largest hydroelectric plants.

32 / Chap. 2: Describing Sets of Data

20,000	5,225	3,409
13,320	5,020	3,300
10,830	4,678	3,200
10,300	4,500	3,200
7,260	4,500	3,000
6,400	4,150	2,715
6,000	4,000	2,700
6,000	3,600	2,700
5,328	3,600	2,700

Source: United States Commission on Large Dams, of the International Commission on Large Dams, Sept., 1989.

Solution

1. The median will be the observation in position $(25 + 1)/2 = 13$. Since the data are ordered from large to small, we see that the 13th measurement is _____.   *4150*

2. The lower hinge will be in position $(13 + 1)/2 = 7$, and the upper hinge will be in position 21. The lower hinge is 3200, and the upper hinge is _____.   *6000*

3. The H-spread is $6000 - 3200 = $ _____. The inner fences are   *2800*

$$3200 - 1.5(2800) = 3200 - 4200 = \underline{}$$ *−1000*

and

$$6000 + 1.5(2800) = 6000 + 4200 = \underline{}.$$ *10,200*

The outer fences are

$$3200 - 3(2800) = 3200 - 8400 = -5200$$

and

$$6000 + 3(2800) = 6000 + 8400 = \underline{}.$$ *14,400*

4. Suspected outliers are those observations between -5200 and _____, or between 10,200 and _____. Unquestionable outliers are those less than -5200 or those greater than _____. By examining the data, we see that 10,300, 10,830, and 13,320 are suspected outliers, while _____ is an unquestionable outlier.   *−1000; 14,400*   *14,400*   *20,000*

5. The box plot is shown below with the ends of the box located at 3200 and 6000. The asterisk (*) designates a suspected outlier and the letter o designates an unquestionable outlier.

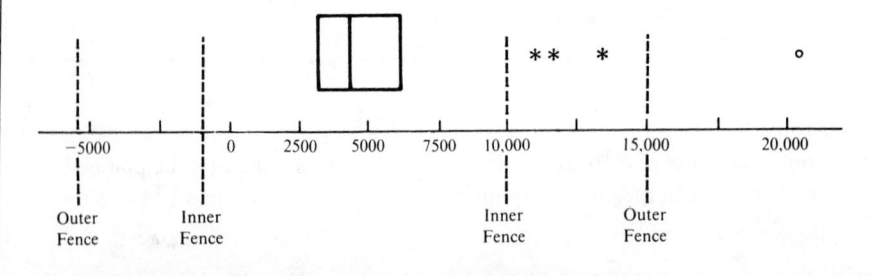

Notice that, in addition to identifying several outliers, the box plot indicates that the set of measurements are skewed to the (right, left); the median is not in the middle of the box, but rather to the left of the middle. Examination of the stem-and-leaf display for these data will confirm this conclusion.

right

Self-Correcting Exercises 2E

1. Consider the following set of data:

   3, 4.5, 7.5, 12, 30, 8.5, 4, 5

   a. Find the median, the lower quartile, and the upper quartile.
   b. Find the z-score for each observation.
   c. Do the z-scores calculated in part (b) indicate that there are any outliers?

2. Use the data given in Exercise 1, Self-Correcting Exercises 2B, to construct a box plot. Are there any outliers? The data are reproduced below in order of ascending magnitude.

18	25	30	34	39	46
19	26	30	34	41	49
21	27	31	35	41	50
21	27	32	35	42	51
23	28	32	36	43	52
23	28	32	37	44	55
23	29	33	38	44	59

3. Construct a box plot for the data given in Exercise 2, Self-Correcting Exercises 2B. The data are reproduced below. Are there any outliers? (*Hint:* Use the stem-and-leaf display constructed in Exercise 2, Self-Correcting Exercises 2B.)

AL	47.0	IN	71.0	NE	57.0	SC	25.5
AK	.6	IA	105.0	NV	2.5	SD	35.0
AZ	8.1	KS	69.0	NH	3.2	TN	91.0
AR	49.0	KY	96.0	NJ	8.3	TX	186.0
CA	84.0	LA	35.0	NM	14.0	UT	13.0
CO	27.3	ME	7.3	NY	39.0	VT	7.1
CT	4.0	MD	15.6	NC	65.0	VA	47.0
DE	3.0	MA	6.9	ND	33.5	WA	38.0
FL	41.0	MI	55.0	OH	87.0	WV	21.0
GA	48.0	MN	90.0	OK	69.0	WI	81.0
HI	4.7	MS	41.0	OR	37.0	WY	8.9
ID	22.3	MO	108.0	PA	54.0		
IL	86.0	MT	24.7	RI	.8		

2.12 The MINITAB Statistical Package

The graphical and numerical descriptive measures for a set of measurements presented in this chapter can be implemented using commands found in the

34 / Chap. 2: Describing Sets of Data

MINITAB statistical package. MINITAB uses a column format for data entry and storage, and a set of commands and subcommands for implementing statistical procedures. The commands to produce a histogram, a stem-and-leaf plot, a set of descriptive measures, and a box plot are illustrated using the data given as the ultimate megawatt capacity of the world's 27 largest hydroelectric plants; this data was stored in column one (C1) of the MINITAB worksheet. MINITAB is but one of a growing list of statistical packages that allows a user to implement many of the commonly used statistical procedures.

MTB > HISTOGRAM C1

Histogram of   C1   N = 27

```
Midpoint   Count
  2000       4    ****
  4000      12    ************
  6000       6    ******
  8000       1    *
 10000       2    **
 12000       0
 14000       1    *
 16000       0
 18000       0
 20000       1    *
```

MTB > STEM AND LEAF C1

Stem-and-leaf of C1   N = 27
Leaf Unit = 1000

```
 11    0  22223333333
 (8)   0  44444555
  8    0  6667
  4    0
  4    1  00
  2    1  3
  1    1
  1    1
  1    1
  1    2  0
```

MTB > DESCRIBE C1

	N	MEAN	MEDIAN	TRMEAN	STDEV	SEMEAN
C1	27	5616	4500	5157	3915	753

	MIN	MAX	Q1	Q3
C1	2700	20,000	3200	6000

MTB > BOXPLOT C1

```
                -------
              -I    +   I- - - -  **      *                        0
                -------
- - - - -+- - - - - -+- - - - - -+- - - - - -+- - - - - -+- - - - -+C1
       3500        7000       10500      14000      17500      21000
```

EXERCISES

1. The following set of data represents the gas mileage for each of 20 cars selected randomly from a production line during the first week in March:

18.1	16.3	18.6	18.7
15.2	19.9	20.3	22.0
19.7	17.7	21.2	18.2
20.9	19.7	19.4	20.2
19.8	17.2	17.9	19.6

   a. What is the range of these data?
   b. Construct a relative frequency histogram for these data by using sub-intervals of width 1.0. (You might begin with 15.15.)
   c. Based on the histogram in part b:
      i. What is the probability that a measurement selected at random from these data will fall in the interval 17.15 to 21.15?
      ii. What is the estimated probability that a measurement taken from the population would be greater than 19.15?
   d. Arrange the measurements in order of magnitude, beginning with 15.2.
   e. What is the median of these data?
   f. The _____ th percentile would be any number lying between 16.3 and 17.2.
   g. The _____ th percentile would be any number lying between 19.9 and 20.2.
   h. Calculate $\bar{x}$, s^2, and s for these data. (Remember to use the shortcut method.)
   i. Do these data conform to Tchebysheff's Theorem? Support your answer by calculating the fractions of the measurements lying in the intervals $\bar{x} \pm ks$ for $k = 1, 2, 3$.
   j. Does the Empirical Rule adequately describe these data?

2. Refer to Exercise 1. Construct a box plot for the data. Are there any outliers?

3. An IRS employee randomly sampled 25 tax returns and for each recorded the number of exemptions claimed by the taxpayer. The data are shown below.

2	0	2	1	3
3	1	1	3	2
1	0	7	0	2
0	0	0	4	1
4	3	2	2	5

   a. What is the range of these data?
   b. Calculate the median, the lower and upper quartiles.
   c. Would a stem-and-leaf display be appropriate for these data? Explain.
   d. Calculate $\bar{x}$, s^2, and s for these data.
   e. Construct a relative frequency histogram for these data using classes of width 1.0. (You might begin with −0.5.)

f. Assuming the above sample is representative of all the taxpayers in the United States,
   i. what is the probability a taxpayer will claim no more than 1 exemption?
   ii. what is the probability a taxpayer will claim 3 or more exemptions?
g. Can the Empirical Rule be applied to these data? Explain.

4. A strain of "long-stemmed roses" was developed with a mean stem length of 15 inches and standard deviation 2.5 inches.
   a. If one accepts as "long-stemmed roses" only those roses with a stem length greater than 12.5 inches, what percentage of such roses would be unacceptable?
   b. What percentage of these roses would have a stem length between 12.5 and 20 inches?
   Hint: Using the symmetry of the normal distribution, 1/2 of 68% of the measurements lie one standard deviation to the left or to the right of the mean, and 1/2 of 95% of the measurements lie two standard deviations to the left or to the right of the mean.

5. The heights of 40 cornstalks ranged from 2.5 feet to 6.3 feet. In presenting these data in the form of a histogram, suppose you had decided to use .5 foot as the width of your class interval.
   a. How many intervals would you use?
   b. Give the class boundaries for the first and the last classes.

6. A machine designed to dispense cups of instant coffee will dispense on the average μ ounces, with standard deviation $\sigma = .7$ ounce. Assume that the amount of coffee dispensed per cup is approximately mound-shaped. If 8-ounce cups are to be used, at what value should μ be set so that approximately 97.5% of the cups filled will not overflow?

7. A pharmaceutical company wishes to know whether an experimental drug being tested in its laboratories has any effect on systolic blood pressure. Fifteen subjects, randomly selected, were given the drug, and the systolic blood pressures recorded (in millimeters) were

172	148	123
140	108	152
123	129	133
130	137	128
115	161	142

   a. Approximate s using the method described in Section 2.9.
   b. Calculate $\bar{x}$ and s for the data.
   c. Find values for the points a and b such that at least 75% of the measurements fall between a and b.
   d. Would Tchebysheff's Theorem be valid if the approximated s were used in place of the calculated s?
   e. Would the Empirical Rule apply to these data?

8. Refer to Exercise 7. Construct a stem-and-leaf display for the data.

9. Refer to Exercise 7. Are there any outliers?
   a. Use the z-score technique.
   b. Use the box plot technique.

10. Toss two coins 30 times, recording for each toss the number of heads observed.
    a. Construct a histogram to display the data generated by the experiment.
    b. Find $\bar{x}$ and s for your data.
    c. Do the data conform to Tchebysheff's Theorem? Empirical Rule?
11. The following data represent the social ambivalence scores for 15 people as measured by a psychological test. (The higher the score, the stronger the ambivalence.)

9	8	15	17	10
14	11	4	12	13
10	13	19	11	9

   a. Using the range, approximate the standard deviation s.
   b. Calculate $\bar{x}$, s^2, and s for these data.
   c. What fraction of the data actually lies in the interval $\bar{x} \pm 2s$?
12. A lumbering company interested in the lumbering rights for a certain tract of slash pine trees is told that the mean diameter of these trees is 14 inches with a standard deviation of 2.8 inches. Assume the distribution of diameters is approximately normal.
    a. What fraction of the trees will have diameters between 8.4 inches and 22.4 inches?
    b. What fraction of the trees will have diameters greater than 16.8 inches?
13. If the mean duration of television commercials on a given network is 1 minute and 15 seconds with a standard deviation of 25 seconds, what fraction of these commercials would run longer than 2 minutes and 5 seconds? Assume that duration times are approximately normally distributed.
14. The following are the prices charged (in cents) for a dozen eggs at 25 supermarkets in a large metropolitan area:

80	81	89	84	78
92	79	82	89	85
78	91	80	76	80
79	90	82	85	79
81	82	79	85	84

   a. Find the range of these data.
   b. Construct a relative frequency histogram. Are the data approximately mound-shaped?
   c. Calculate the mean and standard deviation.
   d. What fraction of the data actually lies in the interval $\bar{x} \pm s$? In the interval $\bar{x} \pm 2s$? Do these results agree with the Empirical Rule?

CHAPTER 3
PROBABILITY AND PROBABILITY DISTRIBUTIONS

3.1 Introduction

We have already stated that our aim is to make inferences about a population based upon sample information. However, in addition to making the inference, we also need to assess how good the inference will be.

Suppose that an experimenter is interested in estimating the unknown mean of a population of observations to within two units of its actual value. If an estimate is produced based upon the sample observations, what is the chance that the estimate is no further than two units away from the true but unknown value of the mean?

If an investigator has formulated two possible hypotheses about a population and only one of these hypotheses can be true, when the sample data are collected she must decide which hypothesis to accept and which to reject. What is the chance that she will make the correct decision?

In both these situations, we have used the term "chance" in assessing the goodness of an inference. But chance is just the everyday term for the concept statisticians refer to as _____. Therefore, some elementary results from the theory of probability are necessary in order to understand how the accuracy of an inference can be assessed.

probability

In the broadest sense, the probability of the occurrence of an event A is a measure of one's belief that the event A will occur in a single repetition of an experiment. One interpretation of this definition that finds widespread acceptance is based upon empirically assessing the probability of the event A by repeating an experiment N times and observing n_A/N, the relative frequency of the occurrence of event A. When N, the number of repetitions, becomes very large, the fraction n_A/N will approach a number we will call $P(A)$, the probability of the occurrence of the event A.

3.2 Probability and the Sample Space

When the probability of an event must be assessed, it is important to be able to visualize under what conditions that event will be realized. An *experiment* is the process by which an observation or measurement is obtained.

An *event* is an outcome of an experiment.

When an experiment is run repeatedly, a population of observations results. A _____ is any set of observations taken from this population. In this context, an observation may be a measurement, such as a person's height, or it may be a description or a category, such as "dead" or "alive," "male" or "female." A simple event is one that cannot be decomposed.

| sample |

A *simple event* is defined as one of the possible outcomes on a single repetition of the experiment. *One and only one* simple event can occur on a single repetition of an experiment.

Simple events are denoted by the letter E with a subscript. *Compound events* consist of a collection of two or more _____ events. Compound events are denoted by capital letters such as A, B, C, and so on.

| simple |

An *event* is collection of one or more simple events.

Example 3.1
An experiment involves ranking three applicants, X, Y, and Z, in order of their ability to perform in a given position. List the possible simple events associated with this experiment.

Solution
Using the notation (X, Y, Z) to denote the outcome that X is ranked first, Y is ranked second, and Z is ranked third, the six possible outcomes or simple events associated with this experiment are

$$E_1: (X, Y, Z),\ E_2: (Y, X, Z),\ E_3: (Z, \underline{\quad\quad}),$$
$$E_4: (X, Z, Y),\ E_5: (Y, Z, X),\ E_6: (\underline{\quad\quad}).$$

| X, Y
| Z, Y, X

If A is the event that applicant X is ranked first, then A will occur if simple event _____ or _____ occurs.

| $E_1; E_4$

Example 3.2
The financial records of two companies are examined to determine whether each company showed a profit (P) or not (N) during the last quarter.
a. List the simple events associated with this experiment.
b. List the simple events comprising the event B, "exactly one company showed a profit."

40 / Chap. 3: Probability and Probability Distributions

Solution
a. The simple events consist of the *ordered* pairs

$$E_1: (P, P), \quad E_2: (P, N),$$
$$E_3: (N, P), \quad E_4: (\underline{\qquad}).$$

NN

b. Event B consists of the simple events _____ and _____ .

$E_2; E_3$

The set of all simple events associated with an experiment is called the *sample space*. A Venn diagram is a pictorial representation of the sample space in which events are depicted as portions of the sample space. The totality of simple events is the sample space, and is denoted by S.

Example 3.3
An oil wildcatter has just enough resources to drill three wells. Each well will either produce oil or will be dry. List the simple events associated with this experiment and construct a Venn diagram to represent the sample space S.

Solution
A typical outcome is (oil, dry, dry), which we will abbreviate to (o, d, d). The simple events are given by the following *ordered triplets*:

$$E_1: (o, o, o), \quad E_5: (d, d, o),$$
$$E_2: (d, o, o), \quad E_6: (d, o, d),$$
$$E_3: (o, d, o), \quad E_7: (\underline{\qquad}),$$
$$E_4: (o, o, d), \quad E_8: (\underline{\qquad}).$$

o, d, d

d, d, d

1. In this experiment, the event "observe exactly two dry wells" is a (simple, <u>compound</u>) event because it is composed of the simple events

compound
$E_5; E_6; E_7$
simple
E_1

_____ , _____ , and _____ . On the other hand, the event "observe no dry wells" is a (simple, compound) event because it is composed of exactly one simple event, namely _____ .

2. Complete the following Venn diagram corresponding to this experiment by assigning the eight simple events to the eight points enclosed by the closed curve:

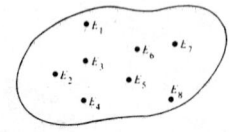

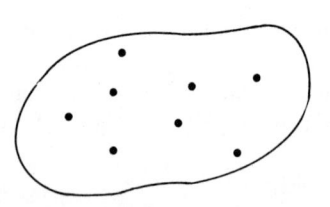

3. Let A be the event "observe no dry wells" and let B be the event "observe at least two wells producing oil." Represent A and B in a Venn Diagram in the space below.

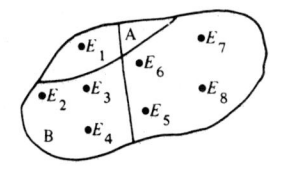

3.3 The Probability of an Event

After the experiment and its resulting sample space have been defined, the next step is to assign probabilities to the simple events. If an experiment is repeated a large number of times N, and the event A is observed n_A times, then the probability that the event will occur on a single repetition of the experiment, denoted by $P(A)$ is approximated by

$$P(A) \approx n_A/N.$$

The actual value of $P(A)$ is defined as the limiting value of n_A/N as N becomes infinitely large.

Using the relative frequency definition of probability, we see that for any event A,

$$\underline{\quad\quad} \leq P(A) \leq \underline{\quad\quad}.$$ 0; 1

If the event A never occurs, then $P(A) = 0$; if the event A always occurs, then $P(A) = \underline{\quad\quad}$. Therefore, the closer its value to 1, the more likely that A is to occur.   1

Two events are *mutually exclusive* if, when one event occurs, the other cannot, and vice versa.

For example, in the toss of a coin, if a head appears on the upper face (event H), then a tail (event T) cannot. Therefore, the events H and T are mutually exclusive. Simple events (are, are not) mutually exclusive; therefore, probabilities associated with simple events satisfy the following conditions.   are

a. $\underline{\quad\quad} \leq P(E_i) \leq \underline{\quad\quad}.$ 0; 1

b. $\sum_{\text{all } i} P(E_i) = \underline{\quad\quad}.$ 1

When it is possible to write down the simple events associated with an experiment and to assess their probabilities, we can find the probability of an event A by summing the probabilities of the simple events contained in _____.

A

The *probability of an event A* is found by summing the probabilities of the simple events in A.

$$P(A) = \sum_{\text{all } E_i \text{ in } A} P(E_i).$$

If there are N simple events, all of which are *equally likely*, and n_A of the simple events are contained in the event A, then $P(A) = n_A/N$. However, in general, simple events will not have equal probabilities.

Example 3.4

Suppose that two coins are tossed and the upper faces recorded. Suppose further that the coins are not fair and the probability that a head results on either coin is greater than a half. The following probabilities are assigned to the simple events:

Simple Event	Outcome	Probability
E_1	HH	.42
E_2	HT	.18
E_3	TH	.28
E_4	TT	.12

a. Verify that this assignment of probabilities satisfies the conditions

$$0 \leq P(E_i) \leq 1 \quad \text{and} \quad \sum_{i=1}^{4} P(E_i) = 1.$$

b. Find the probability of the event A, "the toss results in exactly one head and one tail."

c. Find the probability of the event B, "the toss results in at least one head."

Solution

a. We need but verify the second condition since observation shows that the assigned probabilities satisfy the first condition. Hence,

$$\sum_{i=1}^{4} P(E_i) = P(E_1) + P(E_2) + P(E_3) + P(E_4)$$

.28; .12

$$= .42 + .18 + \underline{\qquad} + \underline{\qquad}$$

1

$$= \underline{\qquad},$$

is

and the second condition (is, is not) satisfied.

b. The event A, "exactly one head and one tail," consists of the simple events E_2 and E_3. Therefore,

$$P(A) = P(E_2) + P(E_3)$$

$= \underline{\hspace{1cm}} + \underline{\hspace{1cm}}$ .18; .28

$= \underline{\hspace{1cm}}.$ .46

c. The event B, "at least one head," consists of the simple events E_1, E_2, and E_3. Therefore,

$$P(B) = P(E_1) + P(E_2) + P(E_3)$$

$= \underline{\hspace{1cm}} + .18 + .28$ .42

$= \underline{\hspace{1cm}}.$ .88

Example 3.5
In a shipment of four radios, R_1, R_2, R_3, and R_4, one radio is defective (say, R_3). If a dealer selects two radios at random to display in his store, what is the probability that exactly one of the radios is defective?

Solution
If we disregard the order of selection of the two radios to be displayed, the possible outcomes are

$E_1: (R_1, R_2), \quad E_4: (R_2, R_3),$

$E_2: (R_1, R_3), \quad E_5: (R_2, \underline{\hspace{1cm}}),$ R_4

$E_3: (R_1, R_4), \quad E_6: (\underline{\hspace{1cm}}).$ R_3, R_4

If the radios are selected at random, all combinations should have the same chance of being drawn. Therefore, we assign $P(E_i) = \underline{\hspace{1cm}}$ to each of the six simple events.   1/6

The event D, "exactly one of the two radios selected is defective," consists of the simple events E_2, E_4, and $\underline{\hspace{1cm}}$. Therefore, E_6

$$P(D) = P(E_2) + P(E_4) + P(E_6)$$

$= 1/6 + 1/6 + 1/6$

$= \underline{\hspace{1cm}}.$ 1/2

Calculating the Probability of an Event: Simple Event Approach

The simple event approach to finding $P(A)$ comprises five steps.
Step 1. Define the experiment.
Step 2. List all the simple events. Test to make certain that none can be decomposed.

44 / Chap. 3: Probability and Probability Distributions

Step 3. Specify which simple events lie in A.
Step 4. Assign appropriate probabilities to the simple events. Make sure that

$$\sum_{\text{all } i} P(E_i) = 1.$$

Step 5. Find $P(A)$ by summing the probabilities for all simple events in A.

Example 3.6
A taste-testing experiment is conducted in a local grocery store. Two brands of soft drinks are tasted by a passing shopper, who is then asked to state a preference for brand C or brand P. Suppose that four shoppers are asked to participate in the experiment and that all four choose brand P. Under the assumption that there is no difference between the two brands, what is the probability of the event A, "all four shoppers choose brand P"?

Solution

16	Step 1. The experiment consists of observing the preferences of each of the four shoppers for either brand P or brand C.
	Step 2. The sample space contains the following _____ simple events.
PPPP; PPPC	E_1: (____), E_5: (____), E_9: (CPPC), E_{13}: (CCPC),
	E_2: (CPPP), E_6: (PPCC), E_{10}: (PCCP), E_{14}: (CPCC),
PCPC	E_3: (PCPP), E_7: (CCPP), E_{11}: (____), E_{15}: (PCCC),
CCCC	E_4: (PPCP), E_8: (CPCP), E_{12}: (CCCP), E_{16}: (____).
PPPP	Step 3. $A = \{(_____)\}$.
1/16	Step 4. The requirement that there be no difference between the two brands implies that a probability of _____ should be assigned to each simple event.
one	Step 5. There is (are) _____ simple event(s) in A. Hence, $P(A)$
1/16	= _____ .

Modified Simple Event Approach
It may be that the list of simple events is quite long. But if equal probabilities are assigned to the simple events, all that is actually required is that you know precisely the number N of events in S and the number n_A of simple events in the event A. Then $P(A) = n_A/N$. If, however, a list is not made of the simple events, you must take care that no simple event in A is overlooked.

Example 3.7
A dealer who buys items in lots of ten selects two of the ten items at random and inspects them thoroughly. He accepts all ten if there are no defectives among the two inspected. Suppose that a lot contains two defective items.
a. What is the probability that the dealer will nonetheless accept all ten?
b. What is the probability that he will find both of the defectives?

Solution

a. Let A be the event that the dealer accepts the lot.

   Step 1. The experiment consists of selecting two items at random from ten items.

   Step 2. The sample space consists of *unordered* pairs of the form (G_1G_2) or (G_7D_1). The number of simple events is $N = 45$; the simple events are listed below.

   (G_1G_2) (G_2G_3) (G_3G_5) (G_4G_8) (G_6G_8)
   (G_1G_3) (G_2G_4) (G_3G_6) (G_4D_1) (G_6D_1)
   (G_1G_4) (G_2G_5) (G_3G_7) (G_4D_2) (G_6D_2)
   (G_1G_5) (G_2G_6) (G_3G_8) (G_5G_6) (G_7G_8)
   (G_1G_6) (G_2G_7) (G_3D_1) (G_5G_7) (G_7D_1)
   (G_1G_7) (G_2G_8) (G_3D_2) (G_5G_8) (G_7D_2)
   (G_1G_8) (G_2D_1) (G_4G_5) (G_5D_1) (G_8D_1)
   (G_1D_1) (G_2D_2) (G_4G_6) (G_5D_2) (G_8D_2)
   (G_1D_2) (G_3G_4) (G_4G_7) (G_6G_7) (D_1D_2)

   Step 3. The event A consists of the simple events containing two good items. In this case, $n_A = 28$.

   Step 4. Since the selection is made at random, each simple event should be assigned the same probability, equal to _____.   1/45

   Step 5. There are 28 simple events in A. Hence $P(A) = $ _____.   28/45

b. Let B be the event that the dealer finds both defective items in the random selection.

   Step 3. The event B consists of the single simple event (_____). D_1D_2
   Hence $n_B = $ _____.   1

   Step 5. Therefore, $P(B) = n_B/N = 1/45$.

Notice that it was very tedious to list the $N = 45$ simple events in this situation. Moreover, it is easy to overlook simple events when their number becomes large. In many instances the counting rules presented in Section 3.7 can be used to find N and n_A without listing the individual simple events. The interested reader is referred to that section.

Self-Correcting Exercises 3A

1. The owner of a camera shop receives a shipment of five cameras from a camera manufacturer. Unknown to the owner, two of these cameras are defective. Suppose that the owner selects two of the five cameras at random and tests them for operability.
   a. Define the experiment.
   b. List the simple events associated with this experiment.
   c. Define the following events in terms of the simple events in part b.
      A: both cameras are defective.
      B: neither camera is defective.

C: at least one camera is defective.
D: the first camera tested is defective.
   d. Find $P(A)$, $P(B)$, $P(C)$, and $P(D)$ by using the simple event approach.

2. A lot containing six items is comprised of four good items and two defective items. Two items are selected at random from the lot for testing purposes.
   a. List the simple events for this experiment.
   b. List the simple events in each of the three following events:

   A: at least one item is defective.
   B: exactly one item is defective.
   C: no more than one item is defective.

   c. Find $P(A)$, $P(B)$, and $P(C)$ using the simple event approach.

3. A public relations office proposed the following experiment to assess public attitudes toward large corporations. A person is shown four photographs (1, 2, 3, and 4) of crimes that have been committed and asked to select what he or she considers to be the two worst crimes. Although all four crimes shown are robberies involving about the same amount of money, 1 and 2 are robberies committed against private citizens, while 3 and 4 are robberies committed against corporations. If the person shows no bias in his or her selection, find the probabilities associated with the following events:

   A: the selection includes pictures 1 and 2.
   B: the selection includes picture 3.
   C: both *A* and *B* occur.
   D: either *A* or *B* or both occur.

4. In quality control of taste and texture, it is common to have a taster compare a new batch of a food product with one having the desired properties. Three new batches are independently tested against the standard and classified as having the desired properties (H) or not having the desired properties (N).
   a. List the simple events for this experiment.
   b. If in fact all three new batches are no different from the standard, all simple events in part a should be equally likely. If this is the case, find the probabilities associated with the following events:

   A: exactly one batch is declared as not having the desired properties.
   B: batch number one is declared to have the desired properties.
   C: all three batches are declared to have the desired properties.
   D: at least two batches are declared to have the desired properties.

3.4 Event Composition and Event Relationships

When attempting to find the probability of an event *A*, it is often useful and convenient to express *A* in terms of other events whose probabilities are known or perhaps easily calculated. Composition of events occurs in one of the two following ways or a combination of these two:

Intersections. The intersection of two events A and B is the event consisting of those simple events that are in both A and B. The intersection of A and B is denoted by AB.

Unions. The union of two events A and B is the event consisting of those simple events that are in either A or B or both A and B. The union of A and B is denoted by $A \cup B$.

Example 3.8

In each of the Venn diagrams that follow, express symbolically the event represented by the shaded area. In each case the sample space S comprises all simple events within the rectangle.

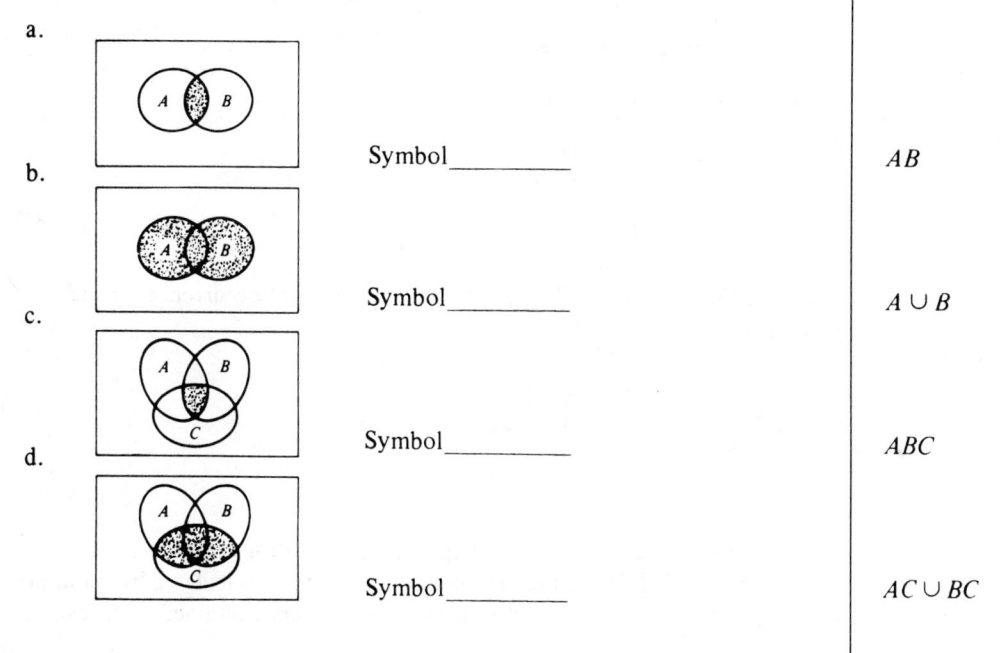

a.

Symbol_____ AB

b.

Symbol_____ $A \cup B$

c.

Symbol_____ ABC

d.

Symbol_____ $AC \cup BC$

Example 3.9

In each of the Venn diagrams below, shade in the event symbolized.
a. Symbol: $A \cup B$.

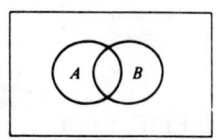

b. Symbol: BC.

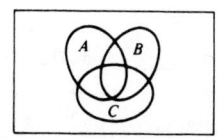

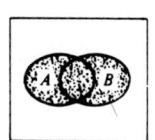

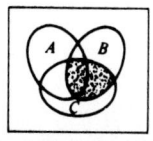

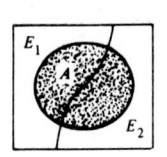

c. Symbol: $AE_1 \cup AE_2$.

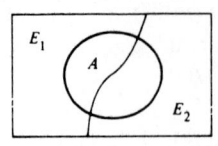

d. Note that $E_1 \cup E_2 = S$ and $AE_1 \cup AE_2 =$ _____.

Example 3.10
Consider an experiment that can result in one of ten simple events with probabilities as given in the table.

E_i	E_1	E_2	E_3	E_4	E_5	E_6	E_7	E_8	E_9	E_{10}
$P(E_i)$	.01	.05	.04	.20	.40	.03	.02	.15	.05	.05

Define the following events:

$$A = \{E_1, E_2, E_3\}$$
$$B = \{E_1, E_3, E_4, E_5\}$$
$$C = \{E_4, E_5, E_6, E_7\}$$

List the simple events and calculate the probability of occurrence for the events shown below.

a. AB.   d. $A \cup C$.   g. C.
b. AC.   e. A.   h. $A \cup B \cup C$.
c. $B \cup C$. f. B.

Solution
a. The event AB consists of the simple events in both A and B. Hence $AB = \{E_1, E_3\}$. Since the probability of an event is calculated by summing the probabilities associated with each simple event contained in the event,

$$P(AB) = P(E_1) + P(\underline{\quad}) = .01 + \underline{\quad} = \underline{\quad}.$$

E_3; .04; .05

b. The events A and C contain _____ common simple events. Hence $P(AC) =$ _____.

no
0

c. $B \cup C = \{E_1, E_3, E_4, E_5, E_6, E_7\}$ and

$$P(B \cup C) = P(E_1) + P(E_3) + P(\underline{\quad}) + P(E_5) + P(\underline{\quad})$$
$$+ P(E_7)$$

E_4; E_6

$$= .01 + .04 + \underline{\quad} + .40 + \underline{\quad} + .02$$

.20; .03

$$= \underline{\quad}.$$

.70

(left margin, top)
A

d. $A \cup C = \{E_1, E_2, E_3, E_4, E_5, E_6, E_7\}$ and

$$P(A \cup C) = \sum_{i=1}^{7} P(E_i) = .75.$$

e. $P(A) = .01 + \underline{} + .04 = \underline{}$.　　　　　　　　　　.05; .10
f. $P(B) = .01 + .04 + \underline{} + .40 = \underline{}$.　　　　　　　　　.20; .65
g. $P(C) = \underline{} + .40 + .03 + .02 = \underline{}$.　　　　　　　　　　.20; .65
h. $A \cup B \cup C = \{E_1, E_2, \underline{}, \underline{}, E_5, E_6, E_7\}$. Hence　　$E_3; E_4$

$$P(A \cup B \cup C) = \sum_{i=1}^{7} P(E_i) = \underline{}.$$

.75

Notice that $A \cup B \cup C = A \cup C$ in this example.

Mutually exclusive events are also called *disjoint* events because they contain no simple events in common. When A and B are mutually exclusive or disjoint, then

1. $P(AB) = \underline{}$　　　　　　　　　　　　　　　　　　　　　　　　　0

2. $P(A \cup B) = P(A) + P(B)$.

For example, in tossing two fair coins, with events A: no heads, B: exactly one head, and C: exactly two heads, let $P(A) = 1/4$, $P(B) = 2/4$, and $P(C) = 1/4$. Since A, B, and C are disjoint, the event $A \cup B$, that there is either zero or one head in the toss has probability $P(A \cup B) = P(A) + P(B)$ = $1/4 + 2/4 = \underline{}$, while the event AB has probability　　　　　　3/4
$P(AB) = \underline{}$.　　　　　　　　　　　　　　　　　　　　　　　　　0

The event relationship that follows often simplifies probability calculations.

The event consisting of all those simple events in the sample space S that are not in the event A is defined as the *complement of* A and is denoted by $\overline{A}$.

It is always true that $P(A) + P(\overline{A}) = \underline{}$. Therefore, $P(A) = 1 - P(\overline{A})$.　　1
If $P(\overline{A})$ can be found more easily than $P(A)$, this relationship greatly simplifies finding $P(A)$.

Example 3.11
If three fair coins are tossed, what is the probability of observing at least one head in the toss?

50 / Chap. 3: Probability and Probability Distributions

Solution

a. Let A be the event that there is at least one head in the toss of three coins. $\overline{A}$ is the event that there are _____no_____ heads in the toss of three coins.

b. There are $N = 8$ possible outcomes for this experiment:

(TTT) (TTH)

(HHT) (THT)

(HTH) (HTT)

(THH) (HHH)

so that each simple event would be assigned a probability of __1/8__.

c. $\overline{A}$ consists of the single simple event (TTT); $P(\overline{A}) =$ __1/8__ and $P(A) = 1 - P(\overline{A}) =$ __7/8__.

When A and B are mutually exclusive, their intersection AB contains no simple events. Notice that the events A and $\overline{A}$ (are, are not) mutually exclusive. [are]

3.5 Conditional Probability and Independence

Two events may be related so that the probability of the occurrence of one event depends upon whether a second event has occurred. Let A be the event that "a defective item is produced" on a production line, and let B be the event that the "production line is not operating within control limits." These two events are related in the sense that $P(A)$ is not the same as the probability of a defective item, given the prior information that the line is not operating within control limits.

Suppose we restrict our attention only to the subpopulation generated when the event B occurs, and look at the fraction of items that are defective. This fraction is the *conditional probability of A, given B,* and in this case would be expected to be larger than $P(A)$. The conditional probability of A, given that B has occurred, is denoted by $P(A|B)$, whereby the vertical bar is read "given" and the event following the bar is the event that has occurred.

The probability that event A will occur, *given that* the event B has occurred, is given by

$$P(A|B) = \frac{P(AB)}{P(B)}$$

for $P(B) > 0$.

Use a Venn diagram with events A and B to see that by knowing the event B has occurred, you effectively exclude any simple events lying outside the event B from further consideration. Since $P(B)$ and $P(AB)$ represent the amounts of probability associated with events B and AB, then

$$P(A|B) = \frac{P(AB)}{P(B)}$$

merely represents the proportion of $P(B)$ that will give rise to the event A.

If $P(A)$ and $P(A|B)$ differ in value, then the probability of A changes, depending on whether it is known that B has occurred.

Two events are said to be *independent* if either

$P(A) = P(A|B)$ when $P(B) > 0$

or

$P(B) = P(B|A)$ when $P(A) > 0$.

Otherwise, the events are said to be *dependent*.

When $P(A|B) = P(A)$, the events A and B are said to be (probabilistically) independent, since the probability of the occurrence of A is not affected by knowledge of the occurrence of B. If $P(A|B) \neq P(A)$, the events A and B are said to be dependent.

Example 3.12

You hold ticket numbers 7 and 8 in an office lottery in which ten tickets numbered 1 through 10 were sold. The winning ticket is drawn at random from those sold. You are told that the winning number is odd. Does this information alter the probability that you have won the lottery or are the two events independent?

Solution

Define the events A and B as follows:
 A: number 7 or 8 is drawn.
 B: an odd number is drawn.

The unconditional probability of winning is $P(A) = 2/10 =$ _____, | 1/5
while the conditional probability of winning is $P(A|B) =$ _____. Your | 1/5
probability of winning remains unchanged; the events A and B are
(dependent, independent). | independent

Example 3.13

Five applicants, all equally qualified, are being considered for a research technician's position. There are three males and two females among the applicants. Define the following events:
 A: female number one is selected.
 B: a female is selected.

If the selection is done at random, find $P(A)$ and $P(A|B)$. Are A and B mutually exclusive? Are A and B independent?

Solution
The unconditional probability, $P(A) = $ _____ and $P(AB) = P(A) = $ _____ . Hence, A and B (are, <u>are not</u>) mutually exclusive. We can find $P(A|B)$ in one of two ways.

a. *Direct enumeration.* If B has occurred, then we need only consider the two female applicants as comprising the new restricted sample space. Hence, $P(A|B) = $ _____ .

b. *Calculation.* By definition,

$$P(A|B) = P(AB)/P(B)$$

$$= \frac{1/5}{\underline{}} = 1/2$$

Since $P(A) \neq P(A|B)$, A and B are (<u>dependent</u>, independent) events.

1/5
1/5; are not

1/2

2/5

dependent

Many events can be viewed as the union or intersection, or both, of simpler events whose probabilities may be known or easily calculated. In such cases, the Additive and Multiplicative Laws of Probability can be used to assess the probability of the event.

The Additive Law of Probability: The probability of the union of events A and B is given by

$$P(A \cup B) = P(A) + P(B) - P(AB).$$

When the events A and B are mutually exclusive, $P(AB) = 0$ and

$$P(A \cup B) = P(A) + P(B).$$

If event A is contained in event B, then $P(A \cup B) = P(B)$.

The Multiplicative Law of Probability: The probability of the intersection of events A and B is given by

$$P(AB) = P(A)P(B|A)$$

or, equivalently, by

$$P(AB) = P(B)P(A|B).$$

If the events are independent, then

$$P(AB) = P(A)P(B).$$

Example 3.14

In the following Venn diagram the ten simple events shown are equally likely. Thus to each simple event is assigned the probability _____.   1/10

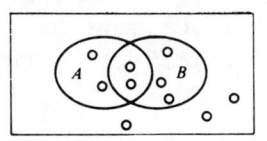

Find the following probabilities:
a. $P(A \cup B) =$ _____.   7/10
b. $P(AB) =$ _____.   2/10
c. $P(B) =$ _____.   5/10
d. $P(A|B) =$ _____.   2/5
e. $P(\bar{B}) =$ _____.   5/10
f. A and B are (independent, dependent). independent

Example 3.15

The personnel files for a large real estate agency lists its 150 employees as follows:

	Years Employed with the Agency		
	0-5 (A)	6-10 (B)	11 or More (C)
Not a college graduate (D)	10	20	20
College graduate (E)	40	50	10

If *one* personnel file is drawn at random from the agency's personnel files, calculate the probabilities requested below.

a. $P(A) =$ _____.   1/3
b. $P(E) =$ _____.   2/3
c. $P(BD) =$ _____.   2/15
d. $P(C|E) =$ _____.   1/10
e. $P(A \cup E) =$ _____.   11/15
f. $P(A|C) =$ _____.   0
g. A and E are (independent, dependent). dependent

Calculating the Probability of an Event: Event Composition Approach

The event composition approach to finding $P(A)$ is comprised of four steps:

Step 1. Define the experiment.

Step 2. Clearly visualize the nature of the simple events. Identify a few to clarify your thinking.

Step 3. Write an equation expressing A as a composition of two or more events. Make certain that the event expressed by the composition is the same set of simple events as the event A.

Step 4. Apply the additive and multiplicative laws of probability as required to the equation found in step 3. It is assumed that the component probabilities are known for the particular composition used.

54 / Chap. 3: Probability and Probability Distributions

Example 3.16
Player A has entered a tennis tournament but it is not yet certain whether player B will enter. Let A be the event that player A will win the tournament and let B be the event that player B will enter the tournament. Suppose that player A has probability 1/6 of winning the tournament if player B enters and probability 3/4 of winning if player B does not enter the tournament. If $P(B) = 1/3$, find $P(A)$.

Solution
Step 1. The experiment is the observation of whether player B enters the tournament and whether player A wins the tournament.
Step 2. The sample space S can be partitioned into four parts, AB, $A\bar{B}$, $\bar{A}B$, and $\overline{AB}$. Recall that $\bar{B}$ is the event that B does not occur, called the _____complement_____ of B, and hence $A\bar{B}$ would be the event that A occurs and B does not occur.
Step 3. Now $B \cup \bar{B} = S$ and $A = A \cap (B \cup \bar{B})$. Therefore $A = AB \cup A\bar{B}$. (Use a Venn diagram to verify this result if you wish.)
Step 4. The events AB and $A\bar{B}$ are mutually exclusive since both events cannot occur simultaneously. Therefore,

$$P(A) = P(AB \cup A\bar{B})$$

$$= P(AB) + P(A\bar{B}).$$

But

$$P(AB) = P(B)P(A|B)$$

$$= (1/3)\ (\underline{1/6}\ )$$

$$= \underline{1/18}\ ,$$

and

$$P(A\bar{B}) = P(\bar{B})P(A|\bar{B})$$

$$= (2/3)\ (\underline{3/4}\ )$$

$$= \underline{1/2}\ .$$

Therefore, $P(A) = \underline{1/18} + \underline{1/2} = \underline{5/9}$.

Example 3.17
The selling style of a temperamental salesman is strongly affected by his success or failure in the preceding attempt to sell. If he has just made a sale, his confidence and effectiveness rise and the probability of selling to his next prospect is 3/4. When he fails to sell, his manner is fearful and the probability that he sells to his next prospect is only 1/3. Suppose that the probability he will sell to his first contact on a given day is 1/2. We will find the probability

of the event A, that he makes at least two sales on his first three contacts on a given day.

Solution

Step 1. The experiment is the observation of this salesman's successes and failures on his first three contacts of the day.

Step 2. As in the preceding example, each simple event is the intersection of compound events. Let S_1 and N_1 denote, respectively, the events "sale on the first contact" and "no sale on the first contact." Similarly, S_2, N_2, S_3, and N_3 denote "sale" and "no sale" on the second and third contacts. Then the eight simple events are $S_1 S_2 S_3$, $S_1 S_2 N_3$, $S_1 N_2 S_3$, $N_1 S_2 S_3$, $S_1 N_2 N_3$, $N_1 S_2 N_3$, $N_1 N_2 S_3$, and $N_1 N_2 N_3$.

Step 3. Since A is the event that at least two sales were made on the first three contacts,

$$A = S_1 S_2 S_3 \cup S_1 S_2 N_3 \cup S_1 N_2 S_3 \cup \underline{\qquad}.$$

$N_1 S_2 S_3$

Step 4. Each of the simple events comprising A is the intersection of three events that are not independent. Notice that for the three events B, C, and D,

$$P(BCD) = P(BC)P(D|BC)$$

$$= P(B)P(C|B)P(D|BC).$$

Therefore,

$$P(S_1 S_2 S_3) = P(S_1)P(S_2|S_1)P(S_3|S_1 S_2)$$

$$= (1/2)(3/4)(3/4) = \underline{\qquad},$$ 9/32

$$P(S_1 S_2 N_3) = (1/2)(3/4)(1/4) = \underline{\qquad},$$ 3/32

$$P(S_1 N_2 S_3) = (1/2)(1/4)(\underline{\qquad}) = \underline{\qquad},$$ 1/3; 1/24

$$P(N_1 S_2 S_3) = (1/2)(\underline{\qquad})(\underline{\qquad}) = \underline{\qquad},$$ 1/3; 3/4; 3/24

and

$$P(A) = 9/32 + 3/32 + 1/24 + 3/24 = \underline{\qquad}.$$ 13/24

Self-Correcting Exercises 3B

1. Refer to Exercise 1, Self-Correcting Exercises 3A.
   a. List the simple events comprising the following events: $A \cup C$, AD, CD and $A \cup D$.
   b. Calculate $P(A \cup C)$, $P(AD)$, $P(CD)$, $P(A \cup D)$.

2. A hospital spokesperson reported that four births had taken place at the hospital during the last twenty-four hours. If we consider only the sex of these four children, recording M for a male child and F for a female child, there are 16 sex combinations possible.
   a. List these 16 outcomes in terms of simple events, beginning with E_1 as the outcome $(FFFF)$.
   b. Define the following events in terms of the simple events $E_1, \ldots, E_{16}$:

   A: two boys and two girls are born.
   B: no boys are born.
   C: at least one boy is born.

   c. List the simple events in the following events

   AB, $B \cup C$, $AC \cup BC$, $\bar{C}$, $\overline{AC}$.

   d. If the sex of a newborn baby is just as likely to be male as female, find the probabilities associated with the eight events defined in parts b and c.
   e. Calculate $P(A|C)$. Are A and C mutually exclusive? Are A and C independent?
   f. Calculate $P(B|C)$. Are B and C independent? Mutually exclusive?

3. Two hundred corporate executives in the Los Angeles area were interviewed. They were classified according to the size of the corporation they represented and their choice as to the most effective method for reducing air pollution in the Los Angeles basin. (Data are fictitious.)

	Corporation Size		
Option	Small (A)	Medium (B)	Large (C)
Car pooling (D)	20	15	20
Bus expansion (E)	30	25	11
Gas rationing (F)	3	8	4
Conversion to natural gas (G)	10	7	5
Antipollution devices (H)	12	20	10

Suppose one executive is chosen at random to be interviewed on a television broadcast.
   a. Calculate the following probabilities and describe each event involved in terms of the problem: $P(A)$, $P(F)$, $P(AF)$, $P(A \cup G)$, $P(AD)$ and $P(\bar{F})$.
   b. Calculate $P(A|F)$ and $P(A|D)$. Are A and F independent? Mutually exclusive? Are A and D independent? Mutually exclusive?

4. An investor holds shares in three independent companies which, according to her business analyst, should show an increase in profit per share with probabilities .4, .6, and .7, respectively. Assume that the analyst's estimates for the probabilities of profit increases are correct.
   a. Find the probability that all three companies show profit increases for the coming year.
   b. Find the probability that none of the companies shows a profit increase.
   c. Find the probability that at least one company shows a profit increase.

5. A marksman is able to hit a small target with probability .9. Assume his shots are independent.
   a. What is the probability that he hits the target with the next three shots?
   b. What is the probability that he hits the target with at least one of the next three shots?
6. Suppose that a person who fails an examination is allowed to retake the examination but cannot take the examination more than three times. The probability that a person passes the exam on the first, second, or third try is .7, .8, or .9, respectively.
   a. What is the probability that a person takes the exam twice before passing it?
   b. What is the probability that a person takes the exam three times before passing it?
   c. What is the probability that a person passes this exam?

3.6 Bayes' Rule (Optional)

A very interesting application of conditional probability is found in Bayes' Rule. This rule assumes that the sample space can be partitioned into k mutually exclusive events (subpopulations) $H_1, H_2, \ldots, H_k$ so that $S = H_1 \cup H_2 \cup H_3 \cup \ldots \cup H_k$. If a single repetition of an experiment results in the event A, with $P(A) > 0$, we may be interested in making an inference as to which event (subpopulation) most probably gave rise to the event A. The probability that the subpopulation H_i was sampled, given that event A occurred, is

$$P(H_i|A) = \frac{P(H_iA)}{P(A)}$$

$$= \frac{P(H_i)P(A|H_i)}{\sum_{j=1}^{k} P(H_j)P(A|H_j)}.$$

The following Venn diagram, with $k = 3$ subpopulations, demonstrates that

$$P(A) = P(AH_1 \cup AH_2 \cup AH_3)$$

$$= P(AH_1) + P(AH_2) + P(AH_3).$$

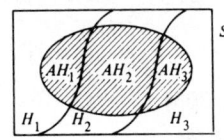

Since $P(AH_i) = P(H_i)P(A|H_i)$, we can write

$$P(A) = P(H_1)P(A|H_1) + P(H_2)P(A|H_2) + P(H_3)P(A|H_3).$$

Hence for k subpopulations,

$$P(A) = \sum_{j=1}^{k} P(H_j)P(A|H_j).$$

In order to apply Bayes' Rule, it is necessary to know the probabilities $P(H_1)$, $P(H_2), \ldots, P(H_k)$. These probabilities are called the __prior_____ probabilities. When the prior probabilities are unknown, it is possible to assume that all subpopulations are equally likely so that $P(H_1) = \ldots = P(H_k) = 1/k$. The conditional probabilities $P(H_i|A)$ are called the __posterior_____ probabilities since these are the probabilities that result after taking account of the sample information contained in the event A.

Example 3.18
A manufacturer of air-conditioning units purchases 70% of its thermostats from company A, 20% from company B, and the rest from company C. Past experience shows that .5% of company A's thermostats, 1% of company B's thermostats, and 1.5% of company C's thermostats are likely to be defective. An air-conditioning unit randomly selected from this manufacturer's production line was found to have a defective thermostat.
a. Find the probability that the defective thermostat was supplied by company A.
b. Find the probability that the defective thermostat was supplied by company B.

Solution
Let A, B, and C be the events that a thermostat selected at random was supplied by company A, B, or C, respectively. Let D be the event that a defective thermostat is observed. (Notice that $A \cup B \cup C = S$ and A, B, and C are mutually exclusive.) The following information is available.

$P(A) = .7$ $P(D|A) = .005$

$P(B) = \underline{.2}$ $P(D|B) = .010$

$P(C) = \underline{.1}$ $P(D|C) = .015$.

a. Using Bayes' Rule to find $P(A|D)$, we have

$$P(A|D) = \frac{P(A)P(D|A)}{P(A)P(D|A) + P(B)P(D|B) + P(C)P(D|C)}$$

$$= \frac{(.7)(.005)}{(.7)(.005) + (.2)(.010) + (.1)(.015)}$$

$$= \frac{.0035}{.0035 + .0020 + .0015}$$

$$= \underline{1/2}.$$

Sect. 3.6: Bayes' Rule (Optional) / 59

b. Using the results of part a to find $P(B|D)$, we have

$$P(B|D) = \frac{P(B)P(D|B)}{P(A)P(D|A) + P(B)P(D|B) + P(C)P(D|C)}$$

$$= \frac{\underline{}}{.0070} \qquad\qquad .0020$$

$$= \underline{}. \qquad\qquad 2/7$$

Further, $P(C|D) = 1 - 1/2 - 2/7 = $ _____ . If one had to make a decision as to which company most probably supplied the defective part, company _____ would be so named.

3/14

A

It is interesting to notice that Bayes' Rule entails deductive rather than inductive reasoning. Usually we are interested in investigating a problem beginning with the cause and reasoning to its effect. Bayes' Rule, however, reasons from the effect (a defective is observed) to the cause (which population produced the defective?). The next example further illustrates this type of logic.

Example 3.19

Suppose that a transmission system, whose input X is either a 0 or a 1 and whose output Y is either a 0 or a 1, mixes up the input according to the following scheme:

$$P(Y = 0 | X = 0) = .90$$

$$P(Y = 1 | X = 0) = .10$$

$$P(Y = 0 | X = 1) = .15$$

$$P(Y = 1 | X = 1) = .85.$$

If $P(X = 0) = .3$, find the following:
a. $P(X = 1 | Y = 1)$.
b. $P(X = 0 | Y = 0)$.

Solution
Since $P(X = 0) = .3$, $P(X = 1) = 1 - .3 = $ _____ .

.7

a. To find $P(X = 1 | Y = 1)$, write

$$P(X = 1 | Y = 1) = \frac{P(X = 1, Y = 1)}{P(Y = 1)}.$$

The numerator is

$$P(X = 1, Y = 1) = P(X = 1)P(Y = 1 | X = 1)$$

$$= (.7)(.85)$$

$$= \underline{}. \qquad\qquad .595$$

The denominator is

$$P(Y = 1) = P(X = 0, Y = 1) + P(X = 1, Y = 1)$$

$$= P(X = 0)P(Y = 1 \mid X = 0) + P(X = 1)P(Y = 1 \mid X = 1)$$

$$= (.3)(.10) + (.7)(.85)$$

.625

$$= \underline{\hphantom{aaaa}}.$$

Therefore,

.952

$$P(X = 1 \mid Y = 1) = .595/.625 = \underline{\hphantom{aaaa}}.$$

b. To find $P(X = 0 \mid Y = 0)$, write

$$P(X = 0 \mid Y = 0) = \frac{P(X = 0, Y = 0)}{P(Y = 0)}.$$

The numerator is

$$P(X = 0, Y = 0) = P(X = 0)P(Y = 0 \mid X = 0)$$

.90

$$= (.3)(\underline{\hphantom{aaaa}})$$

.270

$$= \underline{\hphantom{aaaa}}.$$

The denominator is

$$P(Y = 0) = P(X = 0, Y = 0) + P(X = 1, Y = 0)$$

$$= P(X = 0)P(Y = 0 \mid X = 0) + P(X = 1)P(Y = 0 \mid X = 1)$$

.7; .15

$$= (.3)(.90) + (\underline{\hphantom{aaaa}})(\underline{\hphantom{aaaa}})$$

.375

$$= \underline{\hphantom{aaaa}}.$$

Therefore,

.720

$$P(X = 0 \mid Y = 0) = .270/.375 = \underline{\hphantom{aaaa}}.$$

Self-Correcting Exercises 3C

1. An oil wildcatter must decide whether or not to hire a seismic survey before deciding whether to drill for oil on a plot of land. Given that oil is present, the survey will indicate a favorable result with probability .8; if oil is not present, a favorable result will occur with probability .3. The wildcatter figures that oil is present on the plot of land with probability

equal to .5. Determine the effectiveness of the survey by computing the probability that oil is present on the land, given a favorable seismic survey outcome and comparing this posterior probability with the prior probability of finding oil on the plot of land.

2. Each item coming off a given production line is inspected by either inspector 1 or inspector 2. Inspector 1 inspects about 60% of the production items, while inspector 2 inspects the rest. Inspector 1, who has been at her present job for some time, will not find 1% of the defective items she inspects. Inspector 2, who is newer on the job, misses about 5% of the defective items he inspects. If an item that has passed an inspector is found to be defective, what is the probability that it was inspected by inspector 1?

3.7 Useful Counting Rules (Optional)

There are three basic counting rules that are useful in counting the number of simple events N arising in many experiments. When all the N simple events are equally likely, the probability of an event A can be found without listing the simple events if N, the number of events in S, and n_A, the number of events in A, can be counted, since in this case $P(A) = n_A/N$. This is often important, since N and n_A can become quite large.

> *The mn Rule.* Suppose a procedure can be completed in two stages. If the first stage can be done in m ways and the second stage in n ways after the first stage has been completed, then the number of ways of completing the procedure is mn (m times n).

Example 3.20
An experiment involves ranking three applicants in order of merit. How many ways can the three applicants be ranked?

Solution
The process of ranking three applicants can be accomplished in two stages.
Stage 1: Select the best applicant from the three.
Stage 2: Having selected the best, select the next best from the remaining two applicants.
The ranking of the remaining applicant will automatically be third. The number of ways of accomplishing stage 1 is _____. When stage 1 is completed, there are _____ ways of accomplishing stage 2. Hence there are $(3)(2) =$ _____ ways of ranking three applicants.

| 3 |
| 2 |
| 6 |

Example 3.21
A lot of items consists of four good items (G_1, G_2, G_3, and G_4) and two defective items (D_1 and D_2).
a. How many different samples of size two can be formed by selecting two items from these six?

62 / Chap. 3: Probability and Probability Distributions

b. How many different samples will consist of exactly one good and one defective item?

c. What is the probability that exactly one good and one defective will be drawn?

Solution

a. Selecting two items from six items corresponds to the two-step procedure of (1) picking the first item and (2) picking the second item after picking the first. Hence $m =$ _____ 6; 5 , $n =$ _____ , and the number of ordered pairs is $N = mn = (6)(5) =$ _____ 30 .

b. Selecting one good and one defective item can be done in either of two ways.

1. The *defective* item can be drawn *first* in $m =$ _____ 2 ways and the *good* item drawn *second* in $n =$ _____ 4 ways. Hence there are $mn =$ _____ 8 ways of selecting a defective item on the first draw and a good item on the second draw.

2. However, the *good* item can be drawn *first* in $m =$ _____ 4 ways and the *defective* item drawn second in $n =$ _____ 2 ways, so that there are $mn =$ _____ 8 ways in which a good item is drawn first and a defective item is drawn second

3. Combining the results of Steps 1 and 2, there are exactly $8 + 8 =$ _____ 16 samples that will contain exactly one defective and one good item.

c. Let A be the event that exactly one good and one defective item are drawn. From part a, $N =$ _____ 30 , and from part b, $n_A =$ _____ 16 . Hence

$$P(A) = n_A/N = 16/30 =$$ _____ 8/15 .

An ordered arrangement of r distinct objects is called a *permutation*. The number of permutations consisting of r objects selected from n objects is given by the formula

$$P_r^n = n!/(n-r)! = n(n-1)(n-2)\ldots(n-r+1).$$

Notice that P_r^n consists of r factors, commencing with n.

Example 3.22

In how many ways can three different office positions be filled if there are seven applicants who are qualified for all three positions?

Solution

Notice that assigning the same three people to different office positions would produce different ways of filling the three positions. Hence we need to find the number of permutations (*ordered arrangements*) of three people selected from seven. Therefore,

$$P_3^7 = 7!/4! = (7)(6)(\underline{}) = \underline{}.$$

5; 210

Sect. 3.7: Useful Counting Rules (Optional) / 63

Example 3.23
A corporation will select two sites from ten available sites under consideration for building two manufacturing plants. If one plant will produce flashbulbs and the other cameras, in how many ways can the selection be made?

Solution
We are interested in the number of permutations of two sites selected from ten sites, since if two sites, say 6 and 8, were chosen, and the flashbulb plant was built at site 6 while the camera plant was built at site 8, this would result in a different selection than would occur if the camera plant was built at site 6 and the flashbulb plant at site 8. Therefore, the number of selections is

$$P_2^{10} = (10)(\underline{\hspace{1cm}}) = \underline{\hspace{1cm}}.$$ 9; 90

A selection of r objects from n distinct objects without regard to their ordering is called a *combination*. The number of combinations that can be formed when selecting r objects from n objects is given as

$$C_r^n = \frac{n!}{r!(n-r)!}.$$

Example 3.24
How many different 5-card hands can be dealt from an ordinary deck of 52 cards?

Solution
Since it is the value of the 5 cards and not the order in which they were dealt that will differentiate one 5-card hand from another, the number of distinct 5-card hands is

$$C_5^{52} = \frac{52!}{5!\,47!} = \frac{(52)(51)(50)(49)(48)}{(5)(4)(3)(2)(1)} = \underline{\hspace{1cm}}$$ 2,598,960

Notice that C_5^{52} is the same as C_{47}^{52}. In general,

$$C_r^n = \underline{\hspace{1cm}}.$$ C_{n-r}^n

Example 3.25
An experimenter must select three animals from ten available animals to be used as a control group. In how many ways can the control group be selected?

Solution
Since the order of selection is unimportant, the number of unordered selections is

$$C_3^{10} = \frac{(10)(9)(8)}{(3)(2)(1)} = \underline{\hspace{1cm}}$$ 120

Example 3.26

Refer to Example 3.7. A dealer tests two items randomly chosen from a lot of ten items and accepts the lot if the two items are not defective. Use the counting rules to find the following probabilities if a lot contains two defective items:

a. The probability that the dealer accepts the lot.
b. The probability that both defectives are found.

        A: no defectives are found.
        B: two defectives are found.

Solution

In order to calculate $P(A)$ and $P(B)$ it is necessary to find N, n_A, and n_B.

1. Since a simple event is an unordered pair of the form $(G_1 G_2)$ or $(G_1 D_1)$, the total number of simple events is

$$N = C_2^{10} = \frac{10!}{2!\,8!} = \frac{(10)(9)}{(2)(1)} = 45.$$

2. The number of ways to draw no defectives is the same as the number of ways of drawing _____ good items (from a total of _____ good items). Hence [two; eight]

$$n_A = C_2^8 = \frac{8!}{2!\,6!} = \frac{(8)(7)}{(2)(1)} = \underline{}.$$ [28]

3. The number of ways to draw two defective items (from a total of two defective items) is

$$n_B = C_2^2 = \frac{2!}{0!\,2!} = \underline{}.$$ [1]

4. Using the results of steps 1, 2, and 3,

$$P(A) = n_A/N = 28/45$$

$$P(B) = n_B/N = 1/45.$$

Notice that this method of solution is much less tedious than the solution used in Example 3.7.

Self-Correcting Exercises 3D

1. Refer to Exercise 1, Self-Correcting Exercises 3A.
    a. Using counting rules, count the number of simple events in the experiment.
    b. Count the number of simple events in the events defined in part c of Exercise 1, Self-Correcting Exercises 3A.

c. Find $P(A)$, $P(B)$, $P(C)$, and $P(D)$. Compare your answers with part d of Exercise 1, Self-Correcting Exercises 3A.
2. Five people are being considered for three awards, and no person can receive more than one award.
   a. In how many ways can these awards be given?
   b. If three of these people are city officials, in how many ways could the awards be given to these officials?
   c. If all candidates are equally qualified for the three awards, what is the probability that the awards will be presented to the three city officials?
3. A sociologist is interested in drawing a random sample of six individuals from a group consisting of ten males and ten females.
   a. How many different samples of size six are possible?
   b. How many samples would consist of all males? All females?
   c. How many samples would consist of three males and three females?
   d. If all samples are equally likely, find the probability that the sample would consist of all persons of the same sex.
   e. Find the probability that the sample contains three males and three females.
   f. Would this type of random sampling insure with a high probability that the sample proportion of males and females reflects the 50% proportion in the population?

3.8 Random Variables

Sets of measurements can be classified as either *quantitative* or *qualitative*, according to whether the measurement is a numerical quantity or a descriptive quantity.

Example 3.27
The selling price of 50 homes represents a quantitative set of data, since each measurement is numerical.

Example 3.28
A particular brand of microwave oven is rated by 25 consumers according to overall performance as either excellent, very good, good, fair, or poor. The set of 25 measurements represents a qualitative set of data, since each measurement is one of the five "qualities" given above.

Example 3.29
Identify each of the following sets of data as either quantitative or qualitative.
1. The cost of identical models of a Toyota station wagon was recorded at each of 12 Toyota dealers in Southern California. (quantitative, qualitative)   quantitative
2. In the process of applying for credit, 25 applicants are asked whether or not they currently have an outstanding bank loan. (quantitative, qualitative)   qualitative
3. In 1990, 150 cars were purchased by a local taxi company, and the make of car was recorded for each. There were 45 Fords, 30 Chevrolets, 60 Plymouths, and 15 Dodges. (quantitative, qualitative)   qualitative

quantitative

4. The total sales and the domestic sales were recorded for an electronics firm over the five years 1986 to 1990. (quantitative, qualitative)

quantitative

Recall that an experiment is the process by which an observation (or measurement) is obtained. Most experiments result in numerical outcomes or events. The outcome itself may be a numerical quantity such as height, weight, time, or some rank ordering of a response. That is, the data is (quantitative, qualitative). If the data is qualitative, many times the observations will fall into one of several categories. When categorical observations are made — such as good or defective, color of eyes, income bracket, and so on — we are usually concerned with the number of observations falling into a specified category. Again, the experiment results in a numerical outcome. Each time we observe the outcome of an experiment and assign a numerical value to the event that occurs, we are observing one particular value of a variable of interest. Since the value of this variable is determined by the outcome of a random experiment, we call the variable a *random variable.* Further, when we observe the outcome of an experiment and assign a numerical value to that event, we are in fact defining a functional relationship or a correspondence between events and numerical values. Hence, we choose the following formal definition for a random variable.

> A *random variable* is a numerical-valued function defined on the sample space.

Suppose that x is a random variable. The phrase "defined on a sample space" means that x takes values associated with simple events that are outcomes of an experiment. One and only one value of the random variable x is associated with each event in the sample space. Hence x is a random or chance variable since it takes values according to some probabilistic model. The values that the random variable x may assume form one set and the simple events another. Therefore, the random variable x is said to be a numerical-valued function.

random variable

Suppose that a sample of 100 people was randomly drawn from a population of voters and the number favoring candidate Jones was recorded. This process defines an experiment. The number of voters in the sample favoring candidate Jones is an example of a ______ ______.

Further, suppose that of the 100 voters in the sample, 60 favored Jones. This would not necessarily imply that Jones will win because one could obtain 60 or more in the *sample* favorable to Jones even though only half of the voting *population* favor him. In fact, the crucial question is, "What is the probability that 60 or more of the 100 voters in the sample are favorable to Jones when actually just 50% of the voting population will vote for him?" To answer this question, we need to investigate the probabilistic behavior of the random variable x, the number of favorable voters in a sample of 100 voters. The set of values that the random variable x may assume and the probability $p(x)$ associated with each value of x define a probability distribution. Hence before we can use a random variable to make inferences about a population, we must study some basic characteristics of probability distributions.

Random variables are designated as *discrete* or *continuous* according to the values they may assume in an experiment.

> A *discrete random variable* is one that can assume a countable* number of values.

The following would be examples of discrete random variables:
a. The number of voters favoring a political candidate in a given precinct.
b. The number of defective bulbs in a package of twenty bulbs.
c. The number of errors in an income tax return.

Notice that discrete random variables are basically counts and the phrase "the number of" can be used to identify a discrete random variable.

> A *continuous random variable* can take on the infinitely large number of values associated with the points on a line interval.

The following would be examples of continuous random variables:
a. The time required to complete a medical operation.
b. The height of an experimental strain of corn.
c. The amount of ore produced by a given mining operation.

Classify the following random variables as discrete or continuous:
a. The number of psychological subjects responding to stimuli in a group of thirty. _____   discrete
b. The number of building permits issued in a community during a given month. _____   discrete
c. The number of amoebae in 1 cubic centimeter of water. _____   discrete
d. The juice content of six Valencia oranges. _____   continuous
e. The time to failure for an electronic system. _____   continuous
f. The amount of radioactive iodine excreted by rats in a medical experiment. _____   continuous
g. The number of defects in 1 square yard of carpeting. _____   discrete

It is necessary to make the preceding distinction between the discrete and continuous cases because the probability distributions require different mathematical treatment. In fact, calculus is a prerequisite to any complete discussion of continuous random variables. Arithmetic and elementary algebra are all we need to develop discrete probability distributions.

3.9 Probability Distributions for Discrete Random Variables

The probability distribution for a discrete random variable x consists of the pairs $(x, p(x))$ where x is one of the possible values of the random variable x and $p(x)$ is its corresponding probability. This probability distribution must satisfy two requirements:

---

*Countable means that the values the random variable can assume can be associated with the counting integers, 0, 1, 2, . . . , so that the values can be counted.

68 / Chap. 3: Probability and Probability Distributions

<table><tr><td>1</td><td>1. $\sum_{x} p(x) =$ _____ .</td></tr>
<tr><td>0; 1</td><td>2. _____ $\leq p(x) \leq$ _____ .</td></tr></table>

One can express the probability distribution for a discrete random variable x in any one of three ways:
1. By listing, opposite each possible value of x, its probability $p(x)$ in a table.
2. Graphically as a probability histogram.
3. By supplying a formula together with a list of the possible values of x.

Example 3.30
A businessman has decided to invest $10,000 in each of three common stocks. Four stocks, call them A, B, C, and D, have been recommended to him by a broker and he plans to select three of the four to form an investment portfolio. Unknown to the businessman, stocks A, B, and C will rise in the near future but D will suffer a severe drop in price. If the selection is made at random, find the probability distribution for x, the number of good stocks in the investment portfolio.

Solution
The experiment consists of selecting three of the four stocks. Each of the

<table><tr><td>4
four</td><td>$C_3^4 =$ _____ distinctly different combinations possesses the same chance for selection and these _____ simple events form the sample space. The simple events associated with the four combinations along with their probabilities are shown in the table. The value $x = 3$ is assigned to E_1 because this combination includes all three of the good stocks. Assign values of x to the other three simple events.</td></tr></table>

Simple Events	$P(E_i)$	x
E_1: ABC	1/4	3
E_2: ABD	1/4	____
E_3: ACD	1/4	____
E_4: BCD	1/4	____

2
2
2

1. The probability distribution presented as a table is shown below. When

3/4
1/4

$x = 2$, $p(x)$ is $p(2) = P(E_2) + P(E_3) + P(E_4) =$ _____ . Similarly, $p(3) = P(E_1) =$ _____ .

x	$p(x)$
2	3/4
3	1/4

Note that the two requirements for a discrete probability distribution are satisfied in this example. These are

$\sum_{x} p(x) = 1$

a. _____ ,

$0 \leq p(x) \leq 1$

b. _____ .

2. The probability distribution can also be graphically presented as a probability histogram.

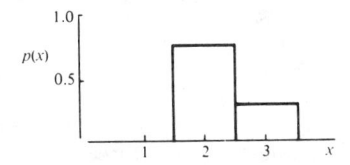

3. A formula appropriate for this probability distribution is

$$p(x) = \frac{C_x^3 C_{3-x}^1}{C_3^4}, \quad x = 2, 3,$$

where

$$C_x^n = \frac{n!}{x!(n-x)!}.$$

(See Section 3.7.) You may verify that the formula does indeed give the correct values for $p(2)$ and $p(3)$.

4. If this experiment were repeated many times, approximately what fraction of the outcomes would result in $x = 2$? _____ What fraction of the total area under the probability histogram lies over the interval associated with $x = 2$? _____

3/4

3/4

Note that the probability distribution provides a theoretical frequency distribution for the hypothetical population associated with the businessman's experiment and thereby relates directly to the content of Chapter 2 in your text.

Example 3.31
A psychological recognition experiment required a subject to classify a set of objects according to whether they had or had not been previously observed. Suppose that a subject can correctly classify each object with probability $p = .7$, that sequential classifications are independent events, and that she is presented with $n = 3$ objects to classify. We are interested in x, the number of correct classifications for the three objects.

Solution
1. This experiment is analogous to tossing three unbalanced coins, where correctly classifying an object corresponds to the observation of a head in the toss of a single coin. Each classification results in one of two outcomes, correct or incorrect. The total number of simple events in the sample space (applying the *mn* rule of Section 3.7) is _____.

8

2. Let *IIC* represent the simple event for which the classification of the first and second objects is incorrect and the third is correct. Complete the listing of all the simple events in the sample space.

CCI

CIC

CCC

E_1: III E_5: CII
E_2: IIC E_6: _____
E_3: ICI E_7: _____
E_4: ICC E_8: _____

3. The simple event E_2 is an *intersection* of three independent events. Applying the Multiplicative Law of Probability,

$$P(E_2) = P(IIC) = P(I)P(I)P(C) = (.3)(.3)(.7) = .063.$$

Similarly, $P(E_1) = .027$ and $P(E_3) = .063$. Calculate the probabilities for all the simple events in the sample space.

.063
.147
.147
.343

$P(E_1) = .027$ $P(E_5) = $ _____
$P(E_2) = .063$ $P(E_6) = $ _____
$P(E_3) = .063$ $P(E_7) = $ _____
$P(E_4) = .147$ $P(E_8) = $ _____.

4. The random variable x, the number of correct classifications for the set of three objects, takes the value $x = 1$ for simple event E_2. Similarly, we would assign the value $x = 0$ to E_1. Assign a value of x to each simple event in the sample space.

Simple Events	x
E_1	0
E_2, E_3, E_5	1
E_4, E_6, E_7	___
E_8	___

2
3

5. The numerical event $x = 0$ contains only the simple event E_1. Summing the probabilities of the simple events in the event $x = 0$, we have $P(x = 0) = P(E_1) = .027$. Similarly, the numerical event $x = 1$ contains three simple events. Summing the probabilities of these simple events, we have $P(x = 1) = p(1) = .189$.

6. The probability distribution $p(x)$ is presented in tabular form.

x	$p(x)$
0	.027
1	___
2	___
3	___

.189
.441
.343

Calculate the probabilities $p(2)$ and $p(3)$ and complete the table.

7. Present $p(x)$ graphically in the form of a probability histogram.

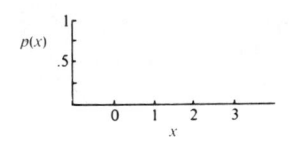

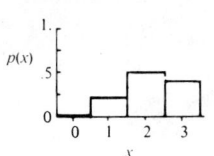

8. After studying Chapter 4, you will be able to express this probability distribution as a formula.

Example 3.32

1. Construct a probability table expressing the probability distribution that is given by the formula

   $p(x) = (1/4)C_x^2$ for $x = 0, 1, 2$.

x	$p(x)$

   0; 1/4
   1; 1/2
   2; 1/4

2. Construct a probability histogram for the probability distribution in part 1.

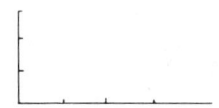

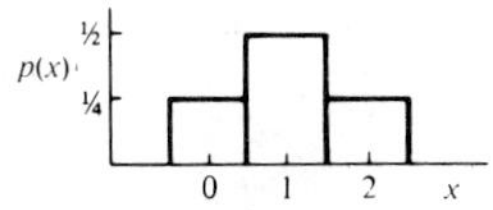

3. If the experiment implied in part 1 were repeated over and over again a large number of times, the frequency histogram would resemble the probability distribution for the random variable x.

Self-Correcting Exercises 3E

1. A subject is shown four photographs, A, B, C, D, of crimes that have been committed and asked to select what she considers the two worst crimes. Although all four crimes shown are robberies involving about the same amount of money, A and B are pictures of robberies committed against private citizens while C and D are robberies committed against corporations. Assuming that the subject does not show discrimination in her selection, find the probability distribution for x, the number of crimes chosen involving private citizens.

2. Someone claims that the following is the probability distribution for a random variable x:

x	-1	0	1	2
$p(x)$	1/10	-2/10	5/10	3/10

   Give two reasons why this is not a valid probability distribution.

3. Suppose that the unemployment rate in a given community is 7%. Four households are randomly selected to be interviewed. In each household,

it is determined whether or not the primary wage earner is unemployed. If the 7% rate is correct, find the probability distribution for x, the number of primary wage earners who are unemployed.

3.10 Mathematical Expectation for Discrete Random Variables

When we develop a probability distribution for a random variable, we are actually proposing a model that will describe the behavior of the random variable in repeated trials of an experiment. For example, when we propose the model for describing the distribution of x, the number of heads in the toss of 2 fair coins, given by

x	$p(x)$
0	1/4
1	1/2
2	1/4

we mean that if the two coins were tossed a large number of times, about one-fourth of the outcomes would result in the outcome "zero heads," one-half would result in the outcome "one head," and the remaining fourth would result in "two heads." A probability distribution is not only a measure of belief that a specific outcome will occur on a single trial but, more important, it actually describes a population of observations on the random variable x. It is reasonable then to talk about and calculate the mean and the standard deviation of a random variable by using the probability distribution as a population model.

The expected value of a random quantity is its average value in the population. In particular, the expected value of x is simply the population mean. The expected value of $(x - \mu)^2$ describes the population variance.

The *mean* or *expected value*, $E(x)$, of a discrete random variable x is

$$\mu = E(x) = \sum_x x\, p(x).$$

The *variance*, σ^2, is given by

$$\sigma^2 = E(x - \mu)^2 = \sum_x (x - \mu)^2 p(x).$$

Example 3.33

Suppose that a psychological experiment is designed in such a way that the patient has two choices for each of three experimental situations into which he is placed. One of the choices is designated as the "correct" choice. The probability distribution for x, the number of correct choices is given below.

x	p(x)
0	.15
1	.35
2	.20
3	.30

Find the mean and the standard deviation of the number of correct choices.

Solution
Before calculating the mean and variance of x, we see that this (is, is not) a valid probability distribution since

is

$$\sum_x p(x) = \underline{} \quad \text{and} \quad \underline{} \leqslant p(x) \leqslant \underline{}.$$

1; 0; 1

1. The expected number of correct choices is calculated as

$$\mu = E(x) = \sum_x x\,p(x)$$

$$= 0(.15) + 1(.35) + 2(.20) + 3(.30)$$

$$= \underline{}.$$

1.65

2. From part 1, $\mu = E(x) = \underline{}$. Then the variance of x is

1.65

$$\sigma^2 = E(x - \mu)^2$$

$$= \sum_x (x - \mu)^2 p(x)$$

$$= (0 - 1.65)^2(.15) + (1 - 1.65)^2(.35)$$

$$\quad + (2 - 1.65)^2(.20) + (3 - 1.65)^2(.30)$$

$$= .408375 + \underline{} + .0245 + \underline{}$$

.147875; .546750

$$= \underline{},$$

1.1275

and $\sigma = \underline{}.$

1.062

If x is a random variable, either continuous or discrete, it can be shown that

$$E(x - \mu)^2 = E(x^2) - \mu^2$$

This result is given in your text (Section 3.10) and can be used as a shortcut computational formula to calculate the variance, σ^2.

Example 3.34
Use the computational formula to calculate σ^2 for Example 3.33.

Solution
For a discrete random variable x,

$$E(x^2) = \sum_x x^2 p(x),$$

which is the average value of x^2 over all its possible values. For this example,

$$E(x^2) = 0^2(.15) + 1^2(.35) + 2^2(.20) + 3^2(.30)$$

3.85

= _____.

Then

1.1275

$$\sigma^2 = 3.85 - (.165)^2 = \underline{}$$

and

1.062

$\sigma = $ _____.

Notice that the numerical results are identical to those found in Example 3.33. Moreover, the computational formula involves fewer steps and in general results in less rounding error than does the definition formula.

Example 3.35
Construct the probability histogram for the distribution of the number of correct choices given in Example 3.33. Visually locate the mean and compare it with the computed value, $\mu = 1.65$.

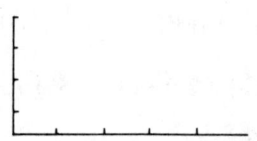

1.5

The approximate value of the mean is _____. Its value is close to that calculated and provides an easy check on the calculated value of $E(x)$.

Example 3.36
A corporation has four investment possibilities: A, B, and C with respective gains of $10, $20, and $50 million and investment D with a loss of $30 million. If one investment will be made and the probabilities of choosing A, B, C, or D are .1, .4, .2, and .3, respectively, find the expected gain for the corporation.

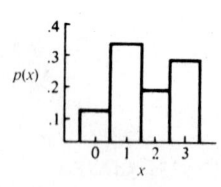

Sect. 3.10: Mathematical Expectation for Discrete Random Variables / 75

Solution
The random variable is x, the corporation's gain, with possible values $10, $20, $50, and −$30 million. The probability distribution for x is given as

x (in millions)	$p(x)$
$ 10	.1
20	.4
50	.2
−30	.3

The expected gain is $E(x) =$ _____ million. $10

Example 3.37
A parcel post service that insures packages against loss up to $200 wishes to reevaluate its insurance rates. If 1 in every 1000 packages had been reported lost during the last several years, what rate should be charged on a package insured for $200 so that the postal service's expected gain is zero? Administrative costs will be added to this rate.

Solution
Let x be the gain to the parcel post service and let r be the rate charged for insuring a package for $200.
1. Complete the probability distribution for x.

x	$p(x)$
r	.999
_____	.001	$(r - 200)$

2. If $E(x)$ is to be zero, we need to solve the equation

$$\sum_x x\,p(x) = 0.$$

Hence, for our problem,

$$r(.999) + (r - 200)(.001) = 0,$$

$r =$ _____. $.20

Self-Correcting Exercises 3F

1. Let x be a discrete random variable with a probability distribution given as

x	−2	−1	0	1
$p(x)$	1/9	1/9	4/9	—

a. Find $p(1)$.

b. Find $\mu = E(x)$.
c. Find σ, the standard deviation of x.

2. A police car visits a given neighborhood a random number of times x per evening. $p(x)$ is given by

x	$p(x)$
0	.1
1	.6
2	.2
3	.1

a. Find $E(x)$.
b. Find σ^2 using both the definition and the computational formulas. Verify that the results are identical.
c. Calculate the interval $\mu \pm 2\sigma$ and find the probability that the random variable x lies within this interval. Does this agree with the results given in Tchebysheff's Theorem?

3. Refer to Exercise 2. What is the probability that the patrol car will visit the neighborhood at least twice in a given evening?

4. You are given the following information. An insurance company wants to insure a $80,000 home against fire. One in every 100 such homes is likely to have a fire; 75% of the homes having fires will suffer damages amounting to $40,000, while the remaining 25% will suffer total loss. Ignoring all other partial losses, what premium should the company charge in order to break even?

3.11 Random Sampling

The objective of the study of statistics is to allow the experimenter to make inferences about a population from information contained in a _____.

sample

Since it is the sample that provides the information that is used in inference making, we must be duly careful about the selection of the elements in the sample so that we do not systematically exclude or include certain elements of the population in our sampling plan. The sample should be representative of the population being sampled.

We call a sample that has been drawn without bias a *random sample*. This is a shortened way of saying that the sample has been drawn in a random manner. Several types of random samples are available for use in a particular situation, depending on the scope of the experiment and the objectives of the experimenter. A commonly employed and uncomplicated sampling plan is called the *simple random sample*.

A *simple random sample* of size n is said to have been drawn if each possible sample of size n in the population has the same chance of being selected.

If a population consists of N elements and we wish to draw a sample of size n from this population, there are

$$C_n^N = \frac{N!}{n!(N-n)!}$$

samples to choose from. A random sample in this situation would be one drawn in such a manner that each sample of size n had the same chance of being drawn, namely, $(C_n^N)^{-1}$ or $1/C_n^N$.

Example 3.38
A medical technician needs to choose four animals for testing from a cage containing six animals. How many samples are available to the technician? List these samples.

Solution
The number of ways to choose four animals from a total of six is

$$C_4^6 = \frac{6!}{4!\,2!} = 15.$$

Designating each animal by a number from 1 to 6, the samples are

(1234)	(1256)	(____)	2345
(1235)	(1345)	(2346)	
(1236)	(1346)	(2356)	
(1245)	(1356)	(2456)	
(____)	(1456)	(____)	1246; 3456

A simple random sampling plan for this experiment would allow each of these 15 possible samples an equal chance of being selected, namely, 1/15.

Although perfect random sampling is difficult to achieve in practice, there are several methods available for selecting a sample that will satisfy the conditions of random sampling when N, the population size, is not too large.
1. *Method A.* List all the possible samples and assign them numbers. Place each of these numbers on a chip or piece of paper and place them in a bowl. Drawing one number from the bowl will select the random sample to be used.
2. *Method B.* Number each of the N members of the population. Write each of these numbers on a chip or slip of paper and place them in a bowl. Now draw n numbers from the bowl and use the members of the population having these numbers as elements to be included in the sample.
3. *Method C.* A useful technique for selecting random samples is one in which a table of random numbers is used to replace the chance device of drawing chips from a bowl.

78 / Chap. 3: Probability and Probability Distributions

population

inferences
cannot

Why is it so important that the sample be randomly drawn? From the practical point of view, one would want to keep the experimenter's biases out of the selection and, at the same time, keep the sample as representative of the _____ as possible. From the statistical point of view, we can assess the probability of observing a random sample and hence make valid _____ about the parent population. If the sample is nonrandom, its probability (can, cannot) in general be determined and hence no valid inferences can be made from it.

EXERCISES

1. Suppose that an experiment requires the ranking of three applicants, A, B, and C, in order of their abilities to do a certain job. The simple events could then be symbolized by the ordered triplets (ABC), (BAC), and so on.
   a. The event A, that applicant A will be ranked first, comprises which of the simple events?
   b. The event B, that applicant B will be ranked third, comprises which of the simple events?
   c. List the events in $A \cup B$.
   d. List the events in AB.
   e. If equal probabilities are assigned to the simple events, show whether or not events A and B are independent.

2. An antique dealer had accumulated a number of small items including a valuable stamp collection and a solid gold vase. To make room for new stock he distributed these small items among four boxes. Without revealing which items were placed in which box, the dealer stated that the stamp collection was included in one box and the gold vase in another. The four boxes were sealed and placed on sale, each at the same price. A certain customer purchased two boxes selected at random from the four boxes. What is the probability that the customer acquired
   a. the stamp collection?
   b. the vase?
   c. at least one of these bonus items?

3. If the probability that an egg laid by an insect hatches is $p = .4$, what is the probability that at least three out of four eggs will hatch?

4. Suppose that on the basis of past experience it is known that a lie detector test will indicate that an innocent person is guilty with probability .08, while the test will indicate that a guilty person is innocent with probability .15. Suppose further that 10% of the population under study has committed a traffic violation. If a lie detector test indicates that a randomly chosen individual from this population has committed a traffic violation, what is the probability that this person is innocent of committing a traffic violation?

5. Eight employees have been found equally qualified for promotion to a particular job. It has been decided to choose five of the employees at random for immediate promotion. How many different groups of five employees are possible?

6. Refer to Exercise 5. Suppose that only one vacancy will occur at a time, and that the five employees must be chosen for assignment sequentially.

These five will then be promoted as vacancies occur in the order they are listed. How many different promotional lists are possible?

7. In the past history of a certain serious disease it has been found that about 1/2 of its victims recover.
   a. Find the probability that exactly one of the next five patients suffering from this disease will recover.
   b. Find the probability that at least one of the next five patients suffering from this disease will recover.

8. The sample space for a given experiment comprises the simple events, E_1, E_2, E_3, and E_4. Let the compound events A, B, and C be defined by these equations:

$$A = E_1 \cup E_2, \quad B = E_1 \cup E_4, \quad C = E_2 \cup E_3.$$

Construct a Venn diagram showing the events, E_1, E_2, E_3, E_4, A, B, and C.

9. Refer to Exercise 8. Probabilities are assigned to the simple events as indicated in the following table:

Simple event	E_1	E_2	E_3	E_4
Assigned probability	1/3	1/3	1/6	—

   a. Supply the missing entry in the table.
   b. Find $P(A)$ and $P(AB)$.
   c. Find $P(A|B)$ and $P(A|C)$.
   d. Find $P(A \cup B)$ and $P(A \cup C)$.
   e. Are A and B mutually exclusive? Independent?

10. An envelope of seeds contains three nonviable seeds and five viable ones. Consider the eight seeds to be distinguishable.
    a. How many different samples of size three can be formed?
    b. How many of these samples of size three comprise two viable seeds and one nonviable seed?
    c. If a sample of size three is selected at random from this envelope, what is the probability that two of these seeds will be viable and the other nonviable?

11. A random sample of size five is drawn from a large production lot with a fraction defective of 10%. The probability that this sample will contain no defectives is .59. What is the probability that this sample will contain at least one defective?

12. A factory operates an 8-hour day shift. Five machines of a certain type are used. If one of these machines breaks down, it is set aside and repaired by a crew operating at night. Suppose the probability that a given machine suffers a breakdown during a day's operation is 1/5.
    a. What is the probability that no machine breakdowns will occur on a given day?
    b. What is the probability that two or more machine breakdowns will occur on a given day?

13. A certain virus disease afflicted the families in 3 adjacent houses in a row of 12 houses. If 3 houses were randomly chosen from a row of 12 houses, what is the probability that the 3 houses would be adjacent? Is there reason to conclude that this virus disease is contagious?

14. A geologist, assessing a given tract of land for its oil content, initially rates the land as having
    i. no oil, with probability 0.7,
    ii. 500,000 barrels, with probability 0.2,
    iii. 1,000,000 barrels, with probability 0.1.

    However, the potential buyer ordered that seismic drillings be performed and found the readings to be "high" based on a "low, medium, high" rating scale. The conditional probabilities, $P(E|H)$, are given in the following table:

	Seismic Readings, E_i		
H_i	E_1, Low	E_2, Medium	E_3, High
H_1: no oil	.50	.30	.20
H_2: 500,000 bbl	.40	.40	.20
H_3: 1,000,000 bbl	.10	.50	.40

    a. Find $P(H_1|E_3)$, $P(H_2|E_3)$, and $P(H_3|E_3)$.
    b. Suppose the seismic readings had been low. Now find $P(H_1|E_1)$, $P(H_2|E_1)$, and $P(H_3|E_1)$.

15. A manufacturer buys parts from a supplier in lots of 10,000 items. The fraction defective in a lot is usually about .1%. Occasionally a malfunction in the supplier's machinery causes the fraction defective to jump to 3%. Records indicate that the probability of receiving a lot with 3% defective is .1. To check the quality of the supplier's lot, the manufacturer selects a random sample of 200 parts from the lot and observes 3 defectives.

    The probability of observing 3 defectives when the fraction defective is .1% is approximated as .0011, and the probability of observing 3 defectives when the fraction defective is 3% is approximated as .0892.

    a. Find the probability that the percentage defective is .1%, given that 3 defectives are observed in the sample.
    b. Find the probability that the percentage defective is 3%, given that 3 defectives are observed in the sample.
    c. Based on your answers to parts a and b, what would you conclude to be the fraction defective in the lot?

16. A car rental agency has three Fords and two Chevrolets left in its car pool. If two cars are needed and the keys are randomly selected from the keyboard, find the probability distribution for x, the number of Fords in the selection.

17. Five equally qualified applicants for a teaching position were ranked in order of preference by the superintendent of schools. If two of the applicants hold master's degrees in education, find the probability distribution for x, the number of applicants holding master's in education ranked first or second.

18. An experiment is run in the following manner: The colors red, yellow, and blue are each flashed on a screen for a short period of time. A subject views the colors and is asked to choose the one he feels was flashed for the longest amount of time. The experiment is repeated three times with the same subject.
    a. If all the colors were flashed for the same length of time, give the

probability distribution for x, the number of times the subject chose the color red. Assume that his three choices are independent.

b. Construct a probability histogram for $p(x)$ found in part a.

19. A publishing company is considering the introduction of a monthly gardening magazine. Advance surveys show the initial market for the magazine will be approximated by the following distribution for x, the number of subscribers:

x	$p(x)$
5,000	.30
10,000	.35
15,000	.20
20,000	.10
25,000	.05

Find the expected number of subscribers and the standard deviation of the number of subscribers.

20. Refer to Exercise 19. Suppose the company expects to charge $20 for an annual subscription. Find the mean and standard deviation of the revenue the company can expect from the annual subscriptions of the initial subscribers.

21. Refer to Exercise 20. Production and distribution costs for the gardening magazine are expected to amount to slightly over $200,000. What is the probability that revenue from initial subscriptions will fail to cover these costs?

22. The following is the probability function for a discrete random variable x:

$$p(x) = (.1)(x + 1), \quad x = 0, 1, 2, 3.$$

a. Find $\mu = E(x)$ and σ^2.
b. Construct a probability histogram for $p(x)$.

23. The probability of hitting oil in a single drilling operation is 1/4. If drillings represent independent events, find the probability distribution for x, the number of drillings until the first success ($x = 1, 2, 3, \ldots$). Proceed as follows:

a. Find $p(1)$.
b. Find $p(2)$.
c. Find $p(3)$.
d. Give a formula for $p(x)$.
Note that x can become infinitely large.
e. Will $\sum_{x=1}^{\infty} p(x) = 1$?

24. The board of directors of a major symphony orchestra has voted to create an employee council for the purpose of handling employee complaints. The council will consist of the president and vice-president of the symphony board and two orchestra representatives. The two orchestra representatives will be randomly selected from a list of 6 volunteers, consisting of 4 men and two women.

a. Find the probability distribution for x, the number of women chosen to be orchestra representatives.
b. Find the mean and variance for x.
c. What is the probability that both orchestra representatives will be women?

25. Given a random variable x with the probability distribution

x	$p(x)$
1	1/8
2	5/8
3	1/4

graph $p(x)$ and make a visual approximation to the mean and standard deviation. (Use your knowledge of Tchebysheff's Theorem to assist in approximating σ.)

26. Refer to Exercise 25 and find the expected value and standard deviation of x. Compare with the answers to Exercise 25.

27. Given the following probability distribution, find the expected value and variance of x.

x	$p(x)$
0	1/2
3	1/3
6	1/6

28. History has shown that buildings of a certain type of construction suffer fire damage during a given year with probability .01. If a building suffers fire damage, it will result in either a 50% or a 100% loss with probabilities .7 and .3, respectively. Find the premium required per $1000 coverage in order that the expected gain for the insurance company will equal zero (break-even point).

29. Experience has shown that a rare disease will cause partial disability with probability .6, complete disability with probability .3, and no disability with probability .1. Only 1 in 10,000 will become afflicted with the disease in a given year. If an insurance policy pays $20,000 for partial disability and $50,000 for complete disability, what premium should be charged in order that the insurance company break even (that is, in order that the expected loss to the insurance company will be zero)?

30. Consider the following situation: A man has an urn containing 20 white and 3 red balls. He asks a little boy to close his eyes and pick 3 balls from the urn. For each red ball selected by the youngster, the man promises him a candy bar. Just as the boy is ready to pick the first ball, the doorbell rings. The man instructs the boy to continue and leaves the room to answer the door. Upon his return he finds the lad has picked 3 red balls. Would you consider this random sampling on the part of the boy?

31. A sidewalk interviewer stopped three men who were walking together, asked their opinions on some topical subjects and found their answers quite similar. Would you consider the interviewer's selection to be random in this case? Is it surprising that similar answers were given by these three men?

CHAPTER 4
SEVERAL USEFUL DISCRETE DISTRIBUTIONS

4.1 Introduction

A random variable that takes a countable number of values corresponding to a countable number of simple events is called a *discrete random variable*. Of the many discrete random variables and their probability distributions found in the sciences and in business and economics, three discrete distributions can be used as a model in many of these situations.

In this chapter, we will learn about the *binomial, Poisson, hypergeometric and several other distributions* used as models for discrete random variables in different settings and contexts.

4.2 The Binomial Probability Distribution

Many experiments in the social, biological, and physical sciences can be reduced to a series of trials resembling the toss of a coin whereby the outcome on each toss will be either a head or a tail. Consider the following analogies:
1. A student answers a multiple-choice question correctly (head) or incorrectly (tail).
2. A voter casts her ballot either for candidate A (head) or against him (tail).
3. A patient having been treated with a particular drug either improves (head) or does not improve (tail).
4. A subject either makes a correct identification (head) or an incorrect one (tail).
5. A licensed driver either has an accident (head) or does not have an accident (tail) during the period his license is valid.
6. An item from a production line is inspected and classified as either defective (head) or not defective (tail).

If any of the above situations were repeated n times and we counted the number of "heads" that occurred in the n trials, the resulting random variable would be a _____ random variable. Let us examine what characteristics these experiments have in common. We will call a head a success (S) and a tail a failure (F). Note that a success does not necessarily denote a desirable outcome but rather identifies the event of interest.

    binomial

The five defining characteristics of a binomial experiment are as follows:

1. The experiment consists of n identical trials.
2. Each trial results in one of two outcomes, success (S) or failure (F).

84 / Chap. 4: Several Useful Discrete Distributions

3. The probability of success on a single trial is equal to p and remains constant from trial to trial. The probability of failure is $q = 1 - p$.
4. The trials are independent.
5. Attention is directed to the random variable x, the total number of successes observed during the n trials.

Although very few real life situations perfectly satisfy all five characteristics, this model can be used with fairly good results if the violations are "moderate." The next several examples will illustrate binomial experiments.

Example 4.1
A procedure (the "triangle test") often used to control the quality of name brand food products utilizes a panel of n "tasters." Each member of the panel is presented three specimens, two of which are from batches of the product known to possess the desired taste while the other is a specimen from the latest batch. Each panelist is asked to select the specimen that is different from the other two. If the latest batch does possess the desired taste, then the probability that a given taster will be "successful" in selecting the specimen from the latest batch is _____ . If there is no communication among the panelists their responses will comprise n independent _____ _____ , with a probability of success on a given trial equal to _____ .

<small>1/3</small>
<small>trials</small>
<small>1/3</small>

Example 4.2
Almost all auditing of accounts is done on a sampling basis. Thus an auditor might check a random sample of n items from a ledger or inventory list comprising a large number of items. If 1% of the items in the ledger are erroneous, then the number of erroneous items in the sample is essentially a _____ random variable with n trials and probability of success (finding an erroneous item) on a given trial equal to _____ .

<small>binomial</small>
<small>.01</small>

Example 4.3
No treatment has been known for a certain serious disease for which the mortality rate in the United States is 70%. If a random selection is made of 100 past victims of this disease in the United States, the number x_1 of those in the sample who died of the disease is essentially a binomial random variable with $n =$ _____ and $p =$ _____ . More important, if observation is made of the next 100 persons in the United States who will in the future become victims of this disease, the number x_2 of these who will die from the disease has a distribution approximately the same as that of x_1 if conditions affecting this disease remain essentially constant for the time period considered.

<small>100; .70</small>

Example 4.4
The continued operation (reliability) of a complex assembly often depends on the joint survival of all or nearly all of a number of similar components. Thus a radio may give at least 100 hours of continuous service if no more than two of its ten transistors fail during the first 100 hours of operation. If the ten transistors in a given radio were selected at random from a large lot of transistors, then each of these (ten) transistors would have the same probability p of failing within 100 hours, and the number of transistors in the radio that will fail within 100 hours is a _____ random variable with

<small>binomial</small>

_____ trials and probability of success on each trial equal to _____. (Success is a word that denotes one of the two outcomes of a single trial and does not necessarily represent a desirable outcome.)

10

p

Three experiments are described below. In each case state whether or not the experiment is a binomial experiment. If the experiment is binomial, specify the number n of trials and the probability p of success on a given trial. If the experiment is not binomial, state which characteristics of a binomial experiment are not met.

1. A fair coin is tossed until a head appears. The number of tosses x is observed. If binomial, $n =$ _____ and $p =$ _____. If not binomial, list characteristic(s) (1, 2, 3, 4, and 5) violated. _____

not binomial
1, 5

2. The probability that an applicant scores above the 90th percentile on a qualifying examination is .10. The examiner is interested in x, the number of applicants (of the 25 taking the examination) that score above the 90th percentile. If binomial, $n =$ _____ and $p =$ _____. If not binomial, list characteristic(s) (1, 2, 3, 4, and 5) violated. _____

25; .10
none

3. A sample of 5 transistors will be selected at random from a box of 20 transistors of which 10 are defective. The experimenter will observe the number x of defective transistors appearing in the sample. If binomial, $n =$ _____ and $p =$ _____. If not binomial, list characteristic(s) (1, 2, 3, 4, and 5) violated. _____

not binomial
3, 4

The probability distribution for a binomial random variable can be derived by considering the toss of n coins with the probability of a head given by $P(\text{Head}) = p$, and the probability of a tail $P(\text{Tail}) = 1 - p = q$. However, rather than derive the binomial distribution in general, we present the *binomial probability distribution* together with its *mean, variance,* and *standard deviation* in the following display.

The probability distribution of x, the number of successes in n trials is given by

$$p(x) = C_x^n p^x q^{n-x}$$

where for $x = 0, 1, 2, \ldots, n$; p is the probability of success on a single trial and

$$C_x^n = \frac{n!}{x!(n-x)!}$$

Mean: $\mu = np$
Variance: $\sigma^2 = npq$
Standard Deviation: $\sigma = \sqrt{npq}$

In the formula for $p(x)$, the quantity $p^x q^{n-x}$ represents the probability associated with a simple event having exactly x successes and $(n - x)$ _____. The combinatorial term defined as $n!/x!(n-x)!$ counts the number of simple events with exactly x successes. The term for $p(x)$ is just one of the terms in the series expansion of $(p + q)^n$, a binomial raised to power n, and hence the name: the binomial distribution. The binomial the-

failures

orem, given in Appendix A.5, shows that this form of a probability distribution sums to one and that its terms lie between 0 and 1.

Example 4.5
The president of an agency specializing in public opinion surveys claims that approximately 70% of all people to whom the agency sends questionnaires respond by filling out and returning the questionnaire. Four such questionnaires are sent out. Let x be the number of questionnaires that are filled out and returned. Then x is a binomial random variable with $n =$ _____ (4) and $p =$ _____ (.70).

1. The probability that no questionnaires are filled out and returned is

$$p(0) = C_0^4 (.7)^0 (.3)^4 = \frac{4!}{0!\,4!} (.7)^0 (.3)^4$$

$$= (.3)^4 = \underline{\hspace{1cm}}.$$ (.0081)

2. The probability that exactly three questionnaires are filled out and returned is

$$p(3) = C_3^4 (.7)^3 (.3)^1$$

$$= 4(.343)(.3) = \underline{\hspace{1cm}}.$$ (.4116)

3. The probability that at least three questionnaires are filled out and returned is

$$P(x \geq 3) = p(3) + p(4)$$

$$= p(3) + C_4^4 (.7)^4 (.3)^0$$

$$= .4116 + \underline{\hspace{1cm}}$$ (.2401)

$$= \underline{\hspace{1cm}}$$ (.6517)

Example 4.6
A marketing research survey shows that approximately 80% of the car owners surveyed indicated that their next car purchase would be either a compact or an economy car. Assume the 80% figure is correct, and five prospective buyers are interviewed.
a. Find the probability that all five indicate that their next car purchase would be either a compact or an economy car.
b. Find the probability that at most one indicates that her next purchase will be either a compact or an economy car.

Solution
Let x be the number of car owners who indicate that their next purchase will be a compact or an economy car. Then $n =$ _____ and $p =$ _____ (5; .8) and the distribution for x is given by

$$p(x) = C_x^5 (.8)^x (.2)^{5-x}, \quad x = 0, 1, \ldots, 5.$$

a. The required probability is $p(5)$, which is given by

$$p(5) = C_5^5 (.8)^5 (.2)^0$$

$$= (.8)^5$$

$$= .32768.$$

b. The probability that at most one car owner indicates that her next purchase will be either a compact or an economy car will be

$$P(x \leq 1) = p(0) + p(1).$$

For $x = 0$,

$$p(0) = C_0^5 (.8)^0 (.2)^5$$

$$= (.2)^5$$

$$= \underline{}.$$ .00032

For $x = 1$,

$$p(1) = C_1^5 (.8)^1 (.2)^4$$

$$= 5 (.8) (.0016)$$

$$= \underline{}.$$ .0064

Hence

$$P(x \leq 1) = .0064 + .00032$$

$$= \underline{}.$$ .00672

As you might expect, the calculation of the binomial probabilities becomes quite tiresome as the number of trials increases. Table 1 of binomial probabilities in Appendix III in your text can be used to find binomial probabilities for values of $p = .01, .05, .10, .20, \ldots, .90, .95, .99$ when $n = 2, 3, \ldots,$ 12, 15, 20, 25.

1. The table entries are not the individual terms for binomial probabilities but rather cumulative sums of probabilities, beginning with $x = 0$ up to and including the value $x = a$. By formula, the entries for n, p, and a are

$$\sum_{x=0}^{x=a} p(x) = p(0) + p(1) + \ldots + p(a).$$

2. By using a table entry, which is $\sum_{x=0}^{a} p(x)$, these tables allow the user to find the following:
a. left-tail cumulative sums,

$$P(x \leq a) = \sum_{x=0}^{a} p(x);$$

b. right-tail cumulative sums,

$$P(x \geq a) = 1 - \sum_{x=0}^{a-1} p(x);$$

c. or individual terms such as

$$P(x = a) = \sum_{x=0}^{a} p(x) - \sum_{x=0}^{a-1} p(x).$$

Example 4.7
Refer to Example 4.6. Find the probabilities asked for by using Table 1, Appendix III.

Solution
For this problem, we will use the table for $n = 5$ and $p = .8$.
1. To find the probability that $x = 5$, we proceed as follows:

$$p(5) = [p(0) + p(1) + p(2) + p(3) + p(4) + p(5)]$$
$$- [p(0) + p(1) + p(2) + p(3) + p(4)]$$

$$= \sum_{x=0}^{5} p(x) - \sum_{x=0}^{4} p(x)$$

$$= 1 - .672$$

$$= .328.$$

2. To find the probability that $x \leq 1$, we need

$$P(x \leq 1) = p(0) + p(1)$$

$$= \sum_{x=0}^{1} p(x)$$

$$= \underline{\qquad}.$$

.007

Sect. 4.2: The Binomial Probability Distribution / 89

3. Let us extend the problem and find the probabilities associated with the terms $x = 2$ and $x = 3$.
   For $x = 2$,

   $$P(2) = \sum_{x=0}^{2} p(x) - \sum_{x=0}^{1} p(x)$$

   $$= .058 - .007$$

   $$= \underline{\qquad}.$$ .051

   For $x = 3$,

   $$p(3) = \sum_{x=0}^{3} p(x) - \sum_{x=0}^{2} p(x)$$

   $$= .263 - .058$$

   $$= \underline{\qquad}.$$ .205

4. Complete the following table:

x	$p(x)$	
0	_____	.000
1	_____	.007
2	_____	.051
3	_____	.205
4	_____	.409
5	_____	.328

   with $\sum_{x=0}^{5} p(x) = \underline{\qquad}.$ 1

5. Graph this distribution as a probability histogram.

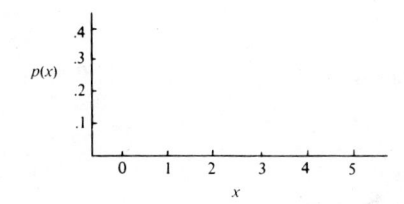

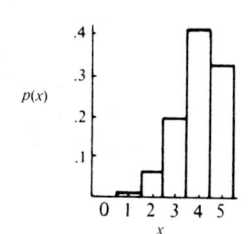

Example 4.8
Using Table 1, Appendix III, find the probability distribution for x if $n = 5$ and $p = 1/2$, and graph the resulting probability histogram.

90 / Chap. 4: Several Useful Discrete Distributions

Solution
1. To find the individual probabilities for $x = 0, 1, 2, \ldots, 5$, we need but subtract successive entries in the table for $n = 5$, $p = .5$.

.031

$$p(0) = \sum_{x=0}^{0} p(x) = \underline{\qquad}$$

.157

$$p(1) = \sum_{x=0}^{1} p(x) - \sum_{x=0}^{0} p(x) = .188 - .031 = \underline{\qquad}$$

.312

$$p(2) = \sum_{x=0}^{2} p(x) - \sum_{x=0}^{1} p(x) = .500 - .188 = \underline{\qquad}$$

.812; .312

$p(3) = \underline{\qquad} - .500 = \underline{\qquad}$

.969; .157

$p(4) = \underline{\qquad} - .812 = \underline{\qquad}$

.969; .031

$p(5) = 1.000 - \underline{\qquad} = \underline{\qquad}$.

2. Using the results of Step 1, we find the probability histogram to be symmetric about the value $x = \underline{\qquad}$.

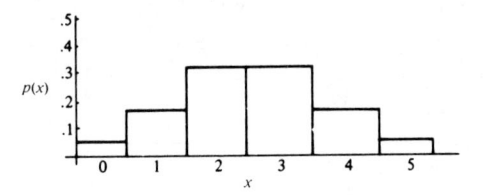

2.5

Example 4.9
Find the probability distribution for x if $n = 5$ and $p = .3$, and graph the probability histogram in this case.

Solution
1. Again, subtracting successive entries for $n = 5$, $p = .3$, we have

.168

$p(0) = \underline{\qquad}$

.360

$p(1) = .528 - .168 = \underline{\qquad}$

.309

$p(2) = .837 - .528 = \underline{\qquad}$

.132

$p(3) = .969 - .837 = \underline{\qquad}$

$p(4) = .998 - .969 = $ _____   .029

$p(5) = 1 - .998 = $ _____   .002

2. Graph the resulting histogram.

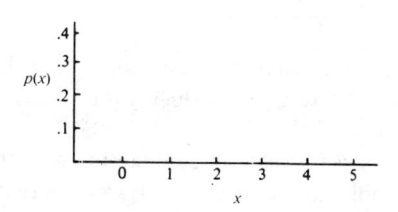

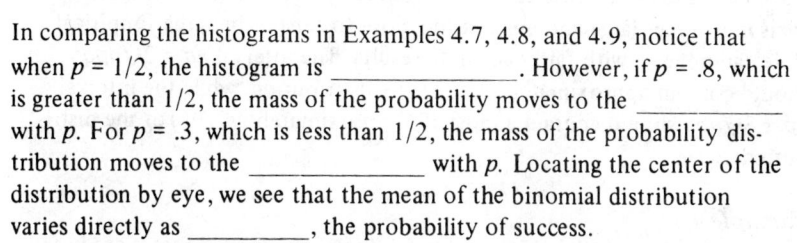

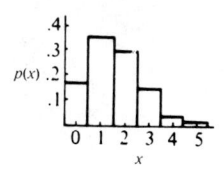

In comparing the histograms in Examples 4.7, 4.8, and 4.9, notice that when $p = 1/2$, the histogram is _____. However, if $p = .8$, which is greater than $1/2$, the mass of the probability moves to the _____ with p. For $p = .3$, which is less than $1/2$, the mass of the probability distribution moves to the _____ with p. Locating the center of the distribution by eye, we see that the mean of the binomial distribution varies directly as _____, the probability of success.

symmetric
right

left

p

Let us consider two more examples. You are now free either to calculate the probabilities by hand or to use the tables when appropriate.

Example 4.10
A preliminary investigation reported that approximately 30% of locally grown poultry were infected with an intestinal parasite which, although not harmful to those consuming the poultry, decreased the usual weight growth rates in the birds and thereby caused a loss in revenue to the growers. A diet supplement believed to be effective against this parasite was added to the birds' rations. During the preparation of poultry that had been fed the supplemental rations for at least two weeks, of 25 birds examined, 3 birds were still found to be infected with the intestinal parasite.

a. If the diet supplement is ineffective, what is the probability of observing 3 or fewer birds infected with the intestinal parasite?

b. If in fact the diet supplement was effective and reduced the infection rate to 10%, what is the probability of observing 3 or fewer infected birds?

Solution
a. With $n = 25$ and $p = .3$, we can use the binomial tables in the text to find $P(x \leq 3)$.

$$P(x \leq 3) = \sum_{x=0}^{3} p(x)$$

$= $ _____ .   .033

.1   b. We can use the same tables with $n = 25$ and $p =$ _____. Hence,

$$P(x \leqslant 3) = \sum_{x=0}^{3} p(x)$$

.764

$$= \underline{\hspace{1cm}}.$$

was

Notice that the sample results are much more probable if the diet supplement (was, was not) effective in reducing the infection rate below 30%.

Tchebysheff's Theorem can be used in conjunction with the distribution of a binomial random variable since *at least* $(1 - 1/k^2)$ of *any* distribution lies within k standard deviations of the mean. However, when the number of trials n becomes large and p is not too close to zero or one, the Empirical Rule can be used with fairly accurate results. The interval $np \pm 2\sqrt{npq}$ should contain approximately 95% of the distribution, while the interval $np \pm 3\sqrt{npq}$ should contain almost all (approximately 99.7%) of the distribution.

Example 4.11
Suppose it is known that 10% of the citizens of the United States are in favor of increased foreign aid. A random sample of 100 United States citizens is questioned on this issue.
  a. Find the mean and standard deviation of x, the number of citizens favoring increased foreign aid.
  b. Within what limits would we expect to find the number favoring increased foreign aid?

Solution
a. With $n = 100$ and $p = .1$,

10      $\mu = np = 100(.1) =$ _____,

9      $\sigma^2 = npq = 100(.1)(.9) =$ _____,

9; 3 $\sigma = \sqrt{npq} = \sqrt{\underline{\hspace{1cm}}} =$ _____.

b. From part a, $\mu = 10$ and $\sigma = 3$. Using two standard deviations we find the interval $\mu \pm 2\sigma$ to be $10 \pm 2(3)$ or 10 ± 6. Since approximately 95% of the distribution lies within this interval, we would expect the number of
4   citizens favoring increased foreign aid to lie between _____ and
16   _____ if, in fact, $p = .1$.

Example 4.12
Each person in a random sample of 64 people was asked to state a preference for candidate A or candidate B. If there is no underlying preference for either candidate, then the probability that an individual chooses candidate A will be
.5 $p =$ _____.

a. What will be the expected number and standard deviation of preferences for candidate A?
b. Within what limits would you expect the number of stated preferences for candidate A to lie?

Solution
Let x be the number of people stating a preference for candidate A. If there really is no preference for either candidate (that is, the voter selects a candidate at random), then x has a binomial distribution with $n = 64$ and $p = $ _____. | .5

a. $\quad \mu = np = 64(.5) = $ _____, | 32

$\sigma^2 = npq = 64(.5)(.5) = $ _____, | 16

$\sigma = \sqrt{npq} = $ _____. | 4

b. From part a, $\mu = 32$ and $\sigma = 4$. Hence, $\mu \pm 2\sigma = 32 \pm 8$. We would expect the number of preferences for candidate A to lie between _____ and _____ if, in fact, $p = .5$. | 24
| 40

Lot Acceptance Sampling
Most manufacturing plants can be thought of as processors that accept raw materials and turn them into finished products. Efficient operation would require that the number of defective items accepted for processing be kept to a minimum and the number of acceptable finished products be kept at a maximum.

These goals can be achieved in different ways. A manufacturer producing television sets would obviously test and adjust *each* set before it leaves the plant but would probably not test each transistor in an incoming lot before accepting the whole shipment. He would probably accept or reject the shipment depending on the number of defective transistors observed in a random sample drawn from that lot. Sometimes the act of testing an item is destructive, so that each item cannot be individually tested. Testing whether a flashbulb produces the required intensity of light obviously destroys the flashbulb.

The process of screening lots is an inferential procedure in which a decision about the proportion defective in a lot (population) is made. The sampling of items from incoming or outgoing lots or the sampling of items from a production line closely approximates the defining characteristics of a binomial experiment. Therefore, the number of defectives in a sample of size n will be distributed as a binomial random variable with parameter p, the proportion of defective items in the population sampled.

A number of sampling schemes are used in industry, the simplest of which is the following:

From the lot select n items at random. Record the number x of defective items found in the sample. If x is less than or equal to a prescribed number a, accept the lot; otherwise reject the lot.

94 / Chap. 4: Several Useful Discrete Distributions

This maximum number a of allowable defectives is called the *acceptance number* for the plan.

The plan above is called a *single sampling plan.* Any such plan is defined by specifying values for the numbers _____ (sample size) and _____ (acceptance number).

$n; a$

How does one decide whether to use plan A ($n = 10, a = 1$), or plan B ($n = 20, a = 2$), or some other plan? One acceptable criterion is that the probability of accepting a good lot should be (high, low) and that the probability of accepting a bad lot should be _____. Thus we might select plan B (with the larger sample size) rather than plan A if good lots have a higher probability of acceptance and bad lots have a lower probability of acceptance when plan B is used. To obtain this comparison, we construct the *operating characteristic curve* for each of these plans. The operating characteristic curve is a graph that shows the probability of acceptance for an incoming lot with fraction defective p. The curve will be shown for values of p ranging from 0 (perfect lot) to 1 (totally defective). If a lot contains no good items ($p = $ _____), then the probability that it will be accepted is _____. If a lot contains no defective items ($p = $ _____), it is certain to be accepted; that is, the probability of acceptance is _____. For intermediate values of p, the operating characteristic curves for two different plans will not, in general, coincide.

high
low

1; 0
0
1

Example 4.13
Construct an operating characteristic curve for the sampling plan $n = 10$, $a = 1$.

Solution
Using this particular plan, we take a sample of size 10 and accept the lot if no more than 1 defective item is found. We assume that the lot is (large, small) enough to justify treating x, the number of defectives found in the sample, as a _____ random variable.

large

binomial

1. Suppose that the lot contains 10% defective items ($p = .1$). The probability of accepting this lot is

$$P(x \leqslant 1) = p(0) + p(1) = C_0^{10}(.1)^0(.9)^{10} + C_1^{10}(.1)^1(.9)^9.$$

It is not necessary to complete this calculation since the result, correct to the nearest thousandth, may be read directly from Table 1, Appendix III, in the text. Thus,

.736

$$C_0^{10}(.1)^0(.9)^{10} + C_1^{10}(.1)^1(.9)^9 = \underline{\qquad}.$$

2. By referring to Table 1, Appendix III, (text), obtain the probabilities of acceptance that are omitted in the following table. Complete the table by filling in the missing entries when $n = 10$ and $a = 1$.

Sect. 4.2 The Binomial Probability Distribution / 95

Fraction Defective, p	Probability of Lot Acceptance
.01	.996
.05	_____ .914
.10	.736
.20	_____ .376
.30	_____ .149
.40	_____ .046
.50	.011

3. A graph may now be constructed showing the probability of acceptance as a function of the fraction defective in the incoming lot. The curve so obtained is called the *operating characteristic curve* for the sampling plan.

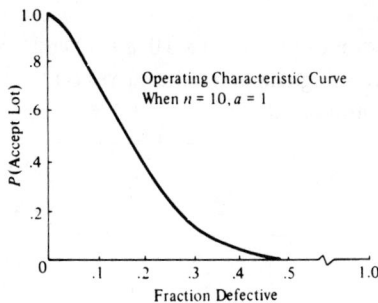

Operating Characteristic Curve When $n = 10$, $a = 1$

Example 4.14
Construct an operating characteristic curve for the plan $n = 20$, $a = 2$.

Solution
The probability of accepting a lot under this plan is the probability of obtaining no more than _____ defective items in a random sample of size 20. 2
Thus the probability of accepting an incoming lot with fraction defective p is

$$P(x \leqslant 2) = \sum_{x=\_\_} C_x^{20} p^x (1-p)^{20-x} \qquad \text{(fill in the summation limits)} \qquad \sum_{x=0}^{2}$$

1. Using Table 1, Appendix III, (text) complete the following table:

Fraction Defective, p	Probability of Lot Acceptance
.05	_____ .925
.10	_____ .677
.20	_____ .206
.30	_____ .035

2. Complete the operating characteristic curve by labeling and scaling the following axes, plotting the points obtained from the table above, and joining the points with a smooth curve.

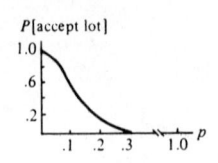

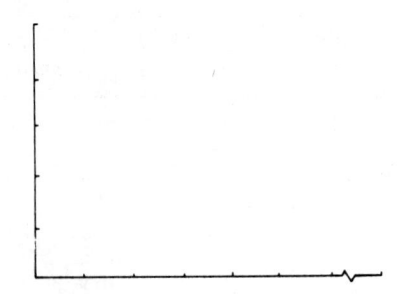

Since an operating characteristic curve falls as one moves to the right, the probability of accepting a lot containing a high fraction defective is (more, less) than the probability of accepting a good lot.

less

Example 4.15
To aid in the comparison of plan A ($n = 10, a = 1$) and plan B ($n = 20, a = 2$), we will show their operating characteristic curves on the same graph, using acceptance values recorded in the following table:

| Fraction Defective, | Probability of Acceptance | |
p	Plan A	Plan B
.00	1.000	1.000
.05	.914	.925
.10	.736	.677
.20	.376	.206
.30	.149	.035
.40	.046	.004
.50	.011	.000

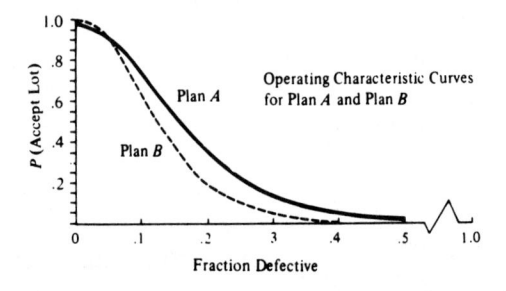

The two curves should cross at about $p = .06$. Thus the probability of accepting a lot with fraction defective less than .06 is (higher, lower) with plan B than with plan A. The probability of accepting a lot with fraction defective more than .06 is _____ with plan B than with plan A. Thus, plan B is more sensitive in discriminating between good and bad lots. The expense of inspecting a larger sample (as in plan B) may be justified by the greater sensitivity of plan B as compared with plan A.

higher

lower

In addition to direct calculation and the use of Table 1 in Appendix III, individual and cumulative binomial probabilities are available in many statistical packages. The following computer output was generated by means of the MINITAB package using the two commands PDF (short for probability density function) and CDF (cumulative distribution function), followed by a semicolon, then the subcommand BINOMIAL N P for which $n = 10$ and $p = .5$ (followed by a period).

```
MTB > PDF;
SUBC > BINOMIAL 10 .5.

       BINOMIAL WITH N =   10   P = 0.500000
            K          P( X  =  K)
            0             0.0010
            1             0.0098
            2             0.0439
            3             0.1172
            4             0.2051
            5             0.2461
            6             0.2051
            7             0.1172
            8             0.0439
            9             0.0098
           10             0.0010


MTB > CDF;
SUBC > BINOMIAL 10 .5.

       BINOMIAL WITH N =   10   P = 0.500000
            K   P( X LESS OR  =  K)
            0             0.0010
            1             0.0107
            2             0.0547
            3             0.1719
            4             0.3770
            5             0.6230
            6             0.8281
            7             0.9453
            8             0.9893
            9             0.9990
           10             1.0000
```

Self-Correcting Exercises 4A

1. A city planner claims that 20% of all apartment dwellers move from their apartments within a year from the time they first moved in. In a particular city, 7 apartment dwellers who had given notice of termination to their landlords are to be interviewed.
   a. If the city planner is correct, what is the probability that 2 of the 7 had lived in the apartment for less than one year?
   b. What is the probability that at least 6 had lived in their apartment for one year or more?
2. Suppose that 70% of the first-class mail from New York to California is delivered within 4 days after being mailed. If 20 pieces of first-class mail are mailed from New York to California:
   a. Find the probability that at least 15 pieces of mail arrive within 4 days of the mailing date.

b. Find the probability that 10 or fewer pieces of mail arrive later than 4 days after the mailing date.

3. In the past history of a certain serious disease it has been found that about 1/2 of its victims recover.
   a. Find the probability that exactly 4 of the next 15 patients suffering from this disease will recover.
   b. Find the probability that at least 4 of the next 15 patients afflicted with this disease will recover.

4. Suppose that 20% of the registered voters in a given city belong to a minority group and that voter registration lists are used in selecting potential jurors. If 80 persons were randomly selected from the voter registration lists as potential jurors, within what limits would you expect the number of minority members on this list to lie?

5. A television network claims that its Wednesday evening prime time program attracts 40% of the television audience. If the 40% figure is correct and if each person in a random sample of 400 television viewers was asked whether he or she had seen the previous show, within what limits would you expect the number of viewers who had seen the previous show to lie? What would you conclude if the interviews revealed that 96 of the 400 had actually seen the previous show?

6. Large lots of portable radios are accepted in accordance with the sampling plan with sample size $n = 4$ and acceptance number $a = 1$.
   a. Complete the following table and construct the operating characteristic curve for this plan. The axes should be properly labeled and scaled.

Fraction defective, p	0	.10	.30	.50	1
Probability of accepting	___	.9477	___	.3125	___

   b. State two essentially different ways in which one might modify the sampling plan above to increase the probability of accepting a lot with fraction defective $p = .10$.

7. To discover the effect on acceptance probabilities of varying the sample size, we study the additional sampling plans ($n = 10, a = 1$) and ($n = 25, a = 1$).
   a. Use Table 1 in the text to complete the following table:

Fraction defective, p	0	.10	.30	.50	1.0
Probability of accepting					
$n = 10, a = 1$	___	.736	___	.011	___
$n = 25, a = 1$	___	___	.002	___	___

   b. Construct the operating characteristic curves for the plan in Exercise 6 and the plans considered in part a.
   c. If the acceptance number is kept the same and the sample size increased, what is the effect on the probability of accepting a given lot?

8. To discover the effect on acceptance probabilities of varying the acceptance number, we study the additional sampling plans ($n = 25, a = 3$) and ($n = 25, a = 5$).
   a. Use Table 1 in the text to complete the following table:

Fraction defective, p	0	.10	.30	.50	1.0
Probability of accepting					
$n = 25, a = 3$	___	___	.033	___	___
$n = 25, a = 5$	___	.967	___	___	___

   b. Construct the operating characteristic curves for the plans considered in part a together with the curve for the plan ($n = 25, a = 1$)(see Exercise 7) on the same set of axes.
   c. If the sample size is kept the same and the acceptance number increased, what is the effect on the probability of accepting a given lot?

4.3 The Poisson Probability Distribution

The Poisson random variable provides a good model for the number of times a specified event occurs in either time or space. The number of weeds in a wheat field, the number of ships entering a harbor on a given day, and the number of telephone calls arriving at a switchboard during a one-minute interval are random variables that can be modeled using the Poisson probability distribution. In these applications, x represents the number of events during a period of time during which an average of μ events can be expected to occur.

Poisson Probability Distribution

$$p(x) = \frac{\mu^x e^{-\mu}}{x!} \qquad x = 0, 1, 2, \ldots$$

where

μ = mean of the random variable x

$\sigma^2 = \mu$

$\sigma = \sqrt{\mu}$

$e = 2.71828 \ldots$ (e is the base of natural logarithms)

The Poisson model is developed using the assumption that the events occur randomly and independently; hence, its use is appropriate when these conditions are met.

Example 4.16

In a food processing and packaging plant, there are, on the average, two packaging machine breakdowns per week. Assuming the weekly machine breakdowns follow a Poisson distribution, what is
1. the probability that there are no machine breakdowns in a given week?
2. the probability that there are no more than two machine breakdowns in a given week?

Solution

Machine breakdowns occur at the average rate of $\mu =$ __2__ breakdowns per week. If the number of breakdowns follows a Poisson distribution, then

$$p(x) = \frac{2^x e^{-2}}{x!} \quad \text{for } x = 0, 1, 2, \ldots$$

1. $P[x = 0] = p(0) = \dfrac{2^0 e^{-2}}{0!}$

   $= \dfrac{e^{-2}}{1}$

   $=$ __.135335__

2. The probability that no more than two machine breakdowns occur in a given week is

$$P[x \leq 2] = p(0) + p(1) + p(2)$$

From part a, we know that $p(0) =$ __.135335__. We need to evaluate $p(1)$ and $p(2)$.

$$p(1) = \frac{2^1 e^{-2}}{1!} = 2(.135335) = \underline{.270670}$$

$$p(2) = \frac{2^2 e^{-2}}{2!} = 2(.135335) = \underline{.270670}$$

Hence,

$$P[x \leq 2] = \underline{.676675}$$

As in the case of the binomial probability distribution, it becomes tedious to calculate individual probabilities by hand. The MINITAB commands CDF and PDF with the subcommand POISSON followed by the appropriate value for μ will generate the individual Poisson probabilities and the cumulative Poisson probabilities, respectively. The MINITAB output generated by the commands PDF; POISSON 2. and CDF; POISSON 2. are shown below.

Sect. 4.3: The Poisson Probability Distribution / 101

```
MTB > PDF;                         MTB > CDF;
SUBC > POISSON 2.                  SUBC > POISSON 2.

  POISSON WITH MEAN = 2.000          POISSON WITH MEAN = 2.000
      K        P(X = K)                K       P(X LESS OR = K)
      0         0.1353                 0            0.1353
      1         0.2707                 1            0.4060
      2         0.2707                 2            0.6767
      3         0.1804                 3            0.8571
      4         0.0902                 4            0.9473
      5         0.0361                 5            0.9834
      6         0.0120                 6            0.9955
      7         0.0034                 7            0.9989
      8         0.0009                 8            0.9998
      9         0.0002                 9            1.0000
     10         0.0000
```

From column _____ (give number), indexing $K = 0, 1$, and 2, we have | 2

$$p(0) = .1353$$

$$p(1) = \text{_____}$$ | .2707

$$p(2) = .2707$$

so that

$$P(x \leq 2) = \text{_____}$$ | .6767

which confirms the results of Example 4.16. Alternatively, from column 4, indexing $K = 2$, we have the result directly:

$$P(x \leq 2) = \text{_____}$$ | .6767

This result could also have been found using Table 2(a) in Appendix III where the Poisson cumulative probabilities, $P(x \leq a) = p(0) + p(1) + \ldots + p(a)$, are given for various values of μ. This table can be used to find right-tailed, left-tailed, or individual Poisson probabilities.

It is important to keep in mind that the Poisson distribution is fixed in time or space. In the last example, the mean number of breakdowns per week was two. The mean number of breakdowns in a three-week period would be 6. The parameter μ in a Poisson distribution is always equal to the *mean* number of rare events observed occurring in a *given unit* of time or space.

The Poisson probability distribution is often used to approximate binomial probabilities in cases where n is _____ and p or q is _____ . | large; small
Generally, the Poisson approximation to binomial probabilities is adequate when the binomial mean, $\mu = np$, is less than or equal to _____ . | seven

Example 4.17
Evidence shows that the probability that a driver will be involved in a serious automobile accident during a given year is .01. A particular corporation employs 100 full-time traveling salesmen. Based upon the above evidence, what is the probability that exactly two of the salesmen will be involved in a serious automobile accident during the coming year?

Solution
This is an example of a binomial experiment with $n =$ __100__ trials and $p =$ __.01__. The exact probability distribution for the number of serious automobile accidents in $n = 100$ trials is

$$p(x) = \frac{100!}{x!(100-x)!}(.01)^x(.99)^{100-x} \quad x = 0, 1, 2, \ldots, 100$$

Since we do not have binomial tables for $n = 100$, we note that the binomial mean $\mu = np = 1$. The Poisson approximation to binomial probabilities can be used in this case with the Poisson mean taken to be $\mu =$ __1__. Therefore,

$$p(2) \approx \frac{(1)^2 e^{-1}}{2!}$$

$$= \frac{.367879}{2}$$

$$= \underline{.1839}$$

Example 4.18
Suppose that past records show that the probability of default on an FHA loan is about .01. If 25 homes in a given area are financed by FHA, use the Poisson approximation to binomial probabilities to find
a. the probability that there will be no defaults among these 25 loans.
b. the probability that there will be two or more defaults.
Compare the values found in parts a and b with the actual binomial probabilities found using Table 1 of Appendix III.

Solution
Although a sample of size $n = 25$ is not usually considered to be large, the value of $p = .01$ is small and $\mu = np = .25$ is less than 7. We will in any case assess the accuracy of the Poisson approximation compared to the actual binomial probabilities. We shall use

$$p(x) = \frac{(.25)^x e^{-.25}}{x!}$$

with $e^{-.25} = .778801$.

1. The probability of $x = 0$ defaults is approximated to be

    $$p(0) \approx \frac{(.25)^0 e^{-.25}}{0!}$$

    = _____ .778801

    The actual probability from Table 1 is _____. .778

2. The probability of two or more defaults can be found by using

    $$P[x \geq 2] = 1 - P[x \leq 1]$$

    = $1 - [p(\underline{\hspace{1cm}}) + p(\underline{\hspace{1cm}})]$ 0; 1

    We need

    $$p(1) \approx \frac{(.25)^1 e^{-.25}}{1!}$$

    = $(.25)(\underline{\hspace{1cm}})$ .778801

    = _____ .194700

    Hence,

    $$P[x \geq 2] \approx 1 - (.778801 + .194700)$$

    = 1 - _____   .973501

    = _____   .026499

    The actual value from Table 1 is

    $P[x \geq 2] = 1 -$ _____   .974

    = _____   .026

Notice that for this problem there is fairly good agreement between the Poisson approximations and the actual binomial probabilities even though n is not large. This is due mainly to the small value of $\mu = np =$ _____.   .25

4.4 The Hypergeometric Probability Distribution

Suppose that we are selecting a sample of n elements from a population containing N elements, some of which are of one type, and the rest of which are of another type. If we designate one type of element as a "success" and the other as "failure", the situation is similar to the binomial experiment described in Section 4.2. However, one of the assumptions required for the application of the binomial probability distribution is that the probability of a success remains _____ from trial to trial. This assumption is violated when-   constant

104 / Chap. 4: Several Useful Discrete Distributions

without — ever the sampling is done (with, **without**) replacement (that is, once an element has been chosen, it cannot be chosen again).

This departure from the conditions required of the ideal binomial experiment is not important when the population is (small, **large**) relative to the _____ size. In such circumstances, the probability p of a success is approximately _____ for each trial or selection.

large
sample
constant

However, if the number of elements in the population is small in relation to the number of elements in the sample, the probability of a success for a given trial is (**dependent on**, independent of) the outcomes of preceding trials. In this case, the number x of successes follows the hypergeometric probability distribution.

dependent on

The probability distribution of a random variable x having the *hypergeometric distribution* is given by the formula

$$p(x) = \frac{C_x^k \, C_{n-x}^{N-k}}{C_n^N}$$

for $x = 0, 1, 2, \ldots,$ _____ if $n < k$ **n**

$x = 0, 1, 2, \ldots,$ _____ if $n \geqslant k$ **k**

where

N = number of elements in the population

k = number of elements in the population that are successes

n = number of elements in the sample which are selected from the population

x = number of successes in the sample

The mean and variance of a hypergeometric random variable is very similar to the mean and variance of a binomial random variable, with $p = k/N$ and $q = (N - k)/N$. The quantity $(N - n)/(N - 1)$ is a correction for the finite population size.

$$\mu = n\left(\frac{k}{N}\right)$$

$$\sigma_2 = n\left(\frac{k}{N}\right)\left(\frac{N-k}{N}\right)\left(\frac{N-n}{N-1}\right)$$

C_b^a defined in Section 3.7 is taken to be zero if $b > a$.

The hypergeometric probability distribution is applicable when one is selecting a sample of elements from a population without _____

replacement

Sect. 4.4: The Hypergeometric Probability Distribution / 105

and one records whether or not each element does or does not possess a certain characteristic.

Example 4.19
An auditor is checking the records of an accountant who is responsible for ten clients. The accounts of two of the clients contain major errors, and the accountant will fail the inspection if the auditor finds even a single erroneous account. What is the probability that the accountant will fail the inspection if the auditor inspects the records of three clients chosen at random?

Solution
Let x be the number of erroneous accounts found in the (population, sample). Then | sample

$N = $ _____ | 10

$k = $ _____ | 2

$N - k = $ _____ | 8

$n = $ _____ | 3

The accountant will fail the inspection if $x = $ _____ or _____. | 1; 2
So P(accountant fails) $= P(x \geq $ _____$) = p($_____$) + p($_____$)$ | 1; 1; 2

$$P(x = 1) = \frac{\left(\frac{2!}{1!\,1!}\right)\left(\right)}{\left(\right)}$$ | $\dfrac{8!}{2!\,6!}$
 | $\dfrac{10!}{3!\,7!}$

$= $ _____ | .467

$$P(x = 2) = \frac{\left(\right)\left(\right)}{\left(\dfrac{10!}{3!\,7!}\right)}$$ | $\dfrac{2!}{2!\,0!}$; $\dfrac{8!}{1!\,7!}$

$= $ _____ | .067

Therefore, the probability that the accountant will fail the inspection is _____ + _____ = _____. | .467; .067; .534

Self-Correcting Exercises 4B

1. The probability of a serious fire during a given year to any one house in a particular city is believed to be .005. A particular insurance company holds fire insurance policies on 1000 homes in this city.

a. Find the probability that the company will not have any serious fire damage claims by the owners of these homes during the next year.
b. Find the probability they will have no more than three claims.

2. In a certain manufacturing plant, wood-grain printed 4' × 8' wall board panels are mass produced and packaged in lots of 100. Past evidence indicates that the number of damaged or imperfect panels per bundle follows a Poisson distribution with mean $\mu = 2$.
   a. Find the probability that there are exactly three damaged or imperfect panels in a bundle of 100.
   b. Find the probability that there are at least two damaged or imperfect panels in a bundle of 100.

3. A home improvement store has purchased two bundles (2 bundles of 100 each) of panels from the manufacturer described in Exercise 2. Find the probability his lot contains no more than four damaged or imperfect panels.

4. The board of directors of a company has voted to create an employee council for the purpose of handling employee complaints. The council will consist of the company president, the vice-president for personnel, and four employee representatives. The four employees will be randomly selected from a list of 15 volunteers. This list consists of nine men and six women.
   a. What is the probability that two or more men will be selected from the list of volunteers?
   b. What is the probability that exactly three women will be selected from the list of volunteers?

5. A bin of 50 parts contains three defective units. A sample of five units is drawn randomly from the bin. What is the probability that no defective units will be selected?

4.5 Other Discrete Distributions

The Discrete Uniform Distribution

Many experiments involving equally likely outcomes are similar to tossing a fair die and observing x, the number of dots on the upper face; $N = 6$ outcomes were possible, each with probability $1/N = 1/6$. For example, in random sampling from a population of N items numbered 1 to N, the probability of selecting one numbered item from this population is $1/N$. In both cases, the resulting probability distribution for the random variable with values $x = 1, 2, \ldots, N$ is called the discrete uniform distribution, because the probability histogram has a constant or uniform height.

Discrete Uniform Distribution

$$p(x) = 1/N \qquad x = 1, 2, \ldots, N$$

$$\mu = \frac{N+1}{2} \qquad \sigma = \sqrt{\frac{N^2 - 1}{12}}$$

Example 4.20
Suppose that a number is randomly drawn from the sample 1 to 100. Let the number drawn be designated as x.

a. Write down the probability distribution for x.
b. Find the mean and standard deviation of x.
c. What proportion of the values of x lie beyond one standard deviation of the mean?

Solution
a. If a number is randomly drawn from the sample 1 to 100, then x has a uniform distribution with $N = 100$ and

$$p(x) = 1/100 \qquad x = 1, 2, \ldots, 100.$$

b. The mean of a uniform random variable is given by

$$\mu = \frac{N + 1}{2}$$

$$= \frac{100 + 1}{2}$$

$$= \underline{} \qquad\qquad 50.5$$

The standard deviation of a uniform random variable is given by

$$\sigma = \sqrt{\frac{N^2 - 1}{12}}$$

$$= \sqrt{\frac{100^2 - 1}{12}}$$

$$= \sqrt{833.25}$$

$$= \underline{} \qquad\qquad 28.9$$

c. We need to find the endpoints of the interval $\mu \pm \sigma$, given by 50.5 ± 28.9, or from 21.6 to 79.4. Therefore, the numbers 1 to 21 and 80 to 100 are outside this interval. Thus, $2(21)/100 = 42\%$ of the values of x lie further than one standard deviation from the mean.

Example 4.21
Consider a game that costs c cents to play. You select a numbered chip at random from a box containing chips numbered 1 to 100. Your prize is the amount in cents of the number on the chip. What should the cost of playing be so that the game is fair; that is, that $E(\text{prize}) = 0$?

Solution
The distribution of x, the number on the chip drawn, is uniform over the numbers 1 to 100. Therefore, the amount of prize money has the same distribution. If the game is to be fair, the cost to play should be set equal to E(prize). But

$$E(x) = \frac{N+1}{2}$$

$$= \frac{100+1}{2}$$

50.5

$$= \underline{}$$

51

Therefore, the cost of the game should be 50.5, or _____ cents per play.

The Geometric Distribution
Another discrete distribution arises as the waiting time for the first success in a series of independent binomial trials, with the probability of success equal to p. The probability that the first success is at trial x, for $x = 1, 2, 3, \ldots$ is

$$P(FFF..FS) = P(F)P(F) \ldots P(F)P(S)$$

$$= [P(F)]^{x-1} P(S)$$

$$= q^{x-1} p.$$

This is summarized in the next display.

The Geometric Distribution

$$p(x) = q^{x-1} p \qquad x = 1, 2, 3, \ldots$$

with

$$\mu = \frac{1}{p} \qquad \sigma = \sqrt{\frac{q}{p^2}}$$

Example 4.22
A fair coin is tossed until the first head is observed. Let x be the number of tosses until the first head is observed.

a. Find the distribution of the random variable x.
b. What is the average number of tosses required until the first head is observed?
c. Find the standard deviation of x.
d. According to Tchebysheff's Theorem, what proportion of tosses until the first head is observed would result in 10 or fewer tosses?

Solution

a. If a fair coin is tossed, then $p = .5$. Therefore,

$$p(x) = (.5)^{x-1}(.5)$$

$$= .5^x$$

for $x = 1, 2, 3, \ldots$

b. Since x has a geometric distribution

$$\mu = 1/p$$

$$= 1/\underline{} \qquad\qquad .5$$

$$= \underline{} \qquad\qquad 2$$

c. The standard deviation of x is given by

$$\sigma = \sqrt{p/q^2}$$

$$= \sqrt{.5/.5^2}$$

$$= \sqrt{2} \quad \text{or} \quad \underline{} \qquad\qquad 1.414$$

d. According to Tchebysheff's Theorem, at least $(1 - 1/k^2)$ of the values of x will lie within the interval $\mu \pm k\sigma$. Therefore, we need to solve for k if the upper bound of the interval $\mu \pm k\sigma$ is equal to 10. Therefore,

$$\mu + k\sigma = 10$$

$$2 + k\sqrt{2} = 10$$

$$k\sqrt{2} = 8$$

$$k = 8/\sqrt{2}$$

$$k = \underline{} \qquad\qquad 5.66$$

Therefore, at least $1 - 1/5.66^2 = 1 - .03 = .97$, or 97% of the trials would not go beyond \underline{} tosses.   10

Self-Correcting Exercises 4C

1. Nonparametric techniques often replace the values of the n observations in a sample of size n by their ranks; i.e., the smallest observation is replaced by 1, the next smallest observation by 2, and so on, with the last observation replaced by n.

a. What is the distribution of r, the rank of an observation drawn at random from the sample of n observations?
b. What is the mean and standard deviation of the random variable r?
c. If $n = 10$, within what bounds would at least 75% of the values of r lie?

2. If x is the number of dots on the upper face in a random throw of a die, use the formulas for μ and σ to verify that $E(x) = 3.5$ and $\sigma = 1.71$.

3. A salesperson has a 20% success rate in selling a new vehicle to potential buyers visiting the new vehicle showroom and lot. Suppose that the sale of a vehicle to one buyer is independent of a sale to another buyer.
   a. What is the distribution of x, the number of potential buyers this salesperson serves until her first sale?
   b. On the average, how many potential buyers would she serve until she makes her first sale during a given week?
   c. At least 75% of the time, her first sale would occur on or before serving how many potential buyers?

EXERCISES

1. A subject is taught to do a task in two different ways. Studies have shown that when subjected to mental strain and asked to perform the task, the subject most often reverts to the method first learned, regardless of whether it was easier or more difficult than the second. If the probability that a subject returns to the first method learned is .8 and six subjects are tested, what is the probability that at least five of the subjects revert to their first learned method when asked to perform their task under mental strain?

2. The taste test for PTC (phenylthiourea) is a favorite exercise for every human genetics class. It has been established that a single gene determines the characteristic, and 70% of the American population are "tasters," while 30% are "nontasters." Suppose 20 people are randomly chosen and administered the test.
   a. Give the probability distribution of x, the number of "nontasters" out of the 20 chosen.
   b. Using appropriate tables, find $P(x \leq 7)$.
   c. Find $P(3 < x \leq 8)$.

3. A multiple-choice test offers four alternative answers to each of 100 questions. In every case there is but one correct answer. Bill responded correctly to each of the first 76 questions when he noted that just 20 seconds remained in the test period. He quickly checked an answer at random for each of the remaining 24 questions without reading them.
   a. What is Bill's expected number of correct answers?
   b. If the instructor assigns a grade by taking 1/3 of the wrong from the number marked correctly, what is Bill's expected grade?

4. On a certain university campus a student is fined $10 for the first parking violation of the academic year. The fine is doubled for each subsequent offense, so that the second violation costs $20, the third $40, and so on. The probability that a parking violation on a given day is detected is .10. Suppose that a certain student will park illegally on each of 20 days during a given academic year.

a. What is the probability he will not be fined?
b. What is the probability that his fines will total no more than $150?

5. Four experiments are described below. Identify which of these might reasonably be treated as a binomial experiment. If a given experiment is clearly not binomial, state what feature disqualifies it. If it is a binomial experiment, write down the probability function for x.
   a. Five percent of the stamps in a large collection are extremely valuable. The stamps are withdrawn one at a time until ten extremely valuable stamps are located. The observed random variable is x, the total number of stamps withdrawn.
   b. There are 15 students in a particular class. The names of these students are written on tags placed in a box. Periodically, a tag is drawn at random from the box and the student with that name is asked to recite. The tag is returned to the box and the proceedings continued. Let x denote the number of times a particular student will be called upon to recite when the teacher draws from the box five times.
   c. This example is conducted in the manner prescribed for part b except that a tag drawn from the box is not returned. Let x denote the number of times the particular student will be called upon to recite when the teacher draws from the box five times.
   d. Sixty percent of the homes in a given country carry fire insurance. A sample of five homes is drawn at random from this country. Let x denote the number of insured homes among the five selected.

6. Shipments of refrigerators are accepted in accordance with the sampling plan ($n = 2, a = 0$).
   a. Find the probability of accepting a lot with fraction defective $p = .01$.
   b. Find the probability of accepting a lot with fraction defective $p = .20$.

7. Refer to Exercise 6.
   a. Find a sampling plan with sample size $n = 20$ that has approximately the same probability of accepting a lot with fraction defective $p = .01$ as the plan ($n = 2, a = 0$).
   b. What is the probability under the plan determined in part a of accepting a lot that has fraction defective $p = .20$?
   c. If you were purchasing refrigerators by the lot, what advantage would there be in using the plan determined in part a rather than the plan ($n = 2, a = 0$)? What disadvantages can you cite for using the plan with sample size $n = 20$?

8. Eastern University has found that about 90% of its accepted applicants for enrollment in the freshman class will actually take a place in that class. In 1991, 1360 applicants to Eastern were accepted. Within what limits would you expect to find the size of the freshman class at Eastern in the fall of 1991?

9. Suppose that the national unemployment rate is 7.1%. A sample of $n = 100$ persons is taken in the Los Angeles area and the number of unemployed persons is recorded.
   a. If the unemployment rate for the Los Angeles area is the same as the national rate, within what limits would you expect the number of unemployed to fall?

b. If 15 of the 100 persons interviewed said that they were unemployed, what would you conclude about the unemployment rate in the Los Angeles area?
10. Suppose that 1 out of 10 homeowners in the state of California have invested in earthquake insurance. If 15 homeowners are randomly chosen to be interviewed,
    a. What is the probability that at least 1 has earthquake insurance?
    b. What is the probability that 4 or more have earthquake insurance?
    c. Within what limits would you expect the number of homeowners insured against earthquakes to fall?
11. Consider 10 management trainees in a firm's rotation program where three of the 10 are members of minority groups. If five of the trainees are randomly assigned to the marketing division, what is the probability that there will be three minority trainees in the group assigned to marketing?
12. Improperly wired control panels were mistakenly installed on two of eight large automated machine tools. It is uncertain which of the machine tools have the defective panels, and a sample of four tools is randomly chosen for inspection.
    a. What is the probability that the sample will include no defective panels? Both defective panels?
    b. Find the binomial approximations for part a. (Optional)
13. A delicatessen has found that the weekly demand for caviar follows a Poisson distribution with a mean of four tins (each tin contains 8 ounces of caviar).
    a. Find the probability that no more than 4 tins are requested during a given week.
    b. As caviar spoils with time, it must be replenished weekly by the delicatessen's owner. How many tins should he buy if it is desired that the probability not exceed .10 that demand cannot be met during a given week?
14. A manufacturer of small mini-computer has found that the average number of service calls per computer each year is 2.2. Assume the number of service calls follows a Poisson distribution.
    a. Find the probability that a particular mini-computer requires no service during a given year.
    b. A small firm has purchased two mini-computers from the manufacturer. Find the probability that the firm requires no service calls during a given year.
    c. Find the probability that the firm requires exactly two service calls during a given year.
15. Refer to Exercise 14. Suppose service calls cost the computer manufacturer an average of $20 each.
    a. What is the expected cost per computer each year?
    b. If the manufacturer has sold 50 mini-computers in the Seattle area, what is the expected annual cost of service in this area?
16. Customers arrive at a certain gasoline filling station at the average rate of one every ten minutes. Assume the arrivals follow a Poisson distribution. The station has only one attendant and he takes an average of

five minutes to service each arrival. What is the probability that two customers arrive while the attendant is servicing an earlier arrival?

17. Suppose the probability that a 24-hour service station has at least 25 customers in a given hour is 0.6.
    a. If x is the number of hours that pass before the service station has at least 25 customers in a given hour, what is the probability distribution of x?
    b. On the average, how long (in hours) will it be before observing a one-hour time period in which 25 or more customers visit the service station?

18. When two fair coins are tossed, each of the four simple events has a probability of 1/4 of occurring. Let x be the number of heads in the toss of the two fair coins, and let y be the number of tosses of these two coins until a head appears on both coins in the toss.
    a. Is the distribution of x the same as the distribution of y?
    b. What is the distribution of the random variable x?
    c. Find and interpret the value of $E(x)$.
    d. What is the distribution of the random variable y?
    e. Find and interpret the value of $E(y)$.
    f. Is it possible for the same basic experiment to give rise to two completely different random variables?

CHAPTER 5
THE NORMAL AND OTHER CONTINUOUS DISTRIBUTIONS

5.1 Probability Distributions for Continuous Random Variables

Not all experiments have sample spaces containing a countable number of simple events. *Continuous random variables,* such as heights, weights, response times, and waiting times, can assume the infinitely many values corresponding to points on a line interval. Since the mathematical treatment of continuous random variables requires the use of calculus, we will merely state some basic concepts. The probability distribution for a continuous random variable can be thought of as the limiting histogram for a very large set of measurements utilizing the smallest possible interval width. In such a case, the outline of the histogram would appear as a continuous curve.

Let us illustrate what happens if we begin with a histogram and allow the interval width to get smaller and smaller while the number of measurements gets larger and larger.

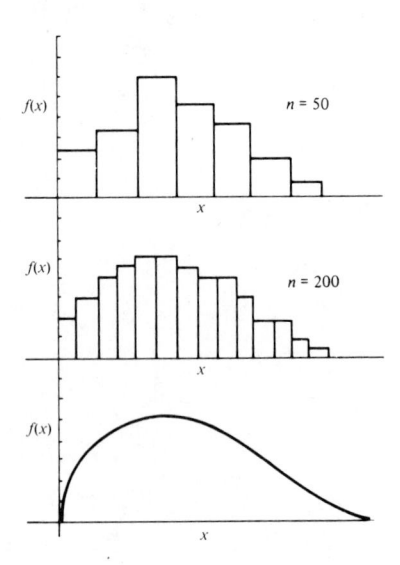

The mathematical function $f(x)$ that traces this curve with varying values of x is called the _____ distribution or the probability density for the random variable x. In the same way that the area under a relative frequency

probability

histogram is _____, the area under the curve $f(x)$ is also equal to _____. The area under the curve between two points, say a and b, represents the _____ that the random variable x will fall into the interval from a to b.

one
one
probability

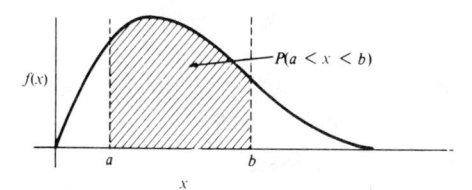

When choosing a model to describe the population of measurements of interest, we must choose $f(x)$ appropriate to our data. Any inferences we may make will only be as valid as the model we are using. It is therefore very important to know as much as possible about the phenomenon under study that will give rise to the measurements that we record.

The idea of modeling the responses that we will record for an experiment may seem strange to you, but you have probably seen models before, though in a different context. For example, when a physicist says, "The distance (s) traversed by a free-falling body is equal to one-half the force of gravity (g) multiplied by the time (t) squared," and writes $s = gt^2/2$, he is merely _____ a physical phenomenon with a mathematical formula. These mathematical models merely provide _____ to reality which further need to be verified by experimental techniques.

modeling
approximations

The Uniform Distribution

The uniform probability distribution is used to model the behavior of a random variable whose values are evenly or uniformly distributed over a given interval. For example, the errors introduced by rounding observations to the nearest inch would probably have a uniform distribution over the interval from -.5 to .5 inch. The time required by lumbering trucks to travel 30 miles from a harvesting area to a lumber mill can be taken to be uniformly distributed over the interval from 30 to 40 minutes.

If x is a uniform random variable over the interval from a to b, its probability density function and its mean and standard deviation are given in the next display.

The Uniform Distribution

$$f(x) = \frac{1}{b-a} \quad a \leq x \leq b$$

$$\mu = (a+b)/2$$

$$\sigma = (b-a)/\sqrt{12}$$

Notice that the mean is the midpoint of the interval and the standard deviation is directly proportional to the length of the interval.

Probabilities associated with a uniform random variable correspond to the rectangular areas under the density function and are found as

$$\text{(length of desired interval)} \times \text{(height given as } 1/(b - a))$$

The uniform distribution is a continuous distribution; hence, $P(c \leq x \leq d)$, $P(c \leq x < d)$, $P(c < x \leq d)$, and $P(c < x < d)$ are all equal, since $P(x = c) = P(x = d) = 0$.

Example 5.1

If rounding errors have a uniform distribution over the interval -.5 to .5,
a. Find the probability that the rounding error is less than .2. Find the probability that the rounding error is less than .2 in magnitude.
b. Find the mean and standard deviation of the rounding error.
c. Within what limits would you expect at least 75% of the possible rounding errors to lie?
d. What is the actual proportion of rounding errors within the limits found in part e?

Solution

Let x represent the rounding error over the interval from -.5 to .5, with probability density function

$$f(x) = 1 \quad \text{for} \quad -.5 < x < .5$$

Note that $a = -.5$, $b = .5$, $b - a = 1$ and therefore, $1/(b - a) = 1$.

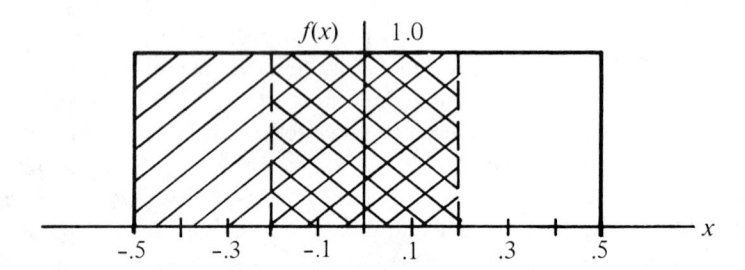

a. The probability that the rounding error is less than .2 corresponds to the area under the interval between $x = -.5$ and $x = .2$. Since the height of $f(x)$ is one,

$$P[x < .2] = [.2 - (-.5)] \times 1 = .7$$

while

$$P[-.2 < x < .2] = [.2 - (-.2)] \times 1 = .4$$

b. With $a = -.5$ and $b = .5$,

$$\mu = (a+b)/2 = [.5 + (-.5)]/2 = 0$$

and

$$\sigma = (a-b)/\sqrt{12} = [.5 - (-.5)]/\sqrt{12}$$
$$= 1/\sqrt{12} = .2887$$

c. From Tchebysheff's Theorem, we know that the interval $\mu \pm 2\sigma$ contains at least 75% of the observations. Therefore the interval

$$0 \pm 2(.2887)$$

$0 \pm$ _____ , or   .5774

from $-.5774$ to $.5774$ contains *at least* _____ of the rounding errors.   75%

d. The actual proportion of rounding errors between $-.5774$ and $.5774$ is equal to the area under the curve between these endpoints. Since these endpoints extend beyond the interval $-.5$ to $.5$, all or 100% of the distribution lies within two standard deviations of the mean.

The Exponential Distribution

The exponential probability distribution is a continuous probability distribution used to model random variables that are waiting times, or random variables that are lifetimes associated with electronic components. The exponential density function, together with the mean and standard deviation of an exponential random variable, is given in the next display.

The Exponential Distribution

$$f(x) = \lambda e^{-\lambda x} \qquad x \geq 0; \lambda \geq 0$$

$$\mu = 1/\lambda$$

$$\sigma = 1/\lambda$$

Notice that for the exponential distribution, $\mu = \sigma$. If the number of events occurring in a given unit of time follows a Poisson distribution with an average of μ events per unit time, then the waiting time for the next event follows an exponential distribution with mean waiting time of $\mu = 1/\lambda$.

Evaluating probabilities associated with an exponential random variable can be simplified by using the following result for *right-tailed* probabilities:

$P[x > a] = e^{-\lambda a}$ for $a > 0$. Once the values of λ and a are known, $e^{-\lambda a}$ can be found from Table 2(b) in Appendix III or by using a calculator that has an exponential function key.

Example 5.2
Suppose that the waiting time at a grocery chain checkout counter follows an exponential distribution with an average waiting time of 10 minutes.
a. What is the probability that you wait longer than 12 minutes at a checkout counter?
b. What is the probability that you wait longer than 15 minutes?
c. With probability .05 you will wait longer than how many minutes at a checkout counter?

Solution
If x, the waiting time at a grocery chain checkout counter follows an exponential distribution with mean 10 minutes, then $\lambda = 1/10$ and

$$P[x > a] = e^{-(a/10)}$$

a. To evaluate

$$P[x > 12] = e^{-(12/10)} = e^{-1.2}$$

from Table 2(b), we see that $e^{-1.2} = .301194$. Therefore,

$$P[x > 12] = .3012$$

b. Similarly,

$$P[x > 15] = e^{-(15/10)} = e^{-1.5} = .223130$$

c. Since we wish to find a value of x, say x_0, with the property that

$$P[x > x_0] = .05,$$

we see that x_0 is the 95th percentile of the distribution. Hence, we need to solve

$$e^{-x_0/10} = .05$$

by using Table 2(b) or by solving the equation algebraically. From Table 2(b), we see that for $a = 2.95$, $e^{-2.95} = .052340$ while for $a = 3.00$, $e^{-3.00} = .049787$. Using linear interpolation, we have

a	$P[x > a]$
2.95	.052340
$x_0/10$	.05
3.00	.049787

.002340

.002553

so that
$$x_0/10 = 2.95 + (.002340/.002553)(.05)$$
$$= 2.9958$$
and $x_0 = 29.96$

Solving the equation directly, we have
$$e^{-x_0/10} = .05$$
$$\ln(e^{-x_0/10}) = \ln(.05)$$
$$-x_0/10 = -2.995732$$
or $x_0 = 29.96$

which agrees with our earlier answer found using Table 2(b).

Self-Correcting Exercises 5A

1. When we round numbers to the nearest tenth, the rounding error is taken to be uniformly distributed between -.05 and .05.
   a. Find the probability that the rounding error will be greater than .025.
   b. Find the probability that the rounding error will be greater than .025 in magnitude.
   c. What is the mean of these rounding errors?
   d. Find the probability that the rounding error lies in the interval $\mu \pm \sigma$.

2. The length of useful life of an electronic component has an exponential distribution with a mean of 8 years.
   a. What is the probability that the component fails before 8 years?
   b. If five percent of these components last longer than x_0, what is the value of x_0?
   c. What is the median lifetime of these electronic components?

3. If calls arriving at a switchboard have a Poisson distribution with an average of 6 calls per minute, then the waiting time between calls has an exponential distribution with a mean waiting time of 10 seconds.
   a. Find the probability that the waiting time between calls exceeds fifteen seconds.
   b. Find the probability that the waiting time is less than 5 seconds or greater than 15 seconds.
   c. Ninety percent of the time, the waiting time between calls is less than x_0. Find the value of x_0.

4. A library book is put on reserve in the library with a loan period of two hours. The time that the book is used by a borrower is uniformly distributed over the interval from 0 to 2 hours.
   a. What is the mean and standard deviation of the time the book is used by a borrower?
   b. What is the probability that a borrower keeps the book at most 1.5 hours?
   c. What proportion of borrowers keep the book longer than 15 minutes, but not longer than the two hour limit?
   d. Ninety-five percent of all borrowers keep the book at least how long?

5.2 The Normal Probability Distribution

Many types of continuous curves are available as models for continuous random variables. However, many continuous random variables that are biological measurements, such as heights and weights or reaction and response times, or measurement errors themselves have mound-shaped distributions that are also bell-shaped. These and other continuous random variables can be modeled by a normal probability distribution whose density function produces the following bell-shaped curve.

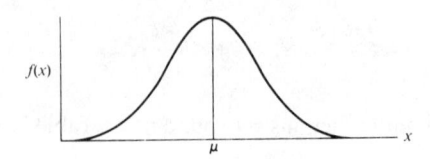

The Normal Probability Distribution

$$f(x) = \frac{1}{\sigma\sqrt{2\pi}} e^{-\frac{1}{2}\left(\frac{x-\mu}{\sigma}\right)^2}, \quad -\infty < x < \infty$$

The symbols used in the function $f(x)$ are defined as follows:
1. π and e are irrational numbers whose approximate values are 3.1416 and 2.7183, respectively.
2. μ and σ are constants which represent the population _____ and _____ _____, respectively.

mean
standard deviation

Encountering a random variable whose values can be extremely small (a large negative value) or extremely large might at first be disconcerting to the student who has heard that heights, weights, response times, and errors of measurements are approximately normally distributed. Surely we do not have heights, weights, or times that are less than zero! Certainly not, but almost all of the distribution of a normally distributed random variable lies within the interval $\mu \pm 3\sigma$. In the case of heights or weights, this interval almost always encompasses positive values. Keep in mind this curve is merely a *model* that approximates an actual distribution of measurements. Its great utility lies in the fact that it *can* be used effectively as a model for so many types of measurements.

The Standard Normal Probability Distribution

Probability is the vehicle through which we are able to make inferences about a population in the form of either estimation or decisions. To make inferences about a normal population, we must be able to compute or otherwise find the probabilities associated with a normal random variable. As explained in Section 5.1, the probability that the normal random variable x lies between two points a and b is equivalent to the area under the normal curve between a and b. However, since the probability distribution for a normal random

variable x depends on the population parameters _____ and _____ . | $\mu; \sigma$
we would be required to recalculate the probabilities associated with x each
time a new value for μ or σ was encountered. We resort to a standardization
process whereby we convert a normal random variable x to a _____ | standard
normal random variable z, which represents the distance of x from its mean μ
in units of the standard deviation σ. To standardize a normal random variable
x, we use the following procedure:
1. From x, subtract its mean μ:

$$x - \mu.$$

This results in the signed distance of x from its mean, a negative sign
indicating that x is to the left of μ while a positive sign indicates x is to the
_____ of μ. | right
2. Now divide by σ:

$$\frac{x - \mu}{\sigma}.$$

Dividing by σ converts the signed distance from the mean to the number of
standard deviations to the right or left of μ.
3. Define

$$z = \frac{x - \mu}{\sigma}.$$

to be the standard normal random variable having the standardized normal distribution with mean 0 and standard deviation 1.

Given the curve representing the distribution of a continuous random variable
x, the probability that $a \leq x \leq b$ is represented by the _____ | area
under the curve between the points _____ and _____ . Hence in | $a; b$
finding probabilities associated with a standardized normal variable z, we
could refer directly to the areas under the curve. These areas are tabulated for
your convenience in Table 3 in Appendix III of the text.

Since the standardized normal distribution is _____ about the | symmetric
mean 0, half of the area lies to the left of 0 and half to the right of 0. Further,
the areas to the left of the mean $z = 0$ can be calculated by using the corresponding and equal area to the right of $z = 0$. Hence Table 3 exhibits areas
only for positive values of z correct to the nearest hundredth. Table 3 gives
the area between $z = 0$ and a specified value of z, say z_0. A convenient notation
used to designate the area between $z = 0$ and z_0 is $A(z_0)$.

For a given value of z, say z_0, this is the area $A(z_0)$, tabulated in Table 3.

a. For $z = 1$, the area between $z = 0$ and $z = 1$ is $A(z = 1) = A(1) = .3413$.
b. For $z = 2$, $A(z = 2) = A(2) = $ _____ . | .4772

122 / Chap. 5: The Normal and Other Continuous Distributions

.4452
.4918

c. For $z = 1.6$, $A(1.6) =$ _____.
d. For $z = 2.4$, $A(2.4) =$ _____.
Now try reading the table for values of z given to two decimal places.

.4951
.2734
.4545
.4979

e. For $z = 2.58$, $A(2.58) =$ _____.
f. For $z = .75$, $A(.75) =$ _____.
g. For $z = 1.69$, $A(1.69) =$ _____.
h. For $z = 2.87$, $A(2.87) =$ _____.

We will now find probabilities associated with the standard normal random variable z by using Table 3.

Example 5.3
Find the probability that z is greater than 1.86, that is, $P(z > 1.86)$.

Solution
Illustrate the problem with a diagram as follows:

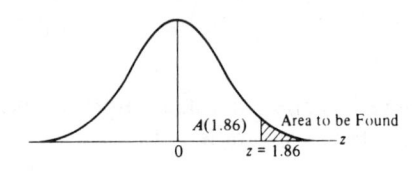

1. The total area to the right of $z = 0$ is equal to .5000.

.4686
.5000

2. From Table 3, $A(1.86) =$ _____.
3. Therefore, the shaded area is found by subtracting $A(1.86)$ from _____.
4. Hence

$$P(z > 1.86) = .5000 - A(1.86)$$

.4686

$$= .5000 - \underline{\qquad}$$

.0314

$$= \underline{\qquad}.$$

Example 5.4
Find $P(z < -2.22)$.

Solution
Illustrate the problem with a diagram.

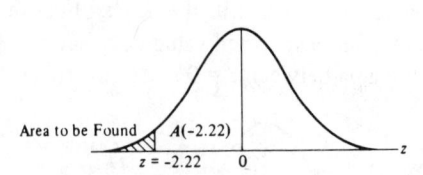

1. Using the symmetry of the normal distribution, $A(-2.22) = A(2.22)$

.4868
left

= _____. The negative value of z indicates that you are to the (left, right) of the mean, $z = 0$.

.4868; .0132

2. $\quad P(z < -2.22) = .5000 - \underline{\qquad} = \underline{\qquad}.$

Example 5.5
Find $P(-1.21 < z < 2.43)$.

Solution
Illustrate the problem with a diagram.

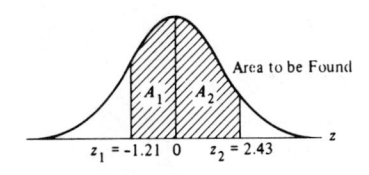

$P(-1.21 < z < 2.43) = P(-1.21 < z < 0) + P(0 < z < 2.43)$

= _____ + _____   .3869; .4925

= _____ .   .8794

A second type of problem that arises is that of finding a value of z, say z_0, such that a probability statement about z will be true. We explore this type of problem with examples.

Example 5.6
Find the value of z_0 such that

$$P(0 < z < z_0) = .3925.$$

Solution
Once again, illustrate the problem with a diagram and list the pertinent information.

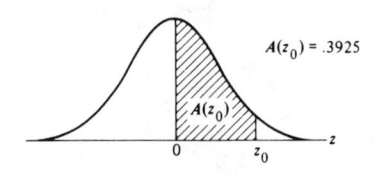

1. Search Table 3 until the area .3925 is found. The value such that

   $A(z_0) = .3925$ is $z_0 =$ _____   1.24

2. $P(0 < z <$ _____ $) = .3925$.   1.24

Example 5.7
Find the value of z_0 such that $P(z > z_0) = .2643$.

Solution
Illustrate the problem and list the pertinent information.

124 / Chap. 5: The Normal and Other Continuous Distributions

[Figure: Normal curve with shaded area to the right of z_0 labeled $A(z_0)$; Shaded Area Equals .2643]

.2357

.2357; .63

.63

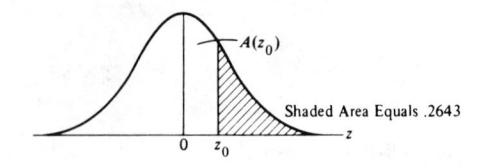

1. $A(z_0) = .5000 - .2643 = $ _____.

2. The value of z_0 such that

    $A(z_0) = $ _____ is $z_0 = $ _____.

3. $P(z > $ _____ $) = .2643$.

Self-Correcting Exercises 5B

1. Find the following probabilities associated with the standard normal random variable z:

    a. $P(z > 2.1)$.   d. $P(-2.75 < z < -1.70)$.

    b. $P(z < -1.2)$.   e. $P(-1.96 < z < 1.96)$.

    c. $P(.5 < z < 1.5)$.   f. $P(z > 1.645)$.

2. Find the value of z, say z_0, such that the following probability statements are true:

    a. $P(z > z_0) = .10$.

    b. $P(z < z_0) = .01$.

    c. $P(-z_0 < z < z_0) = .95$.

    d. $P(-z_0 < z < z_0) = .99$.

3. An auditor has reviewed the financial records of a hardware store and has found that its billing errors follow a normal distribution with mean and standard deviation equal to $0 and $1, respectively.
    a. What proportion of the store's billings are in error by more than $1?
    b. What is the probability that a billing represents an overcharge of at least $1.50?
    c. What is the probability that a customer has been undercharged from $.50 to $1.00?
    d. Within what range would 95% of the billing errors lie?
    e. Of the extreme undercharges, 5% would be at least what amount?

5.3 Use of the Table of Normal Curve Areas for the Normal Random Variable x

We can now proceed to find probabilities associated with any normal random variable x having mean μ and standard deviation σ. This is accomplished by converting the random variable x to the standard normal random variable z, and then working the problem in terms of z.

Since probability statements are written in the form of inequalities, you are reminded of two facts. A statement of inequality is maintained if (1) the same number is subtracted from each member of the inequality and/or (2) each member of the inequality is divided by the same *positive* number.

Example 5.8
The following are equivalent statements about x:

1. $70 < x < 95$.

2. $(70 - 15) < (x - 15) < (95 - 15)$.

3. $\dfrac{70 - 15}{5} < \dfrac{x - 15}{5} < \dfrac{95 - 15}{5}$.

4. $11 < \dfrac{x - 15}{5} < 16$.

Example 5.9
Let x be a normal random variable with mean $\mu = 100$ and standard deviation $\sigma = 4$. Find $P(92 < x < 104)$.

Solution
Recalling that $z = (x - \mu)/\sigma$, we can apply rules 1 and 2 to convert the probability statement about x to one about the standard normal random variable z.

1. $P(92 < x < 104) = P(92 - 100 < x - 100 < 104 - 100)$

$$= P\left(\dfrac{92 - 100}{4} < \dfrac{x - 100}{4} < \dfrac{104 - 100}{4}\right)$$

$$= P(-2 < z < 1).$$

2. The problem now stated in terms of z is readily solved by using the methods of Section 5.2.

$P(92 < x < 104) = P(-2 < z < 1)$

$= A(-2) + A(1)$

$= \underline{\qquad} + \underline{\qquad}$ .4772; .3413

$= \underline{\qquad}$ . .8185

Example 5.10

Let x be a normal random variable with mean 100 and standard deviation 4. Find $P(93.5 < x < 105.2)$.

Solution

$$P(93.5 < x < 105.2) = P\left(\frac{93.5 - 100}{4} < z < \frac{105.2 - 100}{4}\right)$$

1.30

$$= P(-1.63 < z < \underline{\qquad})$$

.4484; .4032

$$= \underline{\qquad} + \underline{\qquad}$$

.8516

$$= \underline{\qquad}.$$

Self-Correcting Exercises 5C

1. If x is normally distributed with mean 10 and variance 2.25, evaluate the following probabilities:

   a. $P(x > 8.5)$.

   b. $P(x < 12)$.

   c. $P(9.25 < x < 11.25)$.

   d. $P(7.5 < x < 9.2)$.

   e. $P(12.25 < x < 13.25)$.

2. An industrial engineer has found that the standard household light bulbs produced by a certain manufacturer have a useful life that is normally distributed with a mean of 250 hours and a variance of 2500.
   a. What is the probability that a randomly selected bulb from this production process will have a useful life in excess of 300 hours?
   b. What is the probability that a randomly selected bulb from this production process will have a useful life between 190 and 270 hours?
   c. What is the probability that a randomly selected bulb from this production process will have a useful life not exceeding 260 hours?
   d. Ninety percent of the bulbs have a useful life in excess of how many hours?
   e. The probability is .95 that a bulb does not have a useful life in excess of how many hours?

3. Scores on a trade school entrance examination exhibit the characteristics of a normal distribution with mean and standard deviation of 50 and 5, respectively.
   a. What proportion of the scores on this examination would be greater than 60?
   b. What proportion of the scores on this examination would be less than 45?
   c. What proportion of the scores on this examination would be between 35 and 65?

d. If to be considered eligible for a place in the incoming class an applicant must score beyond the 75th percentile on this exam, what score must an applicant have to be eligible?

5.4 The Normal Approximation to the Binomial Probability Distribution

The binomial random variable x was defined in Chapter 4 as the number of successes in n independent and identical trials comprising the binomial experiment. The probabilities associated with this random variable were calculated as

$$p(x) = C_x^n p^x q^{(n-x)} \quad \text{for } x = 0, 1, 2, \ldots, n.$$

For large values of n (in fact, if n gets much larger than 10), the binomial probabilities are very tedious to compute. Fortunately, several options are available to us in an effort to avoid lengthy calculations.

1. Table 1 of Appendix III contains the tabulated values $P(x \leq a)$ for n = 1, 2, ..., 12, 15, 20, 25 and for p = .01, .05, .10, .20, ..., .90, .95, .99. However, if the user requires values of n and p for which tables have not been given, Table 1 will not be useful.
2. In Section 4.3, we considered examples in which the Poisson distribution was used to approximate binomial probabilities. This approximation is appropriate when n is large and p (or q) is small, so that $np < 7$. However, there are a great number of situations in which n is large, but $np \geq 7$. In this case, the Poisson approximation (will, will not) be accurate.   will not
3. When n is large and p (or q) is not too small, the binomial probability histogram is fairly symmetric and mound-shaped. This symmetry increases as p gets closer to p = .5. Hence, it would seem reasonable to approximate the distribution of a binomial random variable with the distribution of a normal random variable whose mean and variance are identical to those for the binomial random variable.

When n is *sufficiently large* and p is not too close to zero or one, the random variable x, the number of successes in n trials, is approximately normally distributed with mean np and variance npq.

When can we reasonably apply the normal approximation? For small values of n and values of p close to 0 or 1, the binomial distribution will exhibit a "pile-up" around x = _____ or x = _____. The data will not be _____-shaped and the normal approximation will be poor. For a normal random variable, _____% of the measurements will be within the interval $\mu \pm 2\sigma$. For $\mu = np$ and $\sigma = \sqrt{npq}$, the interval $np \pm 2\sqrt{npq}$ should be within the bounds of the binomial random variable x, or within the interval $(0, n)$, to obtain reasonably good approximations to the binomial probabilities.

0; n
bell
95

To show how the normal approximation is used, let us consider a binomial random variable x with $n = 8$ and $p = 1/2$ and attempt to approximate some binomial probabilities with a normal random variable having the same mean, $\mu = np$, and variance, $\sigma^2 = npq$, as the binomial x. In this case

$$\mu = np = 8(1/2) = \underline{}$$

$$\sigma^2 = npq = 8(1/2)(1/2) = \underline{}$$

4

2

Note that the interval

$$\mu \pm 2\sigma = 4 \pm 2\sqrt{2} = (1.2, 6.8)$$

is contained within the interval $(0, 8)$; therefore our approximations should be adequate. Consider the following diagrammatic representation of the approximation, where $p(x)$ is the frequency distribution for the binomial random variable and $f(x)$ is the frequency distribution for the corresponding normal random variable x.

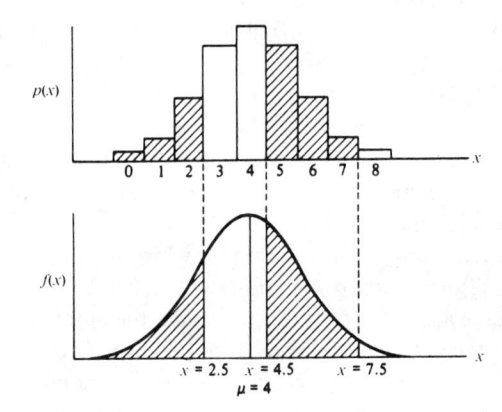

Example 5.11
Find $P(x < 3)$ using the normal approximation.

Solution
$P(x < 3)$ for the binomial random variable with mean $\mu = 4$ and $\sigma = \sqrt{2}$ corresponds to the shaded bars in the histogram over $x = 0, 1,$ and 2. The approximating probability corresponds to the shaded area to the left of $x = 2.5$ in the normal distribution with mean 4 and standard deviation $\sqrt{2}$.
We proceed as follows:

$$P(x < 3) \approx P(x < 2.5)$$

$$= P\left(\frac{x - 4}{\sqrt{2}} < \frac{2.5 - 4}{\sqrt{2}}\right)$$

$$= P(z < -1.06)$$

$$= .5000 - A(-1.06)$$

Sect. 5.4: The Normal Approximation to the Binomial Probability Distribution / 129

= .5000 - _____ .3554

= _____. .1446

Example 5.12
Find $P(5 \leq x \leq 7)$.

Solution
For the binomial random variable with mean 4 and standard deviation $\sqrt{2}$,
$P(5 \leq x \leq 7)$ corresponds to the shaded bars over $x = $ _____,   5
_____, and _____. This corresponds in turn to the shaded area for   6; 7
the approximating normal distribution with mean 4 and standard deviation
$\sqrt{2}$ between $x = $ _____ and $x = $ _____. Therefore,   4.5; 7.5

$$P(5 \leq x \leq 7) \approx P(4.5 < x < 7.5)$$

$$= P\left(\frac{4.5 - 4}{\sqrt{2}} < \frac{x - 4}{\sqrt{2}} < \frac{7.5 - 4}{\sqrt{2}}\right)$$

$$= P(.35 < z < 2.47)$$

$$= A(2.47) - A(.35)$$

= _____ - _____ .4932; .1368

= _____. .3564

Notice that we used $P(x < 2.5)$ to approximate the binomial probability $P(x < 3)$. In like manner we used $P(4.5 < x < 7.5)$ to approximate the binomial probability $P(5 \leq x \leq 7)$. The addition or subtraction of .5 is called the *correction for continuity* since we are approximating a discrete probability distribution with a probability distribution that is continuous. You may become confused as to whether .5 should be added or subtracted in the process of approximating binomial probabilities. A commonsense rule that always works is to examine the binomial probability statement carefully and determine which values of the binomial random variable are included in the statement. (Draw a picture if necessary.) The probabilities associated with these values correspond to the bars in the histogram centered over them. Locating the end points of the bars to be included determines the values needed for the approximating normal random variable.

Example 5.13
Suppose x is a binomial random variable with $n = 400$ and $p = .1$. Use the normal approximation to binomial probabilities to find the following:

a. $P(x > 45)$.

b. $P(x \leq 32)$.

c. $P(34 \leq x \leq 46)$.

Solution

If x is binomial, then its mean and variance are

$$\mu = np = 400(.1) = \underline{40},$$

$$\sigma^2 = npq = 400(.1)(.9) = \underline{36}.$$

a. To find $P(x > 45)$, we need the probabilities associated with the values $46, 47, 48, \ldots, 400$. This corresponds to the bars in the binomial histogram beginning at $\underline{45.5}$. Hence,

$$P(x > 45) \approx P(x > 45.5)$$

$$= P\left(z > \frac{45.5 - 40}{6}\right)$$

$$= P(z > .92)$$

$$= \underline{.1788}.$$

b. To find $P(x \leq 32)$, we need the probabilities associated with the values $0, 1, 2, \ldots,$ up to and including $x = 32$. This corresponds to finding the area in the binomial histogram to the left of $\underline{32.5}$. Hence,

$$P(x \leq 32) \approx P(x < 32.5)$$

$$= P\left(z < \frac{32.5 - 40}{6}\right)$$

$$= P(z < \underline{-1.25})$$

$$= \underline{.1056}.$$

c. To find $P(34 \leq x \leq 46)$, we need the probabilities associated with the values beginning at 34 up to and including 46. This corresponds to finding the area under the histogram between $\underline{33.5}$ and $\underline{46.5}$. Hence,

$$P(34 \leq x \leq 46) \approx P(33.5 < x < 46.5)$$

$$= P\left(\frac{33.5 - 40}{6} < z < \frac{46.5 - 40}{6}\right)$$

$$= P(\underline{-1.08} < z < \underline{1.08})$$

$$= \underline{.7198}.$$

Notice that the interval $\mu \pm 2\sigma$, or 40 ± 12, is well within the binomial range of 0 to 400, so that these approximate probabilities should be reasonably accurate.

Sect. 5.4: The Normal Approximation to the Binomial Probability Distribution / 131

Example 5.14
Past records show that at a given college 20% of the students that began as psychology majors either changed their major or dropped out of school. An incoming class has 110 beginning psychology majors. What is the probability that as many as 30 of these students leave the psychology program?

Solution
1. If x represents the number of students leaving the psychology program, with the probability of losing a student given as $p = .2$, then the required probability for $n = 110$ is

$$P(x \leqslant 30) = \sum_{x=0}^{30} \frac{(110)!}{x!(110-x)!} (.2)^x (.8)^{110-x}$$

2. To use the normal approximation, we need

$$\mu = np = 110(.2) = 22,$$

$$\sigma = \sqrt{npq} = \sqrt{110(.2)(.8)} = \sqrt{17.6} = 4.195.$$

   We now proceed to approximate the probability required in part 1 by using a normal probability distribution with a mean of _____ and a standard deviation of _____.

   22
   4.195

3. $P(x \leqslant 30)$ corresponds to the area to the left of 30.5 for the approximating normal distribution. Hence,

$$P(x \leqslant 30) \approx P(x < 30.5)$$

$$= P\left(\frac{x - 22}{4.195} < \frac{30.5 - 22}{4.195}\right)$$

$$= P(z < \underline{\qquad})$$

2.03

$$= \underline{\qquad}.$$

.9788

4. Hence, we approximate that $P(x \leqslant 30) = $ _____

.9788

Self-Correcting Exercises 5D

1. For a binomial experiment with $n = 20$ and $p = .7$, calculate $P(10 \leqslant x \leqslant 16)$ by
   a. using the binomial tables,
   b. using the normal approximation.
2. If the median income in a certain area is claimed to be $22,000, what is the probability that 37 or fewer of 100 randomly chosen wage earners from this area have incomes less than $22,000? Would the $22,000 figure seem reasonable if your sample actually contained 37 wage earners whose income was less than $22,000?

3. A large number of seeds from a certain species of flower are collected and mixed together in the following proportions according to the color of the flowers they will produce: 2 red, 2 white, 1 blue. If these seeds are mixed and then randomly packaged in bags containing about 100 seeds, what is the probability that a bag will contain the following:
   a. At most 50 "white" seeds?
   b. At least 65 seeds that are not "white"?
   c. At least 25 but at most 45 "white" seeds?
4. Refer to Exercise 3. Within what limits would you expect the number of white seeds to lie with probability .95?

EXERCISES

1. The time required for a garbage truck to make a round-trip to a disposal site and back is uniformly distributed between 0 and 1.5 hours.
   a. What is the probability that a garbage truck makes its round-trip to the disposal site in less than 75 minutes?
   b. What is the probability that the truck requires more than an hour to complete the round-trip?
2. The daily demand for gasoline at a local gas station has an exponential distribution with a mean of 2000 gallons.
   a. What proportion of the time would the demand exceed 4000 gallons?
   b. What proportion of the time would the demand exceed 1000 gallons, but be less than 4000 gallons?
   c. Ninety-five percent of the time, the demand would be less than what amount?
3. The time required to repair an electronic appliance has an exponential distribution with a mean repair time of one hour.
   a. What is the mean and standard deviation of repair times?
   b. Find the probability that the repair time exceeds $\mu + \sigma$?
   c. Fifty percent of all repair times will be shorter than x_0. What is the value of x_0?
4. Find the following probabilities for the standard normal variable z:

   a. $P(z < 1.9)$.   d. $P(-2.8 < z < 1.93)$.

   b. $P(1.21 < z < 2.25)$.   e. $P(-1.3 < z < 2.3)$.

   c. $P(z > -.6)$.   f. $P(-1.62 < z < .37)$.

5. Find the value of z, say z_0, such that the following probability statements are true:

   a. $P(z > z_0) = .2420$.   c. $P(-z_0 < z < z_0) = .90$.

   b. $P(-z_0 < z < z_0) = .9668$.   d. $P(z < z_0) = .9394$.

6. If x is distributed normally with mean 25 and standard deviation 4, find the following:

   a. $P(x > 21)$.

   b. $P(x < 30)$.

   c. $P(15 < x < 35)$.

   d. $P(x < 18)$.

7. A psychological introvert-extrovert test produced scores that had a normal distribution with mean and standard deviation 75 and 12, respectively. If we wish to designate the *highest* 15% as extrovert, what would be the proper score to choose as the cutoff point?

8. A manufacturer's process for producing steel rods can be regulated so as to produce rods with an average length μ. If these lengths are normally distributed with a standard deviation of .2 inch, what should be the setting for μ if one wants at most 5% of the steel rods to have a length greater than 10.4 inches?

9. For a given type of cannon and a fixed range setting, the distance that a shell fired from this cannon will travel is normally distributed with a mean and standard deviation of 1.5 and .1 miles, respectively.
   a. What is the probability that a shell will travel farther than 1.72 miles?
   b. What is the probability that a shell will travel less than 1.35 miles?
   c. What is the probability that a shell will travel at least 1.45 miles but at most 1.62 miles?
   d. If three shells are fired, what is the probability that all three will travel farther than 1.72 miles?

10. In investigating the marital status of adults (persons 14 years and older) in the United States, the U.S. Census Bureau reports that in a recent year, 63.4 percent of all males were married, while 29.3 percent were single, 2.5 percent were widowed, and 4.8 percent were divorced. Suppose for simplicity, that 30% of all males were single in this year. If a random sample of $n = 20$ men were taken in this year, what is the probability that 10 or more are single?
    a. Use the binomial tables, Table I, the Appendix.
    b. Use the normal approximation to the binomial.

11. A controversial issue during the 1981 election in the state of California was a proposition to secure money to build the peripheral water canal, a canal designed to bring water from northern to southern regions of the state. Suppose that 30% of the population favor the canal, while 70% oppose it. If a random sample of $n = 50$ voters is taken, what is the probability that 25 or more favor the canal? That is, what is the probability that the sample will show a majority in favor of the canal when in fact only 30% of the population favor it?

12. A preelection poll taken in a given city indicated that 40% of the voting public favored candidate A, 40% favored candidate B, and 20% were as yet undecided. If these percentages are true, in a random sample of 100 voters, what are the following probabilities:
    a. That at most 50 voters in the sample prefer candidate A?
    b. That at least 65 voters in the sample prefer candidate B?
    c. That at least 25 but at most 45 voters in the sample prefer candidate B?

13. On a well-known college campus, the student automobile registration revealed that the ratio of small to large cars (as measured by engine displacement) is 2 to 1. If 72 car owners are chosen at random from the student body, find the probability that this group includes at most 46 owners of small cars.

14. In introducing a new breakfast sausage to the public, an advertising campaign claimed that 7 out of 10 shoppers would prefer these new sausages over other brands. Suppose 100 people are randomly chosen, and the advertiser's claim is true.
    a. What is the probability that at most 65 people prefer the new sausages?
    b. What is the probability that at least 80 people prefer the new sausages?
    c. If only 60 people stated a preference for the new sausages, would this be sufficient evidence to indicate that the advertising claim is false and that, in fact, less than 7 out of 10 people would prefer the new sausages?

CHAPTER 6
SAMPLING DISTRIBUTIONS

6.1 Introduction

Recall from earlier discussions that a population of measurements results when an experiment is repeated an infinite number of times. This population can be described using numerical descriptive measures called _____ .   parameters
A sample is some subset of the population, and can be described using numerical descriptive measures called _____ . One of our objectives will be to   statistics
use statistics (calculated from the sample) to make inferences about (or to estimate) a population parameter.

Since a statistic is computed from sample measurements, to observe the statistic over and over again, we must sample repeatedly. This repeated sampling will generate a population of possible values of the statistic. The frequency distribution associated with the population of values of the statistic is called its
_____ _____ . Specifically, since it is generated through   probability distribution
repeated sampling, this probability distribution is called a *sampling distribution*.

In this chapter, we will try to clarify the concept of a sampling distribution by deriving the sampling distributions for several statistics when sampling randomly from a finite population. We will then generalize our discussion to include random sampling from an infinite population and consider the sampling distributions of some important sample statistics.

6.2 Sampling Distributions of Statistics

In making inferences about a population based on information contained in a sample, we will use sample statistics to estimate and/or make decisions about population _____ .   parameters

Notice that each sample drawn from a population of interest to the experimenter will contain different elements. Hence, the value of a statistic (such as $\bar{x}$ or s^2) will (change, remain the same) from sample to sample. If one were to   change
draw repeated samples of a constant sample size, many different values of the sample statistic would be obtained. These values could be used to create a rela-

135

136 / Chap. 6: Sampling Distributions

tive frequency distribution to describe the behavior of the statistic in repeated sampling.

Since a sample statistic takes on many different numerical values depending upon the outcome of a random sample, it is classified as a _____

random
variable

_____, and, as such, has a probability distribution associated with it. This probability distribution can be approximated by the relative frequency distribution described above.

The probability distribution for a sample statistic is called its *sampling distribution*. The sampling distribution results when random samples of size n are repeatedly drawn from the population of interest.

Example 6.1
Consider a population of $N = 6$ elements whose values are $x = 3, 3.5, 3.5, 4, 4,$ and 6. If each of the population values are equally likely to be selected in a single random selection, construct the probability distribution for x. Find the mean and the variance of x.

Solution
1. Since each of the $N = 6$ elements has an equal chance of being selected, each

1/6

has probability $1/N =$ _____ . However, some of the elements are identical. Thus,

1/6

$$p(3) = \underline{\hspace{2em}}$$

1/6; 1/3

but $\quad p(3.5) = 1/6 + \underline{\hspace{2em}} = \underline{\hspace{2em}}$.

The probability distribution for x is shown below. Fill in the missing entries.

x	$p(x)$
3	1/6
3.5	_____
4	1/3
6	_____

1/3

1/6

2. Graph the probability distribution for x. The distribution (is, is not) symmetrical.

is not

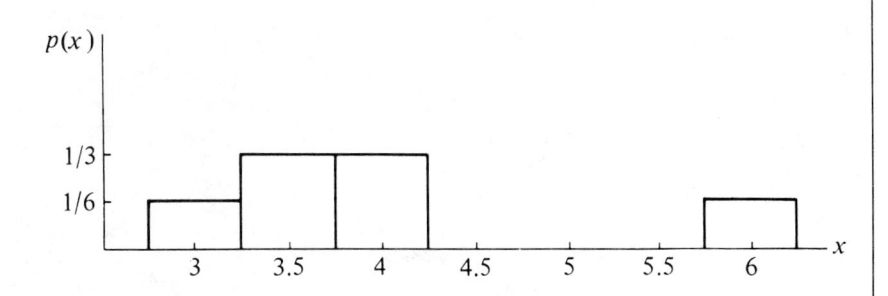

3. The mean value of x is calculated as in Chapter 3.

$$\mu = E(x) = \sum_x x\, p(x) = 3(1/6) + 3.5(1/3) + 4(1/3) + 6(1/6)$$

= _____

4.0

The variance of x is

$$\sigma^2 = \sum (x - \mu)^2 p(x) = \sum x^2 p(x) - \mu^2$$

$$= 3^2(1/6) + 3.5^2(1/3) + 4^2(1/3) + 6^2(1/6) - (4.0)^2$$

= _____ - 16 = _____

16.9166667; .9166667

Example 6.2
Find the sampling distribution for the sample mean $\bar{x}$ when a random sample of size $n = 2$ has been drawn from the population given in Example 6.1.

Solution
When $n = 2$ observations are drawn from $N = 6$, there are $C_2^6 = $ _____ possible samples which can be drawn with equal probability. Since some of the elements in this sample are identical, let us identify the repeated measurements as 3.5 and 3.5*, 4 and 4*, respectively. The 15 possible samples are listed below. Fill in the missing entries in Column 2.

15

138 / Chap. 6: Sampling Distributions

Sample	Sample Values	$\bar{x}$
1	3, 3.5	3.25
2	3, 3.5*	3.25
3	3, 4	_____
4	3, 4*	3.50
5	3, 6	4.50
6	3.5, 3.5*	3.50
7	3.5, 4	_____
8	_____	3.75
9	3.5, 6	4.75
10	3.5*, 4	3.75
11	_____	3.75
12	3.5*, 6	_____
13	4, 4*	4.00
14	4, 6	_____
15	_____	5.00

margin notes: 3.50; 3.75; 3.5, 4*; 3.5*, 4*; 4.75; 5.00; 4*, 6; 2; 4.75

For each possible sample, the sample mean can be different. For example, sample 1 has mean

$$\bar{x} = \frac{3 + 3.5}{2} = 3.75$$

while for sample 9,

$$\bar{x} = \frac{3.5 + 6}{\rule{1cm}{0.15mm}} = \rule{1cm}{0.15mm}$$

Fill in the missing entries in Column 3 of the sample table above. The value of $\bar{x}$ associated with each sample occurs with probability 1/15. The sampling distribution for $\bar{x}$ is shown and graphed below.

$\bar{x}$	$p(\bar{x})$
3.25	2/15
3.50	3/15
3.75	4/15
4.00	_____
4.50	1/15
_____	2/15
5.00	2/15

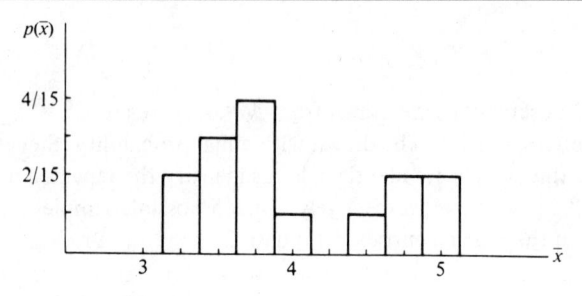

margin notes: 1/15; 4.75

Notice several things in comparing the original population distribution to the sampling distribution of $\bar{x}$.
1. The distribution of $\bar{x}$ is (more, <u>less</u>) variable than the original distribution.   less
2. The distribution of $\bar{x}$ appears (more, <u>less</u>) skewed than the original distribution.   less
3. The mean of $\bar{x}$ is

$$\mu_{\bar{x}} = 3.25(2/15) + 3.5(3/15) + \ldots + 5(2/15)$$

$$= 4.0$$

which is identical to the mean of the original population.
4. The variance of $\bar{x}$ is

$$\sigma_{\bar{x}}^2 = 3.25^2(2/15) + 3.5^2(\underline{\qquad}) + 3.75^2(4/15) + \ldots \qquad 3/15$$

$$+ 5^2(2/15) - (\underline{\qquad})^2 \qquad 4.0$$

$$= \underline{\qquad} - 16 = \underline{\qquad} \qquad 16.366667;\ .36667$$

The variance of $\bar{x}$ is (<u>smaller</u>, larger) than the variance of x, confirming our earlier observation.   smaller

Sampling distributions can be constructed for any finite population using this same method. For example, the experimenter might be interested in the sampling distribution of the sample median, the sample variance, s^2, or the sample standard deviation, s. For each possible sample, he calculates the value of the sample statistic and then uses the results to tabulate the appropriate sampling distribution and to describe the behavior of that statistic in repeated sampling. However, as N increases, it becomes extremely difficult to actually enumerate each sample, unless it can be done empirically on computer. For this reason, we turn to results proven using algebra or higher mathematics to derive the sampling distributions for several commonly used statistics. It can be shown that some of these statistics have desirable properties which will be useful in inference making.

6.3 The Central Limit Theorem and the Sampling Distribution of $\bar{x}$

Consider a meat processing center which as part of its output prepares one-pound packages of bacon. If the weights of the packaged bacon were carefully checked, some weights would be slightly heavier than 16 oz. while others would be slightly lighter than 16 oz. A frequency histogram of these weights would probably exhibit the mound-shaped distribution characteristic of a normally distributed random variable. Why should this be the case? One can think of the weight of each package as differing from 16 oz. due to an error in the weighing process, to a scale that needs adjustment, to the thickness of the slices of bacon so that the package contains either one more or less slice than it should, or perhaps some of the fat has melted, and so on. Hence any one weight would consist of an average weight (hopefully 16 oz.) modified by the addition of random errors that might be either positive or negative.

The Central Limit Theorem loosely stated says that sums or averages are approximately normally distributed with a mean and standard deviation that depend upon the sampled population. If one considers the error in the weight of a one-pound package of bacon as a *sum* of various effects in which small errors are highly likely and large errors are highly improbable, then the Central Limit Theorem helps explain the apparent normality of the package weights.

Equally as important, the Central Limit Theorem assures us that sample means will be approximately normally distributed with a mean and variance that depend upon the population from which the sample has been drawn. This aspect of the Central Limit Theorem will be the focal point for making inferences about populations based upon random samples when the sample size is large.

The Central Limit Theorem (1): If random samples of n observations are drawn from a population with finite mean μ and standard deviation σ, then when n is large, the sample mean $\bar{x}$ will be approximately normally distributed with mean μ and standard deviation $\sigma/\sqrt{n}$.

The quantity $\sigma/\sqrt{n}$ is sometimes called the *standard error* of $\bar{x}$. The Central Limit Theorem could also be stated in terms of the sum of the measurements, Σx_i.

The Central Limit Theorem (2): If random samples of n observations are drawn from a population with finite mean μ and standard deviation σ, then when n is large, Σx_i will be approximately normally distributed with mean $n\mu$ and standard deviation $\sigma\sqrt{n}$.

In both cases the approximation to normality becomes more and more accurate as n becomes large.

The Central Limit Theorem is important for two reasons.

1. It partially explains *why* certain measurements possess approximately a _____ distribution.

normal

estimators
normal

2. Many of the _____ used in making inferences are sums or means of sample measurements and thus possess approximately _____ distributions for large samples. Notice that the Central Limit Theorem

does not
could

(does, does not) specify that the sample measurements come from a normal population. The population (could, could not) have a frequency distribution that is flat or skewed or is nonnormal in some other way. It is the *sample mean* that behaves as a random variable having an approximately normal distribution.

To clarify a point we note that the sample mean $\bar{x}$ computed from a random sample of n observations drawn from any infinite population with mean μ

and standard deviation σ always has a mean equal to μ and a standard deviation equal to $\sigma/\sqrt{n}$. This result is not due to the Central Limit Theorem. The important contribution of the theorem lies in the fact that when n, the sample size, is *large,* we may approximate the distribution of $\bar{x}$ with a *normal* probability distribution.

The Central Limit Theorem relies heavily on the assumption that the sample size n is large. The question of how large n must be is a difficult one to answer and depends upon the characteristics of the underlying population from which we are sampling. Many texts use the rule of thumb that allows application of the Central Limit Theorem if $n > 30$. However, this rule will not always work. If the underlying distribution is (symmetric, skewed), the Central Limit Theorem may be appropriate for $n < 30$. However, if the underlying population is (symmetric, heavily skewed), the distribution of $\bar{x}$ may still be skewed even if $n > 30$. The student will need to use judgment to determine the approximate shape of the underlying population from which he or she is sampling in order to determine whether the Central Limit Theorem will be appropriate. For specific types of applications given in the text, we will provide the sample sizes necessary to insure the applicability of the Central Limit Theorem.

| symmetric

| heavily skewed

Example 6.3
A production line produces items whose mean weight is 50 grams with a standard deviation of 2 grams. If 25 items are randomly selected from this production line, what is the probability that the same mean $\bar{x}$ exceeds 51 grams?

Solution
The underlying population from which we are sampling is the population of weights of items on the production line. Since weights tend to have a mound-shaped distribution, the sampled population is (approximately symmetric, heavily skewed). Thus, the Central Limit Theorem (will, will not) be appropriate.

| approximately symmetric; will

According to the Central Limit Theorem, the sample mean $\bar{x}$ is approximately normally distributed with mean μ and standard deviation $\sigma/\sqrt{n}$. For our problem $\mu = 50$, $\sigma = 2$, and $n = 25$; hence

$$\sigma_{\bar{x}} = \frac{\sigma}{\sqrt{n}} = \frac{2}{\sqrt{25}} = \underline{\qquad}$$

| .4

Therefore,

$$P(\bar{x} > 51) = P\left(\frac{\bar{x} - 50}{.4} > \frac{51 - 50}{.4}\right)$$

$$= P(z > 2.5)$$

$$= .5000 - A(2.5)$$

$$= .5000 - \underline{\qquad}$$

| .4938

$$= \underline{\qquad}$$

| .0062

Example 6.4
A bottler of soft drinks packages cans of soft drink in six-packs.
1. If the fill per can has a mean of 12 fluid ounces and a standard deviation of .2 fluid ounce, what is the distribution of the total fill for a case of 24 cans?
2. What is the probability that the total fill for a case is less than 286 fluid ounces?

Solution
The population from which we are sampling is the population of fills per can, which (should, should not) be an approximately symmetric distribution. [should]

Therefore, for $n = 24$, the Central Limit Theorem (will, will not) be applicable. [will]

1. Using the Central Limit Theorem in its second form, the total fill per case has a mean of $n\mu = 24(12)$ _____ fluid ounces and a standard deviation (or standard error) of $\sigma\sqrt{24} = .2\sqrt{24} =$ _____ fluid ounce. [288; .98]

   The total fill is approximately _____ distributed with mean 288 and standard deviation .98. [normally]

2. Let T represent the total fill per case. We wish to evaluate $P(T < 286)$. Since T is approximately normally distributed with $\mu_T =$ _____ and $\sigma_T =$ _____, [288; .98]

$$P(T < 286) = P\left(\frac{T - 288}{.98} < \frac{286 - 288}{.98}\right) = P(z < -2.04)$$

$$= .5000 - A(-2.04) = .5000 - \underline{\hspace{1cm}} = \underline{\hspace{1cm}}$$

[.4793; .0207]

The Central Limit Theorem applies when a sufficiently (large, small) sample is randomly drawn from a very large or infinite population, regardless of its shape. However, if the population from which we are sampling is itself normal, the sampling distribution of $\bar{x}$ will be _____, regardless of the size of the sample. [large; normal]

If a random sample is drawn from a *normal population* with mean μ and variance σ^2, the sampling distribution of $\bar{x}$ will be normal with mean μ and variance σ^2/n, *regardless of the sample size*.

Example 6.5
Suppose that the bottler of soft drinks in Example 6.4 packages cans of soft drinks in cans whose fill per can is normally distributed with mean 12 fluid ounces and standard deviation .2 fluid ounces. If a six-pack of soda can be considered a random sample of size $n = 6$ from the population, what is the probability that the average fill per can is less than 11.5 fluid ounces?

Solution

1. Since we are sampling from a _____ population, the sampling distribution of $\bar{x}$ will be (approximately, exactly) normal with mean $\mu =$ 12 fluid ounces and variance $\sigma^2/n = .2/6 =$ _____ .

   normal
   exactly
   .033333

2. We wish to evaluate $P($_____$)$, or

   $\bar{x} < 11.5$

$$P(\bar{x} < 11.5) = P\left(\frac{\bar{x} - \mu}{\sigma/\sqrt{n}} < \frac{11.5 - 12}{\sqrt{.03333}}\right)$$

$$= P(z < \underline{\qquad})$$

-2.74

$$= .5 - \underline{\qquad} = \underline{\qquad}$$

.4969; .0031

Self-Correcting Exercises 6A

1. A pharmaceutical company is experimenting with rats to determine if there is a difference in weights for rats fed with and without a vitamin supplement. A frequency distribution of these weights (either with or without the supplement) would probably exhibit the mound-shaped distribution characteristics of a normally distributed random variable. Why should this be the case?

2. An agricultural economist is interested in determining the average diameter of peaches produced by a particular tree. A random sample of $n = 30$ peaches is taken and the sample mean $\bar{x}$ is calculated. Suppose that the average diameter of peaches on this tree is known from previous years' production to be $\mu = 60$ millimeters with $\sigma = 10$ mm. What is the probability that the sample mean, $\bar{x}$, exceeds 65 millimeters?

6.4 The Sampling Distribution of the Sample Proportion

A statistic that is often used to describe a binomial population is $\hat{p}$, the *proportion* of trials in which a success is observed. In terms of previous notation,

$$\hat{p} = \frac{x}{n} = \frac{\text{number of successes in } n \text{ trials}}{n}$$

Recall that, for a binomial population, the proportion of successes in the population is defined as p. The sample proportion, $\hat{p}$, will be used to estimate or make inferences about p.

For a binomial experiment consisting of n trials, let

$x_1 = 1$ if trial one is a success
$x_1 = 0$ if trial one is a failure

$x_2 = 1$ if trial two is a success
$x_2 = 0$ if trial two is a failure

.
.
.

144 / Chap. 6: Sampling Distributions

$x_n = 1$ if trial n is a success
$x_n = 0$ if trial n is a failure

Then x, the number of successes in n trials, can be thought of as a sum of n independent random variables. Using this fact, the statistic $\hat{p} = x/n$ is equivalent to an _____ of these n random variables. The _____ _____ assures us that $\hat{p}$ will be approximately normally distributed when n is large. The mean and standard deviation of $\hat{p}$ can be shown to be

average; Central Limit Theorem

$$\mu_{\hat{p}} = p \quad \text{and} \quad \sigma_{\hat{p}} = \sqrt{\frac{p(1-p)}{n}} = \sqrt{\frac{pq}{n}}$$

Example 6.6
Past records show that at a given college 20% of the students that began as economics majors either changed their major or dropped out of school. An incoming class has 110 beginning economics majors. What is the probability that at most 30% of these students leave the economics program?

Solution
1. If p represents the proportion of students leaving the economics program, with the probability of losing a student given as $p = .2$, then the required probability for $n = 110$ is $P(\hat{p} \leq .30)$.
2. To use the normal approximation, we need

.2

$$\mu_{\hat{p}} = p = \underline{\qquad}$$

.8; .0014545

$$\sigma_{\hat{p}} = \sqrt{\frac{p(1-p)}{n}} = \sqrt{\frac{.2(\underline{\qquad})}{110}} = \sqrt{\underline{\qquad}}$$

.0381

$$= \underline{\qquad}$$

We now proceed to approximate the probability required in part 1 by using a normal probability distribution with a mean of _____ and a standard deviation of _____ .

.2
.0381

3. Using the normal approximation, the value $\hat{p} = .30$ corresponds to a z value of

2.62

$$z = \frac{\hat{p} - p}{\sigma_{\hat{p}}} = \frac{.30 - .20}{.0381} = \underline{\qquad}$$

and

2.62; .4956; .9956

$$P(\hat{p} \leq .30) \approx P(z \leq \underline{\qquad}) = .5000 + \underline{\qquad} = \underline{\qquad}$$

In Chapter 5, we used the normal approximation to the binomial distribution to calculate probabilities associated with x, the number of successes in n trials. In that situation, it was necessary to have the interval $\mu \pm 2\sigma$ or $np \pm 2\sqrt{npq}$

fall within the binomial limits, 0 to n, in order to insure the accuracy of the approximation. Further, a "correction for continuity" was used to make the approximation more accurate. Hence, the standard normal random variable used in this earlier approximation was

$$z = \frac{x \pm \frac{1}{2} - np}{\sqrt{npq}} \approx \frac{x - np}{\sqrt{npq}}$$

If we divide both numerator and denominator of this fraction by n, we see the relationship between the sampling distributions of x and $\hat{p} = x/n$.

$$z = \frac{\frac{x}{n} \pm \frac{1}{2n} - p}{\sqrt{\frac{pq}{n}}} = \frac{\hat{p} \pm \frac{1}{2n} - p}{\sqrt{\frac{pq}{n}}} \approx \frac{\hat{p} - p}{\sqrt{\frac{pq}{n}}}$$

Hence, the sampling distributions of x and $\hat{p}$ (are, are not) equivalent. *The quantity ± 1/2n will be ignored for large values of n, since the value of z changes very little.* The normal approximation will be appropriate if the interval $p \pm 2\sqrt{pq/n}$ falls within the limits 0 to 1.

are

Self-Correcting Exercises 6B

1. For a binomial experiment with $n = 20$ and $p = .5$, calculate $P(.8 \leq \hat{p} \leq .9)$ by
   a. Using the binomial tables, Table 1.
   b. Using the normal approximation to the sampling distribution of $\hat{p}$.
2. A controversial issue in the state of California is the diversion water from northern to southern regions of the state. Suppose that 30% of the population favor the diversion, while 70% oppose it. If a random sample of $n = 50$ voters is taken, what is the probability that 50% or more favor the diversion? That is, what is the probability that the sample will show a majority in favor of the diversion when in fact only 30% of the population favor it?

6.5 The Sampling Distribution of the Difference Between Two Sample Means or Proportions

Chapters 2 through 6 have been concerned with statistics, the science of making inferences about a population based on information contained in a sample, as it relates to a single population of measurements of interest to the experimenter. In particular, Sections 6.3 and 6.4 involved a discussion of the sampling distributions of two important statistics:

1. The sample mean, $\bar{x}$, based on a sample drawn from a population of continuous measurements.
2. The sample proportion, $\hat{p}$, based on a sample drawn from a binomial population.

In practice, an experimenter is often concerned with two populations of measurements; and in general, is interested in knowing whether or not these populations have the same location parameters. In the case of two populations of continuous measurements, this involves a comparison of the two parameters μ_1 and μ_2, which will be made by studying the difference, $\mu_1 - \mu_2$. In the case of two binomial populations, it involves a comparison of the two parameters p_1 and p_2, which will be made by studying the difference, $p_1 - p_2$.

Example 6.7

A pharmaceutical company is interested in the effect of vitamin C on the average cholesterol level in the human body. They design an experiment in which two groups of people will be placed on a specified diet for a given length of time. One group's diet will be supplemented with a particular dose of vitamin C, while the other group's diet will not. The researchers are interested in measuring the difference in the average cholesterol levels for the two groups.

Example 6.8

An agricultural economist is interested in determining the effect of two types of fertilizer on the yield of an apple orchard. He selects 40 trees to be fertilized with one type of fertilizer, while another 40 in a different location within the orchard are fertilized with the second type of fertilizer. The economist is interested in measuring the difference in yield for two fertilizers.

Example 6.9

The political preferences of executives at two large corporations were compared, based on 50 executives from each corporation. The percentage of Democrats were recorded for each, and the difference in the two percentages was recorded.

In all of the above situations, the experimenter is considering two populations, which will be denoted as Population 1 and Population 2. In the first two examples, the populations consist of continuous measurements, although the populations are in fact hypothetical in nature. They involve the conceptual populations of measurements taken on all experimental units (people or trees) that could possibly be treated with the experimental treatment (vitamin C or fertilizer). We are concerned with making inferences about the difference in the two population means, $\mu_1 - \mu_2$. In Example 6.9, the populations are approximately binomial, since the corporations and the number of executives in each corporation are very large. The parameter of interest is $p_1 - p_2$, the difference in the proportion of Democrats in the two corporations.

Consider two populations, 1 and 2, from which we select two independent random samples of size n_1 and n_2, respectively. Sample statistics will be calculated for each of the two samples. In general, denote these sample statistics by x_1 and x_2. We will use the following theorem, which gives the

Sect. 6.5: The Sampling Distribution of the Difference Between Two Sample Means or Proportions

sampling distribution for the difference $(x_1 - x_2)$ in two specific situations. It will be used to obtain the sampling distribution for $\bar{x}_1 - \bar{x}_2$ and to obtain the sampling distribution for $\hat{p}_1 - \hat{p}_2$.

Theorem 6.1: If two independent random variables x_1 and x_2 are normally distributed with means μ_1 and μ_2 and variances σ_1^2 and σ_2^2, respectively, then the difference $(x_1 - x_2)$ will be normally distributed with mean $(\mu_1 - \mu_2)$ and variance $(\sigma_1^2 + \sigma_2^2)$.

---

1. If samples of size n_1 and n_2 are randomly and independently selected from two continuous populations with means μ_1 and μ_2, respectively, a statistic that can be formed in order to make inferences about the difference $(\mu_1 - \mu_2)$ is _____ . If n_1 and n_2 are both larger than _____ , the _____ _____ Theorem allows us to make the following two statements:

    a. $\bar{x}_1$ is approximately _____ distributed with mean $\mu_{\bar{x}_1}$ = _____ and variance

    $\sigma_{\bar{x}_1}^2 = $ _____

    b. $\bar{x}_2$ is approximately _____ distributed with mean $\mu_{\bar{x}_2}$ = _____ and variance

    $\sigma_{\bar{x}_2}^2 = $ _____

    Finally, using Theorem 6.1, $(\bar{x}_1 - \bar{x}_2)$ will be approximately normally distributed with mean

    $$\mu_{\bar{x}_1 - \bar{x}_2} = \mu_1 - \mu_2$$

    and variance

    $$\sigma_{\bar{x}_1 - \bar{x}_2}^2 = \frac{\sigma_1^2}{n_1} + \frac{\sigma_2^2}{n_2}$$

    The standard deviation of $\bar{x}_1 - \bar{x}_2$, sometimes called the *standard error*, is

    $$\sigma_{\bar{x}_1 - \bar{x}_2}^2 = \sqrt{\rule{2cm}{0pt}}$$

2. If samples of size n_1 and n_2 are independently drawn from two binomial populations with parameters p_1 and p_2, respectively, a statistic that can be formed in order to make inferences about the difference $(p_1 - p_2)$ is _____ , where

    $$\hat{p}_1 = \frac{x_1}{n_1} = \frac{\text{number of successes in sample 1}}{\text{number of trials in sample 1}}$$

Answer column:

$(\bar{x}_1 - \bar{x}_2)$; 30
Central Limit

normally
μ_1

$\dfrac{\sigma_1^2}{n_1}$

normally
μ_2

$\dfrac{\sigma_2^2}{n_2}$

$\dfrac{\sigma_1^2}{n_1} + \dfrac{\sigma_2^2}{n_2}$

$(\hat{p}_1 - \hat{p}_2)$

148 / Chap. 6: Sampling Distributions

and

$$\hat{p}_2 = \frac{x_2}{n_2} = \frac{\text{number of successes in sample 2}}{\text{number of trials in sample 2}}$$

As long as $p \pm 2\sqrt{pq/n}$ lies within the range 0 to 1 for both populations, the _____ _____ Theorem allows us to make the following two statements:

Central Limit

a. $\hat{p}_1$ is approximately _____ distributed with mean

normally

$$\mu_{\hat{p}_1} = \underline{\qquad}$$

and variance

p_1

$$\sigma^2_{\hat{p}_1} = \underline{\qquad}$$

$\dfrac{p_1 q_1}{n_1}$

3. $\hat{p}_2$ is approximately _____ distributed with mean

normally

$$\mu_{\hat{p}_2} = \underline{\qquad}$$

p_2

and variance

$$\sigma^2_{\hat{p}_2} = \underline{\qquad}$$

$\dfrac{p_2 q_2}{n_2}$

Using Theorem 6.1, the sampling distribution of $(\hat{p}_1 - \hat{p}_2)$ is approximately normal with mean

$$\mu_{\hat{p}_1 - \hat{p}_2} = \underline{\qquad}$$

$p_1 - p_2$

and variance

$$\sigma^2_{\hat{p}_1 - \hat{p}_2} = \underline{\qquad}$$

$\dfrac{p_1 q_1}{n_1} + \dfrac{p_2 q_2}{n_2}$

The standard deviation of $(\hat{p}_1 - \hat{p}_2)$ is

$$\sigma_{\hat{p}_1 - \hat{p}_2} = \sqrt{\frac{p_1 q_1}{n_1} + \frac{p_2 q_2}{n_2}}$$

The sampling distributions for $\bar{x}_1 - \bar{x}_2$ and $\hat{p}_1 - \hat{p}_2$ can be used to make probability statements about the behavior of $\bar{x}_1 - \bar{x}_2$ or $\hat{p}_1 - \hat{p}_2$ in repeated sampling.

Example 6.10
Random samples of size $n_1 = 50$ and $n_2 = 40$ are drawn from two continuous populations with equal means and equal variances, $\sigma_1^2 = \sigma_2^2 = 10$. What is the probability that the sample means will differ by more than 1?

Sect. 6.5: The Sampling Distribution of the Difference Between Two Sample Means or Proportions / 149

Solution
1. The sampling distribution of $\bar{x}_1 - \bar{x}_2$ is approximately normal with mean

$$\mu_{\bar{x}_1 - \bar{x}_2} = \mu_1 - \mu_2 = \underline{\qquad}$$

0

since $\mu_1 = \mu_2$. The standard deviation of $\bar{x}_1 - \bar{x}_2$ is

$$\sqrt{\frac{\sigma_1^2}{n_1} + \frac{\sigma_2^2}{n_2}} = \sqrt{\frac{\underline{\qquad}}{50} + \frac{10}{40}} = \sqrt{\underline{\qquad}} = \underline{\qquad}$$

10; .45; .6708

2. If $\bar{x}_1$ and $\bar{x}_2$ differ by more than 1, we could have either

$$\bar{x}_1 - \bar{x}_2 > 1 \quad \text{or} \quad \underline{\qquad} > 1$$

$\bar{x}_2 - \bar{x}_1$

The second inequality is equivalent to $\bar{x}_1 - \bar{x}_2$ \underline{\qquad}, so that the probability of interest is

< -1

$$P(\bar{x}_1 - \bar{x}_2 > 1) + P(\bar{x}_1 - \bar{x}_2 < -1)$$

$$= P\left(\frac{(\bar{x}_1 - \bar{x}_2) - 0}{.6708} > \frac{1 - 0}{.6708}\right) + P\left(\frac{(\bar{x}_1 - \bar{x}_2) - 0}{.6708} < \frac{-1 - 0}{.6708}\right)$$

$$= P(z > 1.49) + P(z < \underline{\qquad})$$

-1.49

$$= 2(.5 - \underline{\qquad}) = \underline{\qquad}$$

.4319; .1362

That is, $\bar{x}_1$ and $\bar{x}_2$ will differ by as much as 1 or more approximately \underline{\qquad} of the time, even when the population means are equal.

14%

Example 6.11
A manufacturer of automobiles is conducting research to estimate the difference in the proportion of accidents on the California interstate system and on the New York interstate system that result in fatal injuries to at least one person. He randomly checks the files on 50 automobile accidents on the California interstate system and finds that 8 resulted in fatal injuries. He then randomly checks the files on 50 automobile accidents on the New York interstate system and finds that 9 resulted in fatal injuries. Suppose that the true proportion of fatal injuries is the same for both systems, and that in fact $p_1 = p_2 = .15$. What is the probability of observing a New York percentage at least 2 points higher than the California percentage, as was observed in this experiment?

Solution
1. The probability of interest is

$$P(\hat{p}_1 - \hat{p}_2 > .02)$$

where $\hat{p}_1$ is the (New York, California) sample percentage and $\hat{p}_2$ is the (New York, California) sample percentage.

New York
California

150 / Chap. 6: Sampling Distributions

2. The sampling distribution of $\hat{p}_1 - \hat{p}_2$ is approximately normal with mean

$$\mu_{\hat{p}_1 - \hat{p}_2} = p_1 - p_2 = .15 - .15 = 0$$

and standard deviation

.85
50

$$\sigma_{\hat{p}_1 - \hat{p}_2} = \sqrt{\frac{p_1 q_1}{n_1} + \frac{p_2 q_2}{n_2}} = \sqrt{\frac{.15(.85)}{\rule{0.5cm}{0.15mm}} + \frac{.15(\rule{0.5cm}{0.15mm})}{50}}$$

.00255; .0051; .0714

$$= \sqrt{2(\rule{1cm}{0.15mm})} = \sqrt{\rule{1cm}{0.15mm}} = \rule{1cm}{0.15mm}$$

3. The desired probability is

$$P(\hat{p}_1 - \hat{p}_2 > .02) = P\left(\frac{(\hat{p}_1 - \hat{p}_2) - 0}{.0714} > \frac{.02 - 0}{.0714}\right)$$

.28

$$= P(z > \rule{1cm}{0.15mm})$$

.1103; .3897

$$= .5 - \rule{1cm}{0.15mm} = \rule{1cm}{0.15mm}$$

is not

4. The sample result (is, is not) unlikely, even when the population proportions are identical.

Self-Correcting Exercises 6C

1. Random samples of size $n_1 = n_2 = 35$ are drawn from continuous populations with $\mu_1 = 20$, $\sigma_1^2 = 150$, $\mu_2 = 25$, $\sigma_2^2 = 100$. Find the following probabilities:
   a. $P(\bar{x}_1 - \bar{x}_2 > 1)$
   b. $P(0 \leq \bar{x}_1 - \bar{x}_2 \leq 6)$
   c. $P(|\bar{x}_1 - \bar{x}_2| \geq 2)$
2. Random samples of size $n_1 = n_2 = 100$ are drawn from binomial populations with $p_1 = .3$ and $p_2 = .4$. Find the following probabilities, describing the behavior of $\hat{p}_1 - \hat{p}_2$ in repeated sampling:
   a. $P(|\hat{p}_1 - \hat{p}_2| > .25)$
   b. $P(\hat{p}_1 > \hat{p}_2)$

6.6 Summary

The sampling distribution for a sample statistic is its probability distribution, which results when the behavior of the statistic is examined in repeated sampling. The sampling distribution can be constructed directly for small finite populations. For large or infinite populations, the sampling distribution of a statistic must be found mathematically or approximated empirically.

Central Limit
Theorem

An important result called the \rule{2cm}{0.15mm} \rule{2cm}{0.15mm} \rule{2cm}{0.15mm} allows us to approximate the sampling distribution for statistics which are sums or averages of random variables. Using this theorem, we examined the sampling distributions of four important sample statistics,

$\bar{x}, \bar{x}_1 - \bar{x}_2, \hat{p}, \hat{p}_1 - \hat{p}_2$. All could be approximated by a normal distribution (with the appropriate mean and standard deviation) for large sample sizes. The sampling distributions of these four statistics will serve as the basis for making inferences about their corresponding population parameters. This will be the subject of Chapter 7.

EXERCISES

1. Suppose that an elevator is designed with a permissible load limit of 3000 pounds with a maximum of 20 passengers. If the weights of people using the elevator are normally distributed with a mean of $\mu = 160$ pounds and a standard deviation of 25 pounds, what is the probability that the weight of a group of 20 persons exceeds the permissible load limit?

2. In a past municipal election, a city bond issue passed with 52% of the vote. If a poll involving $n = 100$ people had been taken just prior to the election, what is the probability that the sample proportion favoring the issue would have been 49% or less?

3. Based of 1965-1974 birth records, the probability of a male livebirth in the United States is .513. Of the first $n = 100$ births in January, what is the probability that the proportion of male livebirths exceeds 60%?

4. Suppose that the average assessed values of single family dwellings in a large municipality is $65,000 with a standard deviation of $20,000.
   a. If a random sample of $n = 100$ dwellings is selected and $\bar{x}$, the average assessed value of these dwellings calculated, what is the mean of $\bar{x}$ in repeated sampling?
   b. What is the standard deviation of $\bar{x}$?

5. Suppose that two television networks each allow an average of 5 minutes per half-hour program for advertising with a standard deviation of 2 minutes. What is the probability that the average time devoted to commercials for these two networks differs by less than one minute based on samples of 50 one-half hour programs from each network?

6. Packages of food whose average weight is 16 ounces with a standard deviation of .6 oz are shipped in boxes of 24 packages. What is the probability that a box of 24 packages will weigh more than 392 ounces (24.5 pounds)?

7. The average number of sick days per year in an electronics industry is 7 days with a standard deviation of 3 days. If a sample of $n_1 = 30$ men and $n_2 = 30$ women were selected from among the employees in this industry, what is the probability that the sample mean number of sick days for men exceed that for the women by 2 or more days?

8. Graduate students applying for entrance to many universities must take a Miller Analogies Test. It is known that the test scores have a mean of 75 and a variance of 16. If, during the past year, 100 students applied for graduate admission to a school requiring the Miller Analogies Test, what is the probability that the average score on these 100 tests exceeds 76?

9. The blood pressures for a population of individuals are normally distributed with a mean of 110 and a standard deviation of 7.
   a. If a sample of $n = 10$ individuals is chosen randomly from this population, within what limits would you expect the sample mean to lie with probability .95?
   b. What is the probability that the sample mean will exceed 115?

10. Suppose that the proportion of males and females who favor legalized abortion is the same, and equal to .6. If $n_1 = 50$ males and $n_2 = 100$ females are randomly selected from this population, what is the probability that the proportion of females in the sample favoring abortion exceeds that of the males by at least .2?

CHAPTER 7
LARGE–SAMPLE STATISTICAL ESTIMATION

7.1 Introduction

The objective of statistics is to make _____ about a _____ based on information contained in a _____. Since populations are described by numerical descriptive measures, called _____ of the population, we can make inferences about the population by making inferences about its parameters. For example, lot acceptance sampling involved inferences that resulted in decisions concerning the binomial parameter p. In Chapters 7 and 8, we consider two methods for making inferences concerning populations parameters: estimation and hypothesis testing.

inferences; population
sample
parameters

Example 7.1
Consider the agricultural experimenter interested in the average yield (in tons per acre) of a variety of alfalfa grown for hay. Let μ be the average yield for this variety of alfalfa. If an inference is to be made about μ, two questions could be asked.

1. What is the most likely value of μ for the population of yields from which we are sampling?
2. Is the mean equal to some specified value, say μ_0 or is it not? For example, the experimenter may know that the average yield of this variety of alfalfa in an adjacent field using chemical pest control is at most 2 tons per acre. Can integrated pest management practices produce yields larger than 2 tons per acre?

The first question is one of predicting or estimating the value of the population parameter μ. The second question concerns a test of an hypothesis about μ. If integrated pest management is no more effective than chemical pest control, then the yield will still be 2 tons per acre. However, if integrated pest management increases the yield of alfalfa, then $\mu > 2$ tons per acre. The objective is to determine which of these two hypotheses is correct.

This chapter will be concerned with making inferences about four parameters:
1. μ, the mean of a population of continuous measurements.
2. $\mu_1 - \mu_2$, the difference between the means for two populations of continuous measurements.

154 / Chap. 7: Large-Sample Statistical Estimation

3. p, the parameter of a dichotomous or binomial population.
4. $p_1 - p_2$, the difference in the parameters for two binomial populations.

The quantities to be used in making inferences will be sums or averages of the measurements in a random sample and consequently will possess frequency distributions in repeated sampling that are approximately _____

normal; Central Limit

due to the _____ _____ Theorem, as explained in Chapter 6.

One of the most important concepts to grasp is that estimation as well as hypothesis testing is a two-step procedure. These steps are
1. making the inference, and
2. measuring its goodness.

A measure of the goodness of an inference is essential to enable the person using the inference to measure its reliability. For example, we would wonder how close to the population parameter our estimate is expected to lie.

Rather than follow the section numbers exactly as they appear in your text, we have grouped certain topics together and will consider two more general sections: (1) point estimation and (2) interval estimation. The techniques developed in the process of estimating the four parameters mentioned above will also be used to determine how large the sample must be to achieve the accuracy required by the experimenter.

7.2 Estimation

Using the measurements in a sample to predict the value of one or more parameters of a population is called _____. An _____

estimation; estimator

is a rule that tells us how to calculate an estimate of a parameter based on the information contained in a sample. We can give many different estimators for a particular population parameter. An estimator is often expressed in terms of a mathematical formula in which the estimate is a function of the sample measurements. For example, $\bar{x}$ is an *estimator* of the population parameter μ. If a sample of $n = 20$ pieces of aluminum cable is tested for strength and the mean of the sample is $\bar{x} = 100.7$, then 100.7 is an *estimate* of the population mean strength μ. The estimator of a parameter is usually designated by placing a "hat" (^) over the parameter to be estimated. Thus, an estimator of μ would be $\hat{\mu} = \bar{x}$.

Estimates of a population parameter can be made in two ways. The sample can be used to produce an estimate of μ that is a single number, or to produce an interval, two points intended to enclose the true value of μ.

A *point estimator* of a population parameter is a rule that tells how to calculate a single number based on the sample measurements. The resulting number is called a *point estimate*.

An *interval estimator* of a population parameter is a rule that tells how to calculate two numbers based on the sample measurements that form an interval within which the true value of the parameter is expected to lie. This pair of numbers is called an *interval estimate*.

Sect. 7.3: Point Estimation / 155

The goodness of an estimator is evaluated by observing its behavior in repeated sampling. Let us talk in general about some population parameter which we will denote as θ. An estimator $\hat{\theta}$ for the parameter θ will generate estimates in repeated sampling from the population and will produce a distribution of estimates (numerical values computed from these samples). The distribution of this estimator, $\hat{\theta}$, is called its _____ _____ .   sampling distribution
This estimator would be considered good if the estimates cluster closely about θ. If the mean of the estimates is θ, then $\hat{\theta}$ is said to be an _____   unbiased
estimator for θ and $E(\hat{\theta}) = \theta$. If the spread (variance) of $\hat{\theta}$ is smaller than that of any other estimator, then $\hat{\theta}$ is said to have _____ variance.   minimum
Therefore, a *good estimator* should have the following properties:
1. _____ ,   unbiasedness
2. _____ _____ .   minimum variance
The distributions obtained in repeated sampling are shown below for four different estimators of θ. Which estimator appears to possess the most desirable properties? _____   (d) $\hat{\theta}_4$

(a) $\hat{\theta}_1$ (b) $\hat{\theta}_2$

(c) $\hat{\theta}_3$ (d) $\hat{\theta}_4$

The properties of interval estimators are also determined by repeated sampling. Repeated use of an interval estimator generates a large number of _____ _____ for estimating θ. If an interval estimator   interval estimates
were satisfactory, a large fraction of the interval estimates would enclose the true value of θ. The fraction of such intervals enclosing θ is known as the _____ coefficient. This is not to be confused with the interval   confidence
estimate, called the confidence _____ .   interval

> The probability that a confidence interval will enclose the true value of the estimated parameter is called the *confidence coefficient*.

If the confidence coefficient is .90, and a large number of samples were drawn and each used to calculate a confidence interval, then 90% of these intervals would enclose the true value of the parameter.

7.3 Point Estimation

In this section we will consider point estimation for the following parameters:

156 / Chap. 7: Large-Sample Statistical Estimation

1. μ, the mean of a population of continuous measurements.
2. $(\mu_1 - \mu_2)$, the difference between the means for two populations of continuous measurements.
3. p, a binomial parameter.
4. $(p_1 - p_2)$, the difference in the parameters for two binomial populations.

We assume, in all four cases, that the samples are relatively large so that the estimators possess sampling distributions that are approximately normal due to the _____ _____ Theorem. [Central Limit] The basic estimation problem is the same for all four cases, and therefore we can discuss the problems in general by referring to the estimation of a parameter θ. Thus θ might be any one of the four parameters just mentioned.

To estimate the population parameter θ, a sample of size $n(x_1, x_2, \ldots, x_n)$ is randomly drawn from the population and an estimate of θ is calculated using $\hat{\theta}$. In repeated sampling, a sampling distribution for $\hat{\theta}$ will be generated and will possess the following properties:

1. $E(\hat{\theta}) = $ _____ . [θ]
2. $\hat{\theta}$ is approximately _____ distributed. [normally] Therefore, approximately 95% of the values of $\hat{\theta}$ will lie within 1.96 standard deviations of their mean θ.

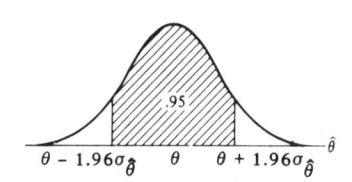

3. The symbol $\sigma_{\hat{\theta}}$ denotes the standard deviation of $\hat{\theta}$. Thus $\sigma_{\hat{\theta}}$ will be the standard deviation of the estimates generated by $\hat{\theta}$ in repeated sampling. The measure of goodness of a particular estimate is the distance that it lies from the target θ. We call this distance _____ _____ _____ _____ . [the error of estimation] When $\hat{\theta}$ possesses the properties stated above, the probability is approximately .95 that the error of estimation will be less than _____ . [$1.96\sigma_{\hat{\theta}}$] We often refer to $1.96\sigma_{\hat{\theta}}$ as the *bound* on the error of estimation. By this we mean that the error will be less than $1.96\sigma_{\hat{\theta}}$ with high probability (say, near .95).

Complete the following table, filling in the estimator and its standard deviation where required:

Parameter	Estimator	Standard Deviation
μ	$\bar{x} = \sum_{i=1}^{n} x_i/n$	$\sigma/\sqrt{n}$
p	$\hat{p} = x/n$	_____ [$\sqrt{pq/n}$]
$\mu_1 - \mu_2$	_____ [$\bar{x}_1 - \bar{x}_2$]	_____ [$\sqrt{\dfrac{\sigma_1^2}{n_1} + \dfrac{\sigma_2^2}{n_2}}$]
$p_1 - p_2$	$\hat{p}_1 - \hat{p}_2$	$\sqrt{\dfrac{p_1 q_1}{n_1} + \dfrac{p_2 q_2}{n_2}}$

Notice that evaluation of the standard deviations given in the table may require values of parameters that are unknown. When the sample sizes are large, the sample estimates can be used to calculate an approximate standard deviation. As a rule of thumb, we will consider samples of size 30 or greater to be large samples.

Example 7.2
The mean length of stay for patients in a hospital must be known in order to estimate the number of beds required. The length of stay, recorded for a sample of 400 patients at a given hospital, produced a mean and a standard deviation equal to 5.7 and 8.1 days, respectively. Give a point estimate for μ, the mean length of stay for patients entering the hospital, and place a bound of error on this estimate.

Solution
1. The point estimate for μ is $\bar{x}$ = _____ .   5.7
2. Since σ is unknown, the *approximate* bound on error is

$$1.96\left(\frac{s}{\sqrt{n}}\right) = 1.96\left(\frac{8.1}{\sqrt{400}}\right) = \underline{\qquad}.$$ .79

Example 7.3
An experimenter is interested in investigating family sizes in an attempt to determine the percentage of families having more than two children under the age of 18. She randomly samples $n = 100$ families and finds 12 families with more than two children under the age of 18. Estimate the true proportion of families with more than two children under 18 and place a bound on the error of estimation.

Solution
1. The point estimate of p is

$$\hat{p} = \frac{x}{n} = \frac{\underline{\quad}}{100} = \underline{\qquad}$$ 12; .12

That is, _____ % of families have more than two children under 18.   12
2. The bound on the error of estimation is $1.96\sigma_{\hat{p}} = 1.96\sqrt{pq/n}$. The quantities p and q are unknown. However, since n is large, $\hat{p}$ and $\hat{q}$ may be substituted for p and q. The *approximate* bound on the error of estimation is

$$1.96\sqrt{\frac{\hat{p}\hat{q}}{n}} = 1.96\sqrt{\underline{\qquad}} = 1.96\sqrt{\underline{\qquad}}$$ $\frac{(.12)(.88)}{100}$; .001056

$$= 1.96(\underline{\qquad}) = \underline{\qquad}$$ .0325; .064

Self-Correcting Exercises 7A

1. In standardizing an examination, the average score on the exam must be known in order to differentiate among examinees taking the examination. The scores recorded for a sample of 93 examinees yielded a mean and a standard deviation of 67.5 and 8.2, respectively. Estimate the true mean score for this examination and place bounds on the error of estimation.
2. Using the following data, give a point estimate with bounds on the error for the difference in mortality rates in breast cancers where radical or simple mastectomy was used as a treatment.

	Radical	Simple
Number died	31	41
Number treated	204	191

3. A physician wishes to estimate the proportion of accidents on the California freeway system that result in fatal injuries to at least one person. He randomly checks the files on 50 automobile accidents and finds that 8 resulted in fatal injuries. Estimate the true proportion of fatal accidents, and place bounds on the error of estimation.
4. In measuring the tensile strength of two alloys, strips of the alloys were subjected to tensile stress and the force (measured in pounds) at which the strip broke recorded for each strip. The data are summarized below.

	Alloy 1	Alloy 2
$\bar{x}$	150.5	160.2
s^2	23.72	36.37
n	35	35

Use these data to estimate the true mean difference in tensile strength by finding a point estimate for $\mu_1 - \mu_2$ and placing a bound on the error of estimation.

7.4 Interval Estimation

An interval estimator is a rule that tells us how to calculate two points based on information contained in a sample. The objective is to form a narrow interval that will enclose the parameter. As in the case of point estimation, we can form many interval estimators (rules) for estimating the parameter of interest. Not all intervals generated by an interval estimator will actually enclose the parameter. The probability that an interval estimate will enclose the parameter is called the _____ _____.

Let $\hat{\theta}$ be an *unbiased* point estimator of θ and suppose that $\hat{\theta}$ generates a normal distribution of estimates in repeated sampling. The mean of this distribution of estimates is _____ and the standard deviation is $\sigma_{\hat{\theta}}$. Then _____% of the point estimates will lie within $1.96\sigma_{\hat{\theta}}$ of the parameter θ. Similarly, _____% will lie in the interval $\theta \pm 1.645\sigma_{\hat{\theta}}$.

---

confidence coefficient

θ
95
90

Suppose we were to construct an interval estimate by measuring the distance $1.645\sigma_{\hat{\theta}}$ on either side of $\hat{\theta}$. *Intervals constructed in this manner will enclose* θ _____ *% of the time* (see below).

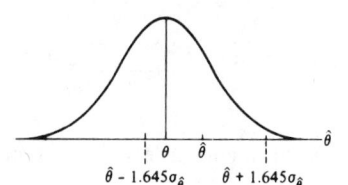

90

Thus for a confidence interval with confidence coefficient $(1 - \alpha)$, we use

$$\hat{\theta} \pm z_{\alpha/2}\sigma_{\hat{\theta}}$$

to construct the interval estimate. The quantity $z_{\alpha/2}$ satisfies the relation $P(z > z_{\alpha/2}) = \alpha/2$, as indicated below:

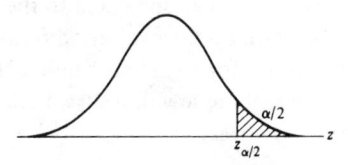

Not all good interval estimators are constructed by measuring $z_{\alpha/2}\sigma_{\hat{\theta}}$ on either side of the best point estimator, but this is true for the parameters μ, p, $(\mu_1 - \mu_2)$, and $(p_1 - p_2)$. These confidence intervals are good for samples that are large enough to achieve approximate normality for the distribution of $\hat{\theta}$ and good approximations for unknown parameters appearing in $\sigma_{\hat{\theta}}$.

A 95% confidence interval for μ is $\bar{x} \pm$ _____ $\sigma/\sqrt{n}$. As a rule of thumb, the sample size n must be greater than or equal to _____ in order that s be a good approximation to σ.

1.96
30

Give the values of z corresponding to the following confidence coefficients:

Confidence Coefficients	$z_{\alpha/2}$
.95	1.96
.90	_____
.99	_____

1.645
2.58

Example 7.4

Refer to Example 7.2. To construct a 90% confidence interval for the mean length of hospital stay, μ, based on the sample of $n = 400$ patients ($\bar{x} = 5.7$ and $s = 8.1$), we calculate

$$\bar{x} \pm z_{\alpha/2}\sigma/\sqrt{n}.$$

Using $z_{.05} = 1.645$ and an estimate for σ given by $s = 8.1$, the interval estimate for the mean length of hospital stay is given as

$$5.7 \pm .67.$$

5.03; 6.37

More properly, we estimate that _____ < μ < _____ with 90% confidence.

The formula for a 95% confidence interval for a binomial parameter p is

$\hat{p} \pm 1.96\sqrt{pq/n}$

_____.

Note that $\hat{p}$ is used to approximate p in the formula for $\sigma_{\hat{p}}$ since its value is unknown.

Example 7.5

An experimental rehabilitation technique employed on released convicts showed that 79 of a total of 121 men subjected to the technique pursued useful and crime-free lives for a three-year period following prison release. Find a 95% confidence interval for p, the probability that a convict subjected to the rehabilitation technique will follow a crime-free existence for at least three years after prison release.

Solution

The sampling described above satisfies the requirements of a binomial experiment consisting of $n = 121$ trials. In estimating the parameter p with a 95% confidence interval, we use the estimator

1.96

$$\hat{p} \pm \text{_____} \sqrt{pq/n}.$$

Since p is unknown, the sample value $\hat{p}$ will be used in the approximation of $\sqrt{pq/n}$. Collecting pertinent information, we have

.65

1. $\hat{p} = x/n = 79/121 = $ _____.

2. $\sqrt{\hat{p}\hat{q}/n} = \sqrt{(.65)(.35)/121} = .04.$

3. The interval estimate is given as

$$.65 \pm 1.96(.04)$$

.08

or $.65 \pm$ _____.

.57; .73

4. We estimate that _____ < p < _____ with 95% confidence.

The exact values for standard deviations of estimators cannot usually be found because they are functions of unknown population parameters. For the following estimators, give the standard deviation and the best approximation of the standard deviation for use in confidence intervals.

Estimator ($\hat{\theta}$)	Standard Deviation ($\sigma_{\hat{\theta}}$)	Best Approximation of Standard Deviation ($\hat{\sigma}_{\hat{\theta}}$)	
$\bar{x}$	$\sigma/\sqrt{n}$	_____	$s/\sqrt{n}$
$\hat{p}$	$\sqrt{pq/n}$	_____	$\sqrt{\hat{p}\hat{q}/n}$
$\bar{x}_1 - \bar{x}_2$	$\sqrt{\dfrac{\sigma_1^2}{n_1} + \dfrac{\sigma_2^2}{n_2}}$	$\sqrt{\dfrac{s_1^2}{n_1} + \dfrac{s_2^2}{n_2}}$	
$\hat{p}_1 - \hat{p}_2$	$\sqrt{\dfrac{p_1 q_1}{n_1} + \dfrac{p_2 q_2}{n_2}}$	$\sqrt{\dfrac{\hat{p}_1 \hat{q}_1}{n_1} + \dfrac{\hat{p}_2 \hat{q}_2}{n_2}}$	

The large-sample confidence intervals for μ, p, $\mu_1 - \mu_2$, and $p_1 - p_2$ will be

$$\hat{\theta} \pm z_{\alpha/2}\, \hat{\sigma}_{\hat{\theta}}$$

where $\hat{\theta}$ is given by $\bar{x}$, $\hat{p}$, $\bar{x}_1 - \bar{x}_2$, and $\hat{p}_1 - \hat{p}_2$, respectively. The table above will determine the appropriate formula for $\sigma_{\hat{\theta}}$ or $\hat{\sigma}_{\hat{\theta}}$.

Example 7.6

An experiment was conducted to compare the mean absorptions of drug in specimens of muscle tissue. Seventy-two tissue specimens were randomly divided between two drugs A and B with 36 assigned to each drug, and the drug absorption was measured for the 72 specimens. The means and variances for the two samples were $\bar{x}_1 = 7.8$, $s_1^2 = .10$ and $\bar{x}_2 = 8.4$, $s_2^2 = .06$, respectively. Find a 95% confidence interval for the difference in mean absorption rates.

Solution

We are interested in placing a confidence interval about the parameter _____ . The confidence interval is $\mu_1 - \mu_2$

$$(\bar{x}_1 - \bar{x}_2) \pm z_{\alpha/2} \sqrt{\dfrac{s_1^2}{n_1} + \dfrac{s_2^2}{n_2}}$$

$(7.8 - 8.4) \pm$ _____ $\sqrt{\dfrac{.10}{36} + \dfrac{.06}{36}}$ 1.96

_____ $\pm .131$ $-.6$

or   _____ $< \mu_1 - \mu_2 <$ _____ . $-.731;\ -.469$

Example 7.7

The voting records at two precincts were compared based on samples of 400 voters each. Those voting Democratic numbered 209 and 263, respectively. Estimate the difference in the fraction voting Democratic for the two precincts, using a 90% confidence interval.

Solution
The confidence interval is

$$(\hat{p}_1 - \hat{p}_2) \pm z_{\alpha/2} \sqrt{\frac{\hat{p}_1 \hat{q}_1}{n_1} + \frac{\hat{p}_2 \hat{q}_2}{n_2}}$$

$$(.5225 - .6575) \pm \underline{} \sqrt{\frac{(.5225)(.4775)}{400} + \frac{(.6575)(.3425)}{400}}$$

1.645

$$\underline{} \pm .057$$

−.135

or $\underline{} < p_1 - p_2 < \underline{}$.

−.192; −.078

Self-Correcting Exercises 7B

1. Suppose it is necessary to estimate the percentage of students on the University of Florida campus who favor constitutional revision of the state constitution. In a random sample of 65 students, 30 stated that they were in favor of revision. Estimate the percentage of students favoring revision with a 98% confidence interval.

2. It is desired to estimate the difference in accident rates between youth and adult drivers. The following data were collected from two random samples, where x is the number of drivers that were involved in one or more accidents:

Youths	Adults
$n_1 = 100$	$n_2 = 200$
$x_1 = 50$	$x_2 = 60$

Estimate the difference $(p_1 - p_2)$ with a 95% confidence interval.

3. The yearly incomes of high school teachers in two cities yielded the following tabulation:

	City 1	City 2
Number of teachers	90	60
Average income	28,520	27,210
Standard deviation	1,510	950

a. If the teachers from each city are thought of as samples from two populations of high school teachers, use the data to construct a 99% confidence interval for the difference in mean annual incomes.

b. Using the results of part a, would you be willing to conclude that these two city schools belong to populations having the same mean annual income?

4. In 39 soil samples tested for trace elements, the average amount of copper was found to be 22 milligrams, with a standard deviation of 4 milligrams. Find a 90% confidence interval for the true mean copper content in the soils from which these samples were taken.

7.5 Choosing the Sample Size

The amount of information contained in a sample depends on two factors:
1. The quantity of information per observation, which depends on the sampling procedure or experimental design.
2. The number of measurements or observations taken, which depends on the sample size.

In this chapter, we will concentrate on choosing the sample size required to obtain the desired amount of information.

One of the first steps in planning an experiment is deciding on the quantity of information that we wish to buy. At first glance it would seem difficult to specify a measure of the quantity of information in a sample relevant to a parameter of interest. However, such a practical measure is available in the bound on the error of estimation; or, alternatively, we could use the half-width of the confidence interval for the parameter.

The larger the sample size, the greater will be the amount of information contained in the sample. This intuitively appealing fact is evident upon examination of the large sample confidence intervals. The width of each of the four confidence intervals described in the preceding section is inversely proportional to the square root of the _____ _____.   sample size

Suppose that $\hat{\theta}$ is an estimator of θ and satisfies the conditions for the large-sample estimators previously discussed. Then the bound B on the error of estimation will be $1.96\sigma_{\hat{\theta}}$. This means that the error (in repeated sampling) will be less than $1.96\sigma_{\hat{\theta}}$ with probability _____ . If B represents the   .95
desired bound on the error, then

a. for a *point* estimator $\hat{\theta}$, the restriction is $1.96\sigma_{\hat{\theta}} = B$;
b. in an interval estimation problem with $(1 - \alpha)$ confidence coefficient, the restriction is $z_{\alpha/2}\sigma_{\hat{\theta}} = B$.

Parts a and b will be equivalent when the confidence coefficient is .95.

Example 7.8
Suppose it is known that $\sigma = 2.25$ and it is desired to estimate μ with a bound on the error of estimation less than or equal to .5 unit with probability .95. How large a sample should be taken?

Solution
The estimator for μ is $\bar{x}$ with standard deviation $\sigma/\sqrt{n}$; $1 - \alpha = .95$, $\alpha/2 = .025$, $z_{.025} = 1.96$. Hence, we solve

$$1.96(\sigma/\sqrt{n}) = B$$

164 / Chap. 7: Large-Sample Statistical Estimation

or $\qquad$ $1.96(2.25/\sqrt{n}) = .5$

$$1.96(2.25/.5) = \sqrt{n}$$

8.82 $\qquad \underline{} = \sqrt{n}$

77.79 $\qquad \underline{} = n.$

78 The solution is to take a sample of size _____ or greater to insure that the bound is less than or equal to .5 unit. Had we wished to have the same

2.58 bound with probability .99, the value $z_{.005} = $ _____ would have been used, resulting in the following solution:

2.58 $\qquad \underline{} (2.25/\sqrt{n}) = .5$

$$\sqrt{n} = 2.58 \, (2.25)/.5$$

134.79 $\qquad n = \underline{}.$

135 Hence, a sample of size _____ or greater would be taken to insure estimation with $B = .5$ unit.

Example 7.9
If an experimenter wished to estimate the fraction of university students that read daily the college newspaper, correct to within .02 with probability .90, how large a sample of students should she take?

Solution
To estimate the binomial parameter with a 90% confidence interval, we would use

1.645 $\qquad \hat{p} \pm \underline{} \sqrt{pq/n}.$

We wish to find a sample size n so that

$$1.645 \sqrt{pq/n} = .02.$$

Since neither p nor $\hat{p}$ is known, we can solve for n by assuming the worst

.5 possible variation, which occurs when $p = q = $ _____. Hence, we solve

$$1.645 \sqrt{(.5)(.5)/n} = .02$$

$$1.645 \, (.5)/.02 = \sqrt{n}$$

1,691.27

or $\qquad \underline{} = n.$

1,692

Therefore, we should take a sample of size _____ or greater to achieve the required bounds, even if faced with the maximum variation possible.

Example 7.10
An experiment is to be conducted to compare two different sales techniques at a number of sales centers. Suppose that the range of sales for the sales centers is expected to be $4,000. How many centers should be included for each of the sales techniques in order to estimate the difference in mean sales correct to within $500?

Solution
We will assume that the two sample sizes are equal, that is, $n_1 = n_2 = n$, and that the desired confidence coefficient is .95. Then

$$\underline{\qquad} = B. \qquad\qquad 1.96\sqrt{\frac{\sigma_1^2}{n} + \frac{\sigma_2^2}{n}}$$

The quantities σ_1^2 and σ_2^2 are unknown but we know that the range is expected to be $4,000. Then we would take $\sigma_1 = \sigma_2 = \underline{\qquad}$ as the best available approximation. Then, substituting into the equation above,   1,000

$$1.96\sqrt{\underline{\qquad}} = 500 \qquad\qquad 2(1{,}000)^2/n$$

or $\quad n = \underline{\qquad}$.   30.73

Thus $n = \underline{\qquad}$ sales centers would be required for each of the two sales techniques.   31

Self-Correcting Exercises 7C

1. A device is known to produce measurements whose errors in measurement are normally distributed with a standard deviation $\sigma = 8$ millimeters. If the average measurement is to be reported, how many repeated measurements should be used so that the error in measurement is no larger than 3 millimeters with probability .95?
2. An experiment is to be conducted to compare the taste threshold levels for each of two food additives as measured by their concentrations in parts per million. How many subjects should be included in each experimental group in order to estimate the mean difference in threshold levels to within 10 units if the range of the measurements is expected to be approximately 80 parts per million for both groups?
3. How many individuals from each of two politically oriented groups should be included in a poll designed to estimate the true difference in proportions favoring a tuition increase at the state university correct to within .01 with probability .95? (In the absence of any prior information regarding the values of p_1 and p_2, solve the problem assuming maximum variation.)
4. If a mental health agency would like to estimate the percentage of local clinic patients that are referred to their counseling center to within 5 percentage points with 90% accuracy, how many patient records should be sampled?

7.6 Summary

Estimation as a method of inference making has been discussed in this chapter. Point estimation and interval estimation for four parameters, μ, p, $\mu_1 - \mu_2$, $p_1 - p_2$, have been presented for large-sample situations.

To provide a brief summary of the preceding sections, complete the following tables.

1. Give the best estimator for each of the following parameters:

Parameter	Estimator
μ	_____
p	_____
$(\mu_1 - \mu_2)$	_____
$(p_1 - p_2)$	_____

[Margin answers: $\bar{x}$, $\hat{p}$, $\bar{x}_1 - \bar{x}_2$, $\hat{p}_1 - \hat{p}_2$]

2. Give the standard deviations for the following estimators:

Estimator	Standard Deviation
$\bar{x}$	_____
$\hat{p}$	_____
$\bar{x}_1 - \bar{x}_2$	_____
$\hat{p}_1 - \hat{p}_2$	_____

[Margin answers: $\sigma/\sqrt{n}$, $\sqrt{\dfrac{pq}{n}}$, $\sqrt{\dfrac{\sigma_1^2}{n_1} + \dfrac{\sigma_2^2}{n_2}}$, $\sqrt{\dfrac{p_1 q_1}{n_1} + \dfrac{p_2 q_2}{n_2}}$]

3. The exact values for standard deviations of estimators cannot usually be found because they are functions of unknown population parameters. Indicate the best approximations of the standard deviations for use in confidence intervals:

Estimator	Best Approximation of Standard Deviation
$\bar{x}$	_____
$\hat{p}$	_____
$\bar{x}_1 - \bar{x}_2$	_____
$\hat{p}_1 - \hat{p}_2$	_____

[Margin answers: $s/\sqrt{n}$, $\sqrt{\dfrac{\hat{p}\hat{q}}{n}}$, $\sqrt{\dfrac{s_1^2}{n_1} + \dfrac{s_2^2}{n_2}}$, $\sqrt{\dfrac{\hat{p}_1 \hat{q}_1}{n_1} + \dfrac{\hat{p}_2 \hat{q}_2}{n_2}}$]

Bounds on error are then given as _____ or its approximation, while large-sample confidence intervals are given as _____ , using the best approximation for unknown parameters in $\sigma_{\hat{\theta}}$.

$1.96\sigma_{\hat{\theta}}$
$\hat{\theta} \pm z\sigma_{\hat{\theta}}$

EXERCISES

1. List the two essential elements of any inference-making procedure.
2. What are two desirable properties of a point estimator $\hat{\theta}$?
3. A bank was interested in estimating the average size of its savings accounts for a particular class of customer. If a random sample of 400 such accounts showed an average amount of $61.23 and a standard deviation of $18.20, place 90% confidence limits on the actual average account size.
4. If 36 measurements of the specific gravity of aluminum had a mean of 2.705 and a standard deviation of .028, construct a 98% confidence interval for the actual specific gravity of aluminum.
5. An appliance dealer sells toasters of two different brands, brand A and brand B. Let p_1 denote the fraction of brand A toasters that are returned to him by customers as defective, and let p_2 represent the fraction of brand B toasters that are rejected by customers as defective. Suppose that of 200 brand A toasters sold, 14 were returned as defective, while of 450 brand B toasters sold, 18 were returned as defective. Provide a 90% confidence interval for $p_1 - p_2$.
6. In a sample of 400 seeds, 240 germinate. Estimate the true germination percentage with a 95% confidence interval.
7. Refer to Exercise 4. Estimate the specific gravity of aluminum and place bounds on the error of estimation. Compare these results to the results obtained in Exercise 4. Can you explain the difference?
8. In a study to establish the absolute threshold of hearing, 70 male college freshmen were asked to participate. Each subject was seated in a soundproof room and a 150 Hz tone was presented at a large number of stimulus levels in a randomized order. The subject was instructed to press a button if he detected the tone. The mean for the group was 21.6 db with $s = 2.1$. Estimate the mean absolute threshold of all men 19-21 years of age and place bounds on the error of estimation.
9. A researcher classified his subjects as innately right-handed or left-handed by comparing thumb-nail widths. He took a sample of 400 men and found that 80 men could be classified as left-handed according to his criterion. Estimate p for all males with a 95% confidence interval, where p represents the probability a man tests to be left-handed.
10. An entomologist wishes to estimate the average development time of the citrus red mite correct to within .5 day. From previous experiments it is known that σ is in the neighborhood of 4 days. How large a sample should the entomologist take to be 95% confident of her estimate?
11. A manufacturer of dresses believes that approximately 20% of his product contains flaws. If he wishes to estimate the true percentage to within 8%, how large a sample should he take?
12. It is desired to estimate $\mu_1 - \mu_2$ from information contained in independent random samples from populations with variances $\sigma_1^2 = 9$ and $\sigma_2^2 = 16$. If the two sample sizes are to be equal ($n_1 = n_2 = n$), how large should n be

in order to estimate $\mu_1 - \mu_2$ with an error less than 1.0 (with probability equal to .95)?

13. To compare the effect of stress in the form of noise upon the ability to perform a simple task, 70 subjects were divided into two groups. The first group of 30 subjects were to act as a control, while the second group of 40 were to be the experimental group. Although each subject performed the task in the same control room, each of the experimental group subjects had to perform the task while loud rock music was being played in the room. The time to finish the task was recorded for each subject and the following summary was obtained:

	Control	Experimental
n	30	40
$\bar{x}$	15 minutes	23 minutes
s	4 minutes	10 minutes

Find a 99% confidence interval for the difference in mean completion times for these two groups.

14. In an experiment to assess the strength of the hunger drive in rats, 30 previously trained animals were deprived of food for 24 hours. At the end of the 24-hour period each animal was put into a cage where food was dispensed if the animal pressed a lever. The length of time the animal continued pressing the bar (although he was receiving no food) was recorded for each animal. If the data yielded a sample mean of 19.3 minutes with a standard deviation of 5.2 minutes, estimate the true mean time and place bounds on the error of estimation.

15. Last year's records of auto accidents occurring on a given section of highway were classified according to whether the resulting damage was $200 or more and to whether or not a physical injury resulted from the accident. The tabulation follows:

	Under $200	$200 or more
Number of accidents	32	41
Number involving injuries	10	23

a. Estimate the true proportion of accidents involving injuries and damage of $200 or more for similar sections of highway and place bounds on the error of estimation.

b. Estimate the true difference in proportion of accidents involving injuries for accidents involving less than $200 in damage and those involving $200 or more with a 95% confidence interval.

CHAPTER 8

LARGE-SAMPLE TESTS OF HYPOTHESES

8.1 Introduction

In Chapter 7 we presented estimation as an inferential technique. We developed both point and interval estimators for the binomial parameter p, and the mean μ of a population of continuous measurements, as well as the differences $\mu_1 - \mu_2$ and $p_1 - p_2$.

We now turn our attention to a decision-making form of inference, hypothesis testing. In hypothesis testing, we formulate an hypothesis about a population in terms of its _____, and then, after observing a _____ drawn from this population, we decide whether our sample value could have come from the hypothesized population. We then accept or reject the hypothesized value. Hypothesis testing will be discussed for the case of a binomial parameter p, and for μ, the mean of a population of continuous measurements.

<small>parameters; sample</small>

8.2 A Statistical Test of an Hypothesis

A statistical test of an hypothesis consists of four parts:
1. _____ _____, (H_0): This is the hypothesis to be tested and gives hypothesized values for one or more population parameters.
2. _____ or _____ _____, (H_a): This is the hypothesis against which H_0 is tested. We look for evidence in the sample that will cause us to reject H_0 in favor of H_a.
3. Test statistic: This function of the sample values extracts the information about the parameter contained in the sample. The observed value of the test statistic leads us to reject one hypothesis and accept the other.
4. Rejection region: Once the test statistic to be used is selected, the entire set of values that the statistic may assume is divided into two regions. The acceptance region consists of those values most likely to have arisen if H_0 were true. The rejection region consists of those values most likely to have arisen if H_a were true. If the observed value of the test statistic falls in the rejection region, H_0 is rejected; if it falls in the acceptance region, H_0 is accepted.

<small>Null hypothesis</small>

<small>Research alternative hypothesis</small>

Our approach will be to draw a random sample from the population of interest and decide whether we will accept or reject an hypothesized value of the specified parameter. The decision will be made on the basis of whether the sample

170 / Chap. 8: Large-Sample Tests of Hypotheses

probable
improbable

evidence

results are highly _____ and support the hypothesized value or are _____ and fail to support the hypothesized value. Our procedure is very similar to a court trial in which the accused is assumed innocent until proven guilty. In fact, our sample acts as the _____ for or against the accused. What we do is to compare the hypothesized value with reality. Let us illustrate how a test of an hypothesis is conducted.

Example 8.1
A large orchard has averaged 140 pounds of apples per tree per year. A new fertilizer is tested to try to increase yield. Forty trees are randomly selected and the mean and standard deviation of yield are $\bar{x} = 143.2$ and $s = 9.4$. Do the data indicate a significant increase in yield?

Solution
If there is no increase in the average yield for fertilized trees, then the average value μ of the hypothetical population of fertilized trees from which we have obtained a random sample is $\mu = 140$. If, on the other hand, this is not true and the fertilizer tends to increase yield, then μ, the mean yield for fertilized trees, would be greater than 140. Since the researcher is interested in detecting an increase in yield. the statement $\mu > 140$ is the _____ or

research
alternative; null
hypothesis

_____ hypothesis, while the statement $\mu = 140$ is the _____ _____.

Assuming that there is no increase in the average yield for fertilized trees, it would be extremely unlikely in a sample of $n = 40$ trees to observe certain values of $\bar{x}$. Since $\bar{x}$ has an approximate normal distribution with $\mu_{\bar{x}} = 140$ and

1.486

$$\sigma_{\bar{x}} \approx \frac{s}{\sqrt{n}} = \frac{9.4}{\sqrt{40}} = \underline{\qquad},$$

we can calculate the probability of observing a particular value of $\bar{x}$ or something even more extreme. For example, suppose that $\bar{x} = 150$. This is a highly unlikely event since

$$P(\bar{x} > 150) = P\left(z > \frac{150 - 140}{1.486}\right)$$

6.73; 0

$$= P(z > \underline{\qquad}) \approx \underline{\qquad}.$$

The occurrence of $\bar{x} = 141$, however, is not so unlikely, since

$$\frac{141 - 140}{1.486}$$

$$P(\bar{x} > 141) = P\left(z > \underline{\qquad}\right)$$

.67

$$= P(z > \underline{\qquad})$$

.2486; .2514

$$= .5 - \underline{\qquad} = \underline{\qquad}.$$

In conducting a test of hypothesis, the possible outcomes for $\bar{x}$ are divided into those for which we agree to reject the null hypothesis [those values of $\bar{x}$ that are

much (greater, less) than $\mu = 140$] and those for which we accept the null hypothesis [those values of $\bar{x}$ that are (close to, far away from) $\mu = 140$.] | greater
close to

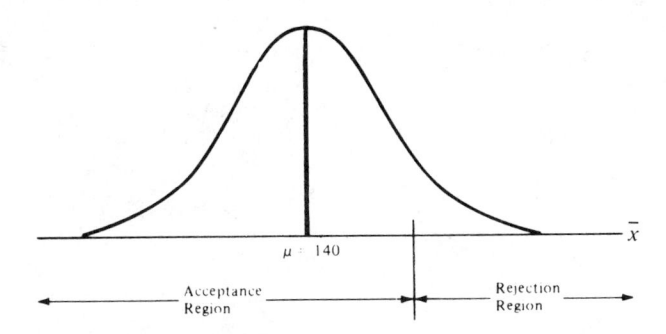

There are many possible rejection regions available to the experimenter. A sound choice among various reasonable rejection regions can be made after considering the possible errors that can be made in a test of an hypothesis.

The following table is called a *decision table* in which we look at the two possible states of nature (H_0 and H_a) and the two possible decisions in a test of an hypothesis. Fill in the missing entries as either "correct" or "error":

	Decision	
Null Hypothesis	Reject H_0	Accept H_0
True	_____	correct
False	correct	_____

error
error

An error of type I is made when we reject H_0 when H_0 is (true, false). An error of type II is made when we fail to reject H_0 when H_a is (true, false). In considering a statistical test of any hypothesis, it is essential to know the probabilities of committing errors of type I and type II when the test is used in order to assess the goodness of the test. | true
true

A *type I error* for a statistical test is the error made by rejecting the null hypothesis when it is true. The probability of committing a type I error is denoted by α and

$$\alpha = P(\text{type I error}) = P(\text{reject } H_0 \text{ when } H_0 \text{ is true}).$$

A *type II error* for a statistical test is the error made by accepting the null hypothesis when it is false and the alternative hypothesis is true. The probability of committing a type II error is denoted by β and

$$\beta = P(\text{type II error}) = P(\text{accept } H_0 \text{ when } H_a \text{ is true}).$$

172 / Chap. 8: Large-Sample Tests of Hypotheses

For this example, suppose we set the region $\bar{x} > 142$ as the rejection region.

1. If H_0 is true and $\mu = 140$,

$$\alpha = P(\text{reject } H_0 \text{ when } H_0 \text{ true})$$

$$= P(\bar{x} > 142) = P\left(z > \frac{142 - 140}{1.486}\right)$$

1.35; .4115

$$= P(z > \underline{\qquad}) = .5 - \underline{\qquad}$$

.0885

$$= \underline{\qquad}.$$

2. If H_0 is false and $\mu = 145$,

$$\beta = P(\text{accept } H_0 \text{ when } H_a \text{ true})$$

$$= P(\bar{x} < 142 \text{ when } \mu = 145)$$

$$= P\left(z < \frac{142 - 145}{1.486}\right)$$

−2.02; .4783

$$= P(z < \underline{\qquad}) = .5 - \underline{\qquad}$$

.0217

$$= \underline{\qquad}.$$

Notice that β is a function of H_a, since by definition,

$$\beta = P(\text{accepting } H_0 \text{ when } H_a \text{ true}).$$

By saying H_a is true, we mean that the true value of the population parameter is that given by H_a, and β is computed using that value.

Consider a second rejection region, $\bar{x} > 144$. This rejection region is (larger, smaller

smaller) than the first region.

1. If H_0 is true, and $\mu = 140$,

$$\alpha = P(\text{reject } H_0 \text{ when } H_0 \text{ true})$$

$$= P(\bar{x} > 144 \text{ when } \mu = 140)$$

$\dfrac{144 - 140}{1.486}$

$$= P\left(z > \underline{\qquad}\right)$$

$$= P(z > 2.69)$$

$$= .5 - .4964 = .0036.$$

2. If H_0 is false and $\mu = 145$,

$$\beta = P(\text{accept } H_0 \text{ when } H_a \text{ true})$$

$$= P(\bar{x} < 144 \text{ when } \mu = 145)$$

$$= P\left(z < \underline{\hspace{1cm}}\right) \qquad\qquad \frac{144 - 145}{1.486}$$

$$= P(z < -.67)$$

$$= .5 - \underline{\hspace{1cm}} = \underline{\hspace{1cm}}. \qquad\qquad .2486;\ .2514$$

Notice that both α and β depend on the rejection region employed and that when the sample size n is fixed, α and β are inversely related: as one increases, the other _____. Increasing the sample size provides more information on which to make the decision and will reduce the probability of a type II error. Since these two quantities measure the risk of making an incorrect decision, the experimenter chooses reasonable values for α and β and then chooses the rejection region and sample size accordingly. Since experimenters have found that a 1-in-20 chance of a type I error is usually tolerable, common practice is to choose $\alpha \leq$ _____ and a sample size n large enough to provide the desired control of the type II error.

decreases

.05

A graph of β, the probability of a type II error, as a function of the true value of the parameter under test is called the *operating characteristic curve* for the test. The operating characteristic curves that illustrated how a lot acceptance plan operated as a function of the sample size n and the acceptance number a and the binomial parameter p were actually plots of β for various values of p.

For any statistical test, the value of α is usually selected and fixed in advance, often at values reflecting a 1-in-20 or a 1-in-100 chance of committing a type I error. The value of β, the probability of incorrectly accepting the null hypothesis, depends on the sample size n and the true value of the parameter under test.

For example, if the null hypothesis specifies $H_0: \mu = 140$, which alternative hypothesis would be more easily detected, $H_a: \mu = 141$ or $H_a: \mu = 200$? The answer, of course, is $H_a: \mu = 200$. In fact, a very large sample would be required to detect $H_a: \mu = 141$, that the mean is one unit larger than that specified by H_0. The ability of a test to reject H_0 when H_a is true is called the power of the test, because a statistical test of an hypothesis is set up with the intention of disproving, and hence rejecting, H_0 when H_a is true.

The *power of a statistical test* is given by

$$1 - \beta = P(\text{rejecting } H_0 \text{ when } H_a \text{ is true})$$

and serves as a measure of the ability of a test to perform as required.

8.3 A Large-Sample Test of an Hypothesis

As in previous sections, the parameter of interest (μ, $\mu_1 - \mu_2$, p, or $p_1 - p_2$) will be referred to as θ. If a point estimator $\hat{\theta}$ exists which is normally distributed with average value θ, we can employ $\hat{\theta}$ as a test statistic to test the hypothesis $H_0: \theta = \theta_0$.

If H_a states that $\theta > \theta_0$, that is, the value of the parameter is greater than that given by H_0, then the sample value for $\hat{\theta}$ should reflect this fact and be (larger, smaller) than a value of $\hat{\theta}$ when sampling from a population whose mean is θ_0. [larger]
Hence we would reject H_0 for large values of $\hat{\theta}$. "Large" can be interpreted as too many standard deviations to the right of the mean θ_0. The value of $\hat{\theta}$ selected to separate the acceptance and rejection regions is called the _____ _____ of the test statistic. [critical value]

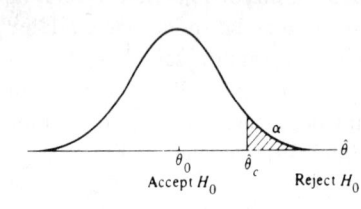

$\hat{\theta}_c$ in the diagram represents the critical value of $\hat{\theta}$ and the shaded area to the right of $\hat{\theta}_c$ is equal to _____. This is a one-tailed statistical test. [α]

A similar picture could have been used with the critical value of $\hat{\theta}$ to the left of the mean for testing $H_0: \theta = \theta_0$ against $H_a: \theta$ _____ θ_0. Then we [<] would reject for values of $\hat{\theta}$ lying too many standard deviations to the left of θ_0 (resulting in a _____-_____ test in the left tail) and [one-tailed] would reject H_0 for small values of $\hat{\theta}$.

A third type of alternative hypothesis would be $H_a: \theta \neq \theta_0$, where we seek departures either greater or less than θ_0. This results in a _____-tailed [two] statistical test.

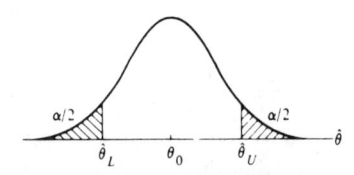

In order that the probability of a type I error be equal to α, two critical values of $\hat{\theta}$ must be found, one having area $\alpha/2$ to its right ($\hat{\theta}_U$) and one having area $\alpha/2$ to its left ($\hat{\theta}_L$). H_0 will be rejected if $\hat{\theta} \geq \hat{\theta}_U$ or $\hat{\theta} \leq \hat{\theta}_L$.

Since the estimator $\hat{\theta}$ is normally distributed, we can standardize the normal variable $\hat{\theta}$ by converting the distance that $\hat{\theta}$ departs from θ_0 to z (the number of standard deviations to the left or right of the mean). Thus we will use z as the test statistic. The four elements of the test are as follows:

1. $H_0: \theta = \theta_0$.

2. One of the three alternatives:

a. $H_a: \theta > \theta_0$. (right-tailed)

b. $H_a: \theta < \theta_0$. (left-tailed)

c. $H_a: \theta \neq \theta_0$. (two-tailed)

3. Test statistic z, where

$$z = \frac{\hat{\theta} - \theta_0}{\sigma_{\hat{\theta}}}.$$

4. Rejection region: Let z_a be a value of z having area a to its right.
   a. For $H_a: \theta > \theta_0$,

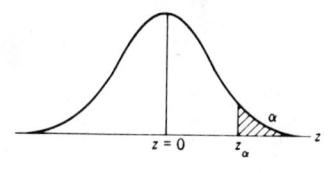

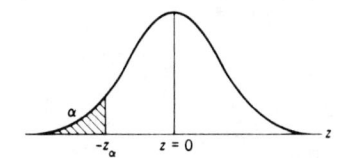

   Reject H_0 if $z > z_\alpha$.

   b. For $H_a: \theta < \theta_0$,

   Reject H_0 if $z < -z_\alpha$.

   c. For $H_a: \theta \neq \theta_0$,

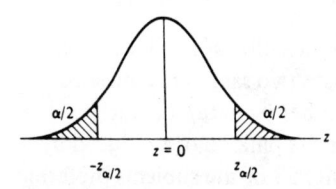

   Reject H_0 if $z > z_{\alpha/2}$
   or $z < -z_{\alpha/2}$
   ($|z| > z_{\alpha/2}$)

We now apply this test of an hypothesis.

Example 8.2

Test of a population mean μ. Test the hypothesis at the $\alpha = .05$ level, that a population mean $\mu = 10$ against the hypothesis that $\mu \neq 10$ if, for a sample of 81 observations, $\bar{x} = 12$ and $s = 3.2$. Note the following:

a. $\theta = \mu$.

b. $\theta_0 = \mu_0 = 10$.

c. $\hat{\theta} = \bar{x}$.

d. $\sigma_{\hat{\theta}} = \sigma/\sqrt{n}$.

e. $\alpha = .05$.

Solution
Since σ is unknown, use s, the sample standard deviation, as its approximation. Then the elements of the test are as follows:

1. $H_0: \mu = 10$.

2. $H_a: \mu \neq 10$.

3. Test statistic:

$$z = \frac{\bar{x} - \mu_0}{\sigma/\sqrt{n}}. \quad \text{(using } s \text{ if } \sigma \text{ is unknown)}$$

4. Rejection region:

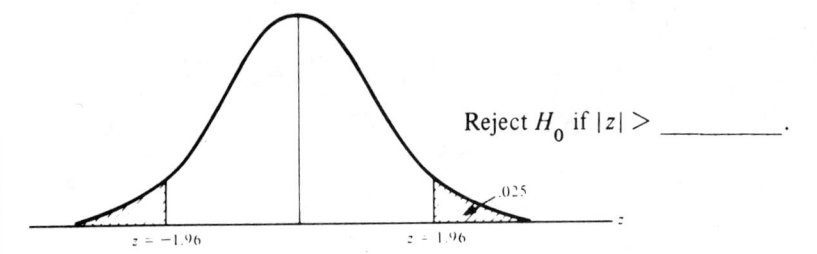

1.96   Reject H_0 if $|z| > $ _____.

.025

$z = -1.96$ $z = 1.96$

Having defined the test, calculate z.

12
3.2

$$z = \frac{\bar{x} - \mu_0}{s/\sqrt{n}} = \frac{\_\_\_ - 10}{\_\_\_/\sqrt{81}} = \frac{2}{.356} = 5.62.$$

reject

Since $z = 5.62 > 1.96$, the decision is (reject, do not reject) H_0 with $\alpha = .05$.

Example 8.3
Test of a binomial p. In assessing the effect of the color of food on taste preferences, 65 subjects were each asked to taste two samples of mashed potatoes, one of which was colored pink, the other its natural color. Although both samples were identical except for the pink color produced by the addition of a drop of tasteless food coloring, 53 of the subjects preferred the taste of the sample possessing its natural color. Does this indicate that subjects tend to be adversely affected by the pink color? Test at the $\alpha = .01$ level of significance.

Solution
In this problem, with $p = P$ (choose natural color), we have the following:

a. $\theta = p$ d. $\sigma_{\hat{\theta}} = \sqrt{p_0 q_0/n}$

½   b. $\theta_0 = p_0 = $ _____

.01   c. $\hat{\theta} = \hat{p} = x/n$.   e. $\alpha = $ _____

Note that p_0 and q_0 are used in $\sigma_{\hat{p}}$ since we are testing $H_0: p = p_0$. The elements of the test are as follows:

1. $H_0: p = \frac{1}{2}$

2. $H_a: p > \frac{1}{2}$

3. Test statistic:

$$z = \frac{\hat{p} - p_0}{\sqrt{p_0 q_0 / n}}$$

4. Rejection region:

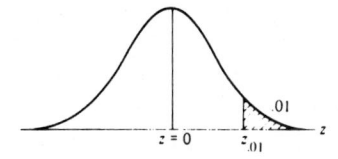

Reject H_0 if $z >$ _____ .   2.33

To calculate z, we need $\hat{p} = x/n = 53/65 = .815$. Then,

$$z = \frac{.815 - \underline{}}{\sqrt{.5(.5)/65}} = \frac{.315}{.062} = 5.08$$ .500

Since $z = 5.08 > 2.33$ we (reject, do not reject) H_0 and conclude that color (does, does not) adversely affect a subject's choice.

reject
does

Example 8.4

Test of an hypothesis concerning $(\mu_1 - \mu_2)$. Suppose an educator is interested in testing whether a new teaching technique is superior to an old teaching technique. The criterion will be a test given at the end of a six-week period and the technique resulting in a significantly higher score will be judged superior. Test the hypothesis that the test means for the techniques are the same against the alternative hypothesis that the new technique is superior at the $\alpha = .05$ level, based on the following sample data:

	New Technique	Old Technique
	$\bar{x}_1 = 69.2$	$\bar{x}_2 = 67.5$
	$s_1^2 = 49.3$	$s_2^2 = 64.5$
	$n_1 = 50$	$n_2 = 80$

Solution

1. $H_0: \mu_1 - \mu_2 =$ _____ .   0

2. $H_a: \mu_1 - \mu_2 > 0$.

178 / Chap. 8: Large-Sample Tests of Hypotheses

3. Test statistic:

$$z = \frac{(\bar{x}_1 - \bar{x}_2) - 0}{\sqrt{\dfrac{s_1^2}{n_1} + \dfrac{s_2^2}{n_2}}}$$

(since σ_1^2 and σ_2^2 are unknown).

4. Rejection region:

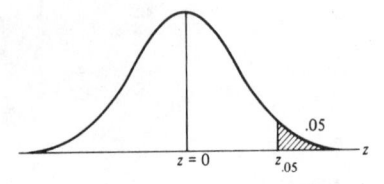

1.645   Reject H_0 if $z > $ _____ .

Computing the value of the test statistic,

69.2; 67.5

$$z = \frac{(\underline{} - \underline{}) - 0}{\sqrt{\dfrac{49.3}{50} + \dfrac{64.5}{80}}}$$

1.7; 1.27

$$= \frac{\underline{}}{\sqrt{716.9/400}} = \frac{1.7}{1.34} = \underline{} .$$

does not; will not

Since the value of z (does, does not) fall in the rejection region, we (will, will not) reject H_0. Before deciding to accept H_0 as true, we may wish to evaluate the probability of a type II error for meaningful values of $\mu_1 - \mu_2$ described by H_a. Until this is done, we will state our decision as "do not reject H_0."

Example 8.5

Test of an hypothesis concerning $(p_1 - p_2)$. To investigate possible differences in attitude about a current political problem, 100 randomly selected voters between the ages of 18 and 25 were polled and 100 randomly selected voters over age 25 were polled. Each was asked if he or she agreed with the government's position on the problem. Forty-five of the first group agreed, while 63% of the second group agreed. Do these data represent a significant difference in attitude for these two groups?

Solution
This problem involves a test of the difference between two binomial proportions, $p_1 - p_2$. The relevant data are given in the table.

	Group 1	Group 2
n	100	100
$\hat{p}$	.45	.63

1. $H_0: p_1 - p_2$ _____. $= 0$

2. $H_a: p_1 - p_2$ _____. $\neq 0$

3. For testing the hypothesis of *no difference* between proportions, the test statistic is

$$z = \frac{(\hat{p}_1 - \hat{p}_2) - 0}{\sqrt{\hat{p}\hat{q}\left(\frac{1}{n_1} + \frac{1}{n_2}\right)}}$$

with

$$\hat{p} = \frac{x_1 + x_2}{n_1 + n_2}.$$

4. Rejection region:

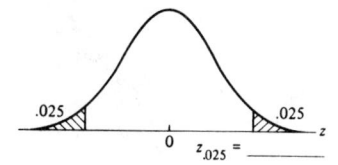

    $z_{.025} =$ _____   1.96

For a two-tailed test with $\alpha = .05$, we will reject H_0 if $|z| >$ _____.   1.96
To calculate the test statistic, we need

$$\hat{p} = \frac{x_1 + x_2}{n_1 + n_2} = \frac{45 + 63}{200} = \underline{\qquad}.$$ .54

Then

$$z = \frac{(.45 - .63) - 0}{\sqrt{(.54)(.46)(2/100)}}$$

$$= \underline{\qquad} / .0705$$ $-.18$

$$= \underline{\qquad}.$$ -2.55

Since $|-2.55| = 2.55 > 1.96$, we (reject, do not reject) H_0 and conclude that   reject
there (is, is not) a significant difference in opinion between these two age   is
groups with respect to this issue.

Example 8.6
Refer to Example 8.1. Using $\alpha = .05$, set up a formal test of hypothesis to test the researcher's claim.

Solution

1. $H_0: \mu = 140$.

2. $H_a: \mu > 140$.

Chap. 8: Large-Sample Tests of Hypotheses

3. Test statistic:

$$z = \frac{\bar{x} - \mu_0}{\sigma/\sqrt{n}}. \quad \text{(using } s \text{ if } \sigma \text{ is unknown)}$$

4. Rejection region:

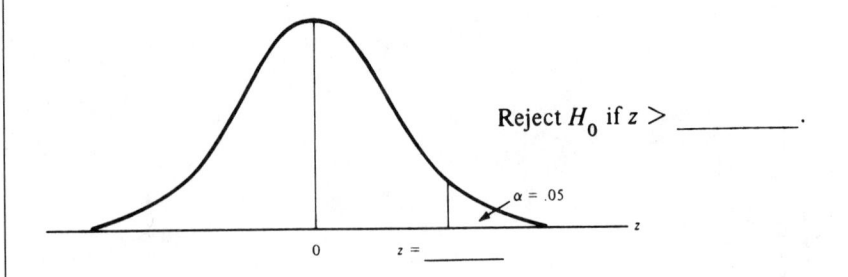

1.645   Reject H_0 if $z > $ _____ .

1.645 $\alpha = .05$, $z = $ _____

Having defined the test, calculate z.

143.2

$$z = \frac{\bar{x} - \mu_0}{\sigma/\sqrt{n}} \approx \frac{\underline{} - 140}{9.4/\sqrt{40}}$$

3.2
   2.15
1.486

$$= \frac{\underline{}}{\underline{}} = \underline{}.$$

2.15; reject
does

Since $z = $ _____ > 1.645, the decision is to (reject, not reject) H_0 with $\alpha = .05$. The new fertilizer (does, does not) increase yield.

Example 8.7
Refer to Example 8.6. Calculate β and the power of the test when $\mu_a = 145$.

Solution
1. The acceptance region for the test in Example 8.6 consists of values of $\bar{x}$ which were greater than

$$\mu + 1.645\,\sigma/\sqrt{n} = 140 + 1.645(9.4/\sqrt{n})$$

$$= 140 + 1.645(1.486)$$

2.44

$$= 140 + \underline{}$$

142.44

$$= \underline{}$$

2. The value of β is the area under the frequency distribution of $\bar{x}$ to the (left, right) of 142.44. If the value of μ is 145 when H_a is true, then $\bar{x}$ is normally distributed with mean $\mu = 145$ and standard deviation $\sigma\sqrt{n} = 1.486$.

right

3. Therefore,

$$\beta = P(\bar{x} < 142.22)$$

$$= P\left(\frac{\bar{x} - 145}{1.486} < \frac{142.44 - 145}{1.486}\right)$$

$$= P(z < -1.72)$$

so that

$$\beta = .5 - .4573 = .0427.$$

Various values of β and $1 - \beta$ corresponding to values of μ are given below. Fill in any missing entries.

μ	z	β	$1 - \beta$	
140	1.645	.9500	.0500	
141	.97	_____	.1185	.8815
142	.30	.6179	.3821	
143	−.38	.3520	_____	.6480
144	−1.05	.1469	.8531	
145	−1.72	_____	.9573	.0427
146	−2.40	.0088	.9912	
147	−3.07	.0011	_____	.9989
148	−3.74	.0000	1.0000	

Examine the graph of the power curve which follows.

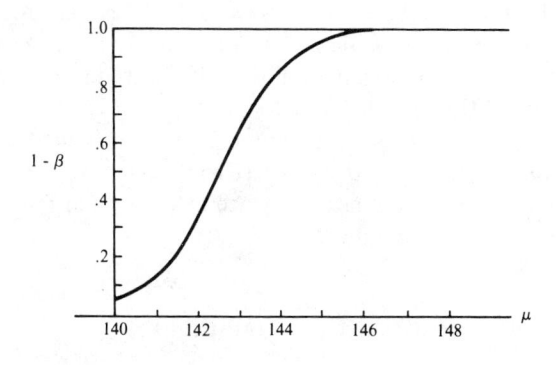

Notice that the power increases to _____ as the value of μ increases in value beginning with $\mu = 140$. Although not calculated, the power decreases to _____ as the value of the mean becomes increasingly smaller than $\mu = 140$. If the test had been two-tailed, the power curve would have been U-shaped with power at its minimum value of .05 when μ = _____, and then increasing symmetrically for values of μ equally spaced to the left and right of $\mu =$ _____.

one

zero

140

140

Self-Correcting Exercises 8A

1. To investigate a possible "built-in" sex bias in a graduate school entrance examination, 50 male and 50 female graduate students who were rated as above-average graduate students by their professors were selected to participate in the study by actually taking this test. Their test results on this examination are summarized in the following table.

	Males	Females
$\bar{x}$	720	693
s^2	104	85
n	50	50

   Do these data indicate that males will, on the average, score higher than females of the same ability on this exam? Use $\alpha = .05$.

2. A machine shop is interested in determining a measure of the current year's sales revenue in order to compare it with known results from last year. From the 9,682 sales invoices for the current year to date, the management randomly selected $n = 400$ invoices and from each recorded x, the sales revenue per invoice. Using the following data summary, test the hypothesis that the mean revenue per invoice is $6.35, the same as last year, versus the alternative hypothesis that the mean revenue per invoice is different from $6.35, with $\alpha = .05$.

   Data Summary

   $n = 400$ $\qquad \sum_{i=1}^{400} x_i = \$2464.40 \qquad \sum_{i=1}^{400} x_i^2 = 16156.728$

3. A physician found 480 men and 420 women among 900 patients admitted to a hospital with a certain disease. Is this consistent with the hypothesis that in the population of patients hospitalized with this disease, half the cases are male? Use $\alpha = .10$.

4. Random samples of 100 shoes manufactured by machine A and 50 shoes manufactured by machine B showed 16 and 6 defective shoes, respectively. Do these data present sufficient evidence to suggest a difference in the performances of the machines? Use $\alpha = .05$.

8.4 The Level of Significance of a Statistical Test

The structure of a statistical test of hypothesis can be summarized as follows:
1. State the null and _____ (or _____) hypotheses. *research; alternative*
2. Choose a test statistic.
3. Choose a value of α and, depending on the nature of the alternative hypothesis, establish a one- or two-tailed rejection region for which the α level is approximately the level chosen.
4. Perform the experiment, calculate the test statistic and come to a conclusion based on the observed value of the test statistic. If the value of the test

statistic falls in the rejection region we _____ the null hypothesis in favor of the alternative hypothesis. If the value of the test statistic is not in the rejection region we (can, cannot) reject the null hypothesis in favor of the alternative hypothesis.

One difficulty in utilizing the approach outlined above is that the choice of the α level is to some extent subjective. Another researcher may disagree with your conclusions regarding the research hypothesis because he or she does not agree with your choice of the α level. The _____ _____ or p-value for an observed value of the test statistic is the smallest α value for which the null hypothesis could be rejected. In using the rare-event philosophy, we reject the null hypothesis in favor of the alternative whenever the total probability of the observed value of the test statistic and any rarer event is _____.

Consider the problem discussed in Example 8.6. In that problem we tested $H_0: \mu = 140$ versus $H_a: \mu > 140$ based on an observed value of $\bar{x} = 143.2$. Since this is a one-tailed test and "large" values of $\bar{x}$ belong in the rejection region, the smallest rejection region that contains $\bar{x}$ is the region $\bar{x} \geq 143.2$. The p-value associated with this observation is therefore

$$p\text{-value} = P(\bar{x} \geq \underline{\qquad})$$

$$= P\left(z \geq \frac{143.2 - 140}{9.4/\sqrt{40}}\right)$$

$$= P(z > 2.15)$$

$$= .5 - .4842$$

$$= \underline{\qquad}.$$

Thus, any researcher who would specify an α value greater than or equal to _____ would conclude that there (is, is not) sufficient evidence to accept the research hypothesis. Since this is a relatively small α value, it would probably be acceptable, and the researcher (would, would not) reject H_0.

The significance-level approach to hypothesis testing does not require specifying an α level before the analysis is undertaken. Rather, the degree of disagreement with the null hypothesis is quantified by the calculation of the _____ _____, which is used to make decisions regarding the research hypothesis.

	reject
	cannot
	level of significance
	small
	143.2
	.0158
	.0158; is
	would
	p-value

Example 8.8
Calculate the level of significance for the data in Example 8.2.

Solution
The hypothesis to be tested is

$$H_0: \mu = 10$$

$$H_a: \mu \neq 10$$

184 / Chap. 8: Large-Sample Tests of Hypotheses

two; 12

5.62

upper

less than

.4990; .001

doubling

.001; .002

⩾

less

null
one
twice

This is (one, two) -tailed test and the observed value of $\bar{x}$ was $\bar{x} = $ _____, or equivalently,

$$z = \frac{\bar{x} - \mu_0}{\sigma/\sqrt{n}} \approx \underline{}$$

This value of the test statistic is in the (upper, lower) tail of the distribution if the null hypothesis is true. The collection of all values of z as rare or rarer than $z = 5.62$ has probability $P(z \geq 5.62)$. Since the value $z = 5.62$ is not given in Table 1 of Appendix III, we can say only that the probability of interest is (less than, greater than) that given for the last tabled entry. That is,

$$P(z \geq 5.62) \text{ is less than } P(z \geq 3.09)$$

$$= .5 - \underline{} = \underline{}$$

This probability accounts for only the values of $\bar{x}$ in the upper tail of the distribution. There would be values of $\bar{x}$ which are equally as rare in the lower tail of the distribution. In order to account for these realizations, the significance level is calculated by _____ the value found above, so that

$$p\text{-value} < 2(\underline{}) = \underline{}$$

Thus, for any researcher specifying an α value (⩾, ⩽) .002, the conclusion would be to accept the alternative hypothesis, $\mu \neq 10$.

In summary, the procedure used to determine levels of significance is as follows:
1. Look at the alternative hypothesis. If it is an upper-tail alternative, list all values of the test statistic greater than or equal to the value observed. If it is a lower-tail alternative, list all values of the test statistic _____ than or equal to the value observed. If it is a two-tailed alternative, determine which tail of the distribution contains the observed value and list all values as rare or rarer than the observed value that are in that tail.
2. Calculate the probability of getting a value of the test statistic in the collection of values given in part (1) assuming the _____ hypothesis is true. If the test is _____ -tailed, this is the p-value.
3. If the test is two-tailed, the p-value is _____ the probability calculated in (2).

Self-Correcting Exercises 8B

1. Refer to Exercise 2, Self-Correcting Exercises 8A. Give the level of significance for the test and interpret your results.
2. Refer to Exercise 4, Self-Correcting Exercises 8A. Give the level of significance for the test and interpret your results.
3. Suppose it is hypothesized that $p = .1$. Test this hypothesis against the alternative that $p < 0.1$ if the number of successes is $x = 8$ in a sample of $n = 100$. Give the level of significance for the test and make a decision based on the p-value.

EXERCISES

1. What are the four essential elements of a statistical test of an hypothesis?
2. Assume that a certain set of "early returns" in an election is actually a random sample of size 400 from the voters in that election. If 225 of the voters in the sample voted for candidate A, could we assert with $\alpha = .01$ that candidate A has won?
3. Refer to Exercise 2 and give the level of significance for the test. Interpret your results.
4. A grocery store operator claims that the average waiting time at a checkout counter is 3.75 minutes. To test this claim, a random sample of 30 observations was taken. Test the operator's claim at the 5% level of significance, using the sample data shown below.

		Waiting Time in Minutes		
3	4	3	4	1
1	0	5	3	2
4	3	1	2	0
3	2	0	3	4
1	3	2	1	3
2	4	2	5	2

5. Refer to Exercise 4 and give the level of significance for the test. Interpret your results.
6. In a maze running study, a rat is run in a T maze and the result of each run recorded. A reward in the form of food is always placed at the right exit. If learning is taking place, the rat will choose the right exit more often than the left. If no learning is taking place, the rat should randomly choose either exit. Suppose that the rat is given $n = 100$ runs in the maze and that he chooses the right exit $x = 64$ times. Would you conclude that learning is taking place? Give the level of significance of the test and make a decision based on the p-value.
7. To test the effectiveness of a vaccine, 150 experimental animals were given the vaccine and 150 were not. All 300 were then infected with the disease. Among those vaccinated, 10 died as a result of the disease. Among the control group (i.e., those not vaccinated), there were 30 deaths. Can we conclude that the vaccine is effective in reducing the mortality rate? Use a significance level of .025.
8. Two diets were to be compared. Seventy-five individuals were selected at random from a population of overweight people. Forty of this group were assigned diet A and the other 35 were placed on diet B. The weight losses in pounds over a period of 1 week were found and the following quantities recorded:

	Sample Size	Sample Mean (pounds)	Sample Variance
Diet A	40	10.3	7
Diet B	35	7.3	3.25

   a. Do these data allow the conclusion that the expected weight loss under diet A (μ_A) is greater than the expected weight loss under diet B (μ_B)? Test at the .01 level. Draw the appropriate conclusion.

b. Construct a 90% confidence interval for $\mu_A - \mu_B$.

9. In a manufacturing plant employing a double inspection procedure, the first inspector is expected to miss an average of 25 defective items per day with a standard deviation of 3 items. If the first inspector has missed an average of 29 defectives per day based upon the last 30 working days, is he working up to company standards? (Use $\alpha = .01$.)

10. Refer to Exercise 9 and give the level of significance for the test. Interpret your results.

11. To compare the effect of stress in the form of noise upon the ability to perform a simple task, 70 subjects were divided into two groups: the first group of 30 subjects was to act as a control while the second group of 40 was to be the experimental group. Although each subject performed the task in the same control room, each of the experimental-group subjects had to perform the task while loud rock music was being played in the room. The time to finish the task was recorded for each subject and the following summary was obtained:

	Control	Experimental
n	30	40
$\bar{x}$	15 minutes	23 minutes
s	4 minutes	10 minutes

Does there appear to be a difference in the average completion times for the two groups? Find the approximate level of significance and interpret its value.

12. An experimenter has prepared a drug-dose level which he claims will induce sleep for at least 80 percent of people suffering from insomnia. After examining the dosage we feel that his claims regarding the effectiveness of his dosage are inflated. In an attempt to disprove his claim we administer his prescribed dosage to 50 insomniacs and observe that $x = 37$ have had sleep induced by the drug dose. Is there enough evidence to refute his claim?

13. During the 1990–91 academic year, approximately 93% of all students enrolled at a state university were residents of that state. Since the total student population in this university system is assumed to be quite large, the number x of state residents in a random sample of size n is approximately a binomial random variable. Suppose that the administrators of the university are interested in determining whether the percentage of state residents has changed for the academic year 1991–92. A random sample of $n = 150$ students consists of 142 state residents. Does the data provide sufficient evidence to prove that the percentage has changed?
a. Test at the $\alpha = .05$ level of significance.
b. Find the observed significance level for the test.

14. Polychlorinated biphenols (PCBs) have been found to be dangerously high in some game birds found along the marshlands of the southeastern coast of the United States. The Federal Drug Administration (FDA) considers a concentration of PCBs higher than 5 parts per million (ppm) in these game birds to be dangerous for human consumption. A sample of $n = 38$ game birds produced an average of $\bar{x} = 7.2$ ppm with a standard deviation of $s = 6.2$ ppm.

a. Find a 90% confidence interval estimate of the true average ppm of PCBs in the population of game birds sampled.
b. Use a statistical test of hypothesis to determine whether the mean ppm of PCBs differs from the FDA's recommended limit of $\mu = 5$ ppm.
c. What are β and $1 - \beta$ if the true mean ppm of PCBs is 6 ppm? If the true mean is 7 ppm?
d. Find $1 - \beta$ when $\mu = 8, 9, 10$, and 12. Use these values to construct a power curve for the test under consideration.
e. For what values of μ does this test have power greater than or equal to .90?

CHAPTER 9
INFERENCE FROM SMALL SAMPLES

9.1 Introduction

Central
Limit

Large-sample methods for making inferences about a population were considered in the preceding chapters. When the sample size was large, the _____ _____ Theorem assured the approximate normality of the distribution of the estimators $\bar{x}$ or $\hat{p}$. However, time, cost, or other limitations may prevent an investigator from collecting enough data to feel confident in using large-sample techniques. When the sample size is small, $n < 30$, the Central Limit Theorem may no longer apply. This difficulty can be overcome if the investigator is reasonably sure that his or her measurements constitute a sample from a _____ population.

normal

randomly

The results presented in this chapter are based on the assumption that the observations being analyzed have been _____ drawn from a normal population. This assumption is not as restrictive as it sounds, since the normal distribution can be used as a model in cases where the underlying distribution is mound-shaped and fairly symmetrical.

9.2 Student's *t* Distribution

When the sample size is large, the statistic

$$\frac{\bar{x} - \mu}{\sigma/\sqrt{n}}$$

is approximately distributed as the standard normal random variable z. What can be said about this statistic when n, the sample size, is small and the sample variance s^2 is used to estimate σ^2?

If the parent population is not normal (nor approximately normal), the behavior of the statistic given above is not known in general when n is small. Its distribution could be empirically generated by repeated sampling from the population of interest. If the parent population *is* normal, we can rely upon the results of W. S. Gosset, who published under the pen name Student. He drew repeated samples from a normal population and tabulated the distribution of a statistic which he called *t*, where

$$t = \frac{\bar{x} - \mu}{s/\sqrt{n}}.$$

The resulting distribution for t has the following properties:

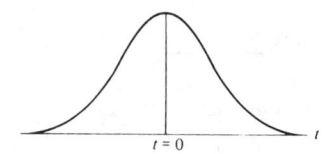

[graph of symmetric mound-shaped distribution centered at $t = 0$]

1. The distribution is _____-shaped.   mound
2. The distribution is _____ about the value $t = 0$.   symmetrical
3. The distribution has more flaring tails than z; hence t is (more, less) variable than the z statistic.   more
4. The shape of the distribution changes as the value of _____, the sample size, changes. n
5. As the sample size n becomes large, the t distribution becomes identical to the _____ _____ distribution.   standard normal

These results are based on the following two assumptions:

1. The parent population has a _____ distribution. The t statistic is, however, relatively stable for nonnormal _____-shaped distributions.   normal; mound
2. The sample is a _____ sample. When the population is normal, this assures us that $\bar{x}$ and s^2 are independent.   random

For a fixed sample size, n, the statistic

$$z = \frac{\bar{x} - \mu}{\sigma/\sqrt{n}}$$

contains exactly _____ random quantity, the sample mean _____.   one; $\bar{x}$
However, the statistic

$$t = \frac{\bar{x} - \mu}{s/\sqrt{n}}$$

contains _____ random quantities, _____ and _____.   two; $\bar{x}$; s
This accounts for the fact that t is (more, less) variable than z. In fact, $\bar{x}$ may be large while s is small or $\bar{x}$ may be small while s is large. Hence it is said that $\bar{x}$ and s are _____, which means that the value assumed by $\bar{x}$ in no way determines the value of s.   more; independent

As the sample size changes, the corresponding t distribution changes so that each value of n determines a different probability distribution. This is due to the variability of s^2, which appears in the denominator of t. Large sample sizes produce (more, less) stable estimates of σ^2 than do small sample sizes. These different probability curves are identified by the degrees of freedom associated with the estimator of σ^2.   more

The term "degrees of freedom" can be explained in the following way. The sample estimate s^2 uses the sum of squared deviations in its calculation.

190 / Chap. 9: Inference from Small Samples

0	Recall that $\sum_{i=1}^{n} (x_i - \bar{x}) = $ _____. This means that if we know the values of $n - 1$ deviations, we can determine the last value uniquely since their sum must be zero. Therefore, the sum of squared deviations,
$n - 1$	$\sum_{i=1}^{n} (x_i - \bar{x})^2$, contains only _____ independent deviations and not n independent deviations, as one might expect. Degrees of freedom refer to the number of independent deviations that are available for estimating σ^2. When n observations are drawn from one population, we use the estimator
	$$\hat{\sigma}^2 = s^2 = \sum_{i=1}^{n} (x_i - \bar{x})^2 / (n - 1).$$
$n - 1$ $n - 1$	In this case, the degrees of freedom for estimating σ^2 are _____ and the resulting t distribution is indexed as having _____ degrees of freedom.
	### The Use of Tables for the t Distribution
right left	We define t_α as that value of t having an area equal to α to its _____, and $-t_\alpha$ is that value of t having an area equal to α to its _____. Consider the following diagram:
	[diagram of t-distribution showing Area = .01 at $-t_{.01}$ on left and Area = .05 at $t_{.05}$ on right, with $t = 0$ in center]
symmetrical	The distribution of t is _____ about the value $t = 0$; hence, only the positive values of t need be tabulated. Problems involving left-tailed values of t can be solved in terms of right-tailed values, as was done with the z statistic.
left	A negative value of t simply indicates that you are working in the (left, right) tail of the distribution.
	Table 4 of the text tabulates *commonly used* critical values, t_α, based on $1, 2, \ldots, 29, \infty$ degrees of freedom for $\alpha = .100, .050, .025, .010, .005$. Along the top margin of the table you will find columns labeled t_α for the various values of α, while along the right margin you will find a column marked degrees of freedom, d.f. By cross-indexing you can find the value t having an area equal to α to its right and having the proper degrees of freedom.
	### Example 9.1
	To find the critical value of t for $\alpha = .05$ with 5 degrees of freedom, find 5 in the right margin. Now by reading across, you will find $t = 2.015$ in the $t_{.05}$ column. In the same manner, we find that for 12 degrees of freedom,
2.179	$t_{.025} = $ _____. In using Table 4, think of your problem in terms of α, the area to the right of the value of t, and the degrees of freedom used to estimate σ^2.

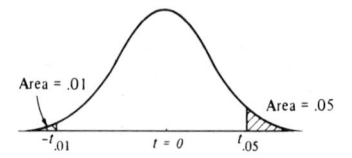

Compare the different values of t based on an infinite number of degrees of freedom with those for a corresponding z. You can perhaps see the reason for choosing a sample size greater than _____ as the dividing point for using the z distribution when the standard deviation s is used as an estimate for _____.

30

σ

Example 9.2

Find the critical values for t when t_α is that value of t with an area of α to its right, based on the following degrees of freedom:

	α	d.f.	t	
a.	.05	2	_____	2.920
b.	.005	10	_____	3.169
c.	.10	28	_____	1.313
d.	.01	16	_____	2.583
e.	.025	20	_____	2.086

Students taking their first course in statistics usually ask the following questions at this point: "How will I know whether I should use z or t? Is sample size the only criterion I should apply?" No, sample size is not the only criterion to be used. When the sample size is *large*, both

$$T_1 = \frac{\bar{x} - \mu}{\sigma/\sqrt{n}} \quad \text{and} \quad T_2 = \frac{\bar{x} - \mu}{s/\sqrt{n}}$$

behave as a standard normal random variable z, regardless of the distribution of the parent population. When the sample size is *small* and the sampled population is *not normal*, then, in general, neither T_1 nor T_2 behaves as z or t. In the special case when the parent population is *normal*, then T_1 behaves as z and T_2 behaves as t.

Use this information to complete the following table when the sample is drawn from a *normal* distribution:

	Sample Size	
Statistic	$n < 30$	$n \geq 30$
$\dfrac{\bar{x} - \mu}{s/\sqrt{n}}$	_____	t or app. z
$\dfrac{\bar{x} - \mu}{\sigma/\sqrt{n}}$	_____	_____

t

z; z

9.3 Small-Sample Inferences Concerning a Population Mean

Small-Sample Test Concerning a Population Mean μ

A test of an hypothesis concerning the mean μ of a *normal* population when $n < 30$ and σ is unknown proceeds as follows:

1. $H_0: \mu = \mu_0$.

2. H_a: appropriate one- or two-tailed alternative.
3. Test statistic:

$$t = \frac{\bar{x} - \mu}{s/\sqrt{n}}.$$

4. Rejection region with $\alpha = P(\text{falsely rejecting } H_0)$:
   a. For $H_a: \mu > \mu_0$, reject H_0 if $t > t_\alpha$ based on $n - 1$ degrees of freedom.
   b. For $H_a: \mu < \mu_0$, reject H_0 if $t < -t_\alpha$ based on $n - 1$ degrees of freedom.
   c. For $H_a: \mu \neq \mu_0$, reject H_0 if $|t| > t_{\alpha/2}$ based on $n - 1$ degrees of freedom.

Example 9.3
A new electronic device that requires 2 hours per item to produce on a production line has been developed by company A. While the new product is being run, profitable production time is used. Hence, the manufacturer decides to produce only six new items for testing purposes. For each of the six items, the time to failure is measured, yielding the measurements 59.2, 68.3, 57.8, 56.5, 63.7, and 57.3 hours. Is there sufficient evidence to indicate that the new device has a mean life greater than 55 hours at the $\alpha = .05$ level?

Solution
To calculate the sample mean and standard deviation we need

$$\Sigma x_i = 362.8 \quad \text{and} \quad \Sigma x_i^2 = 22{,}043.60.$$

$$\bar{x} = \Sigma x_i/n = 362.8/6 = \underline{}.$$

$$s^2 = \frac{1}{n-1}\left[\Sigma x_i^2 - \frac{(\Sigma x_i)^2}{n}\right]$$

$$= \frac{1}{5}\left[22{,}043.60 - \frac{(362.8)^2}{6}\right]$$

$$= 21.2587,$$

with $\quad s = \sqrt{21.2587} = 4.61073$

The test proceeds as follows:

1. $H_0: \mu \underline{}$.
2. $H_a: \mu \underline{}$.
3. Test statistic:

$$t = \frac{\bar{x} - 55}{s/\sqrt{n}}.$$

4. Rejection region:

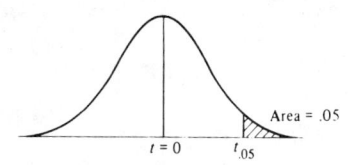

Based on 5 degrees of freedom, reject H_0 if $t >$ _____. | 2.015
Now calculate the value of the test statistic:

$$t = \frac{\bar{x} - 55}{s/\sqrt{n}} = \frac{\_\_\_\_\_ - 55}{\_\_\_\_\_/\sqrt{6}} = \frac{\_\_\_\_\_}{1.8823} = \_\_\_\_\_.$$

60.4667; 5.4667; 2.90
4.61073

Since the observed value (is, is not) larger than 2.015, we (reject, do not reject) H_0. There (is, is not) sufficient evidence to indicate that the new device has a mean life greater than 55 hours at the 5% level of significance.

is; reject
is

Confidence Interval for a Population Mean μ

In estimating a population mean, we can use either a point estimator with bounds on the error of estimation or an interval estimator having the required level of confidence.

Small sample estimation of the mean of a *normal* population with σ unknown involves the statistic

$$\frac{\bar{x} - \mu}{s/\sqrt{n}},$$

which has a _____ distribution with _____ degrees of freedom. | t; $n - 1$
The resulting $(1 - \alpha)100\%$ confidence interval estimator is given as

$$\bar{x} \pm t_{\alpha/2} s/\sqrt{n},$$

where $t_{\alpha/2}$ is that value of t based on $n - 1$ degrees of freedom having an area of $\alpha/2$ to its right. The lower confidence limit is _____ and the upper confidence limit is _____. The point estimator of μ is _____ and the bound on the error of estimation can be taken to be $t_{\alpha/2} \, s/\sqrt{n}$. A proper interpretation of a $(1 - \alpha)100\%$ confidence interval for μ would be stated as follows: In repeated sampling, (_____)100% of the _____ _____ so constructed would enclose the true value of the mean μ.

$\bar{x} - t_{\alpha/2} s/\sqrt{n}$
$\bar{x} + t_{\alpha/2} s/\sqrt{n}$; $\bar{x}$

$1 - \alpha$
confidence intervals

Example 9.4
Using the data from Example 9.3 of this section, find a 95% confidence interval estimate for μ, the mean life in hours for the new device.

Solution
The pertinent information from Example 9.3 follows.

$\bar{x} = 60.4667$ \qquad d.f. = _____ | 5

194 / Chap. 9: Inference from Small Samples

$s/\sqrt{n} = 1.8823$ $\alpha/2 = .025$

2.571

$t_{.025} = $ _____ .

The confidence interval will be found by using

$$\bar{x} \pm t_{.025}\, s/\sqrt{n}\, .$$

Substituting $\bar{x}$, $s/\sqrt{n}$, and $t_{.025}$, we have

2.571

60.4667 ± _____ (1.8823)

4.84

60.4667 ± _____

or

55.63; 65.31

(_____ , _____).

Example 9.5
In a random sample of ten cans of corn from supplier B, the average weight per can of corn was $\bar{x} = 9.4$ oz. with standard deviation, $s = 1.8$ oz. Does this sample contain sufficient evidence to indicate the mean weight is less than 10 oz. at the $\alpha = .01$ level? What p-value would you report?

Solution
1. The following information is needed:

10

$n = $ _____

9.4

$\bar{x} = $ _____

1.8

$s = $ _____

.01

$\alpha = $ _____ .

2. Set up the test as follows:

10

$H_0: \mu = $ _____

10

$H_a: \mu < $ _____ .

3. Test statistic:

$$t = \frac{\bar{x} - \mu}{s/\sqrt{n}}.$$

Sect. 9.3: Small-Sample Inferences Concerning a Population Mean

4. Rejection region:

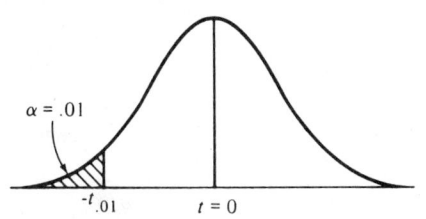

Based on 9 degrees of freedom, reject H_0 if $t <$ _____ .   −2.821

Calculate:

$$t = \frac{\bar{x} - \mu}{s/\sqrt{n}}$$

$$= \frac{9.4 - (\underline{\qquad})}{1.8/\sqrt{10}} \qquad\qquad 10$$

$$= \frac{(\underline{\qquad})}{.57} \qquad\qquad -.6$$

$$= \underline{\qquad} . \qquad\qquad -1.05$$

Since the calculated value of t (does, does not) fall in the rejection   does not
region we conclude that the data (do, do not) present sufficient   do not
evidence to indicate that the mean weight per can is less than
10 oz.

The level of significance for this one-tailed test is defined to be

$$p\text{-value} = P(t \leqslant -1.05) = P(t \geqslant 1.05),$$

where t has a Student's t distribution with $n - 1 = 9$ degrees of freedom. From Table 4, with d.f. = 9, the critical value of $t = 1.05$ falls to the left of the smallest value given. That value is $t = 1.383$ with area .10 to its right.

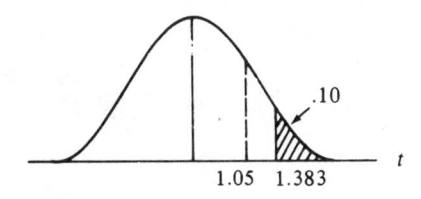

Hence, the area to the right of $t = 1.05$ is (greater, less) than .10, and the   greater
p-value is _____ than .10.   greater

Example 9.6
Refer to Example 9.5. Find a 98% confidence interval for μ.

196 / Chap. 9: Inference from Small Samples

Solution

.02 $\alpha =$ _____

9   d.f. = _____

2.821 $t_{\alpha/2} =$ _____ .

Calculate:

$$\bar{x} \pm t_{\alpha/2} s/\sqrt{n}$$

9.4; .57   (_____) ± 2.821 (_____)

9.4; 1.6   (_____) ± (_____) .

7.8   Therefore, the 98% confidence interval required is (_____,
11.0   _____) .

The MINITAB package can be used to perform a test of hypothesis concerning a population mean μ, and to calculate a confidence interval estimate of μ for any specified level of confidence. The command TTEST, followed by the column number in which the data are stored, will implement a one-sample t-test. The subcommand ALTERNATIVE followed by a -1, 0, or 1 will implement a left-tailed, two-tailed, or right-tailed test procedure, respectively. If this subcommand is not used, the procedure implements a two-tailed test. The command TINTERVAL, followed first by the confidence coefficient expressed as a decimal, and then by the column number in which the data are stored, will produce the appropriate upper and lower confidence limits for μ. If the confidence coefficient is not specified, a 95% confidence interval estimate is calculated.

In the displays that follow, the data from Example 9.3 were entered into column one (C1), and the commands DESCRIBE, TTEST and TINTERVAL, together with their appropriate arguments and subcommands, were implemented using the data in column one.

MTB > DESCRIBE C1

	N	MEAN	MEDIAN	TRMEAN	STDEV	SEMEAN
C1	6	60.47	58.50	60.47	4.61	1.88

	MIN	MAX	Q1	Q3
C1	56.50	68.30	57.10	64.85

MTB > TTEST 55 C1;
SUBC > ALTERNATIVE 1.

TEST OF MU = 55.00 VS MU G.T. 55.00

	N	MEAN	STDEV	SE MEAN	T	P VALUE
C1	6	60.47	4.61	1.88	2.90	0.017

```
MTB > TINTERVAL C1

           N      MEAN     STDEV    SE MEAN    9.50 PERCENT C.I.
   C1      6      60.47    4.61     1.88       ( 55.63,  65.31)
```

The DESCRIBE command allows us to verify our calculated values of $\bar{x}$ = _____ and s = _____. In addition, the estimated standard deviation of $\bar{x}$, called the standard error of the mean (SEMEAN), is given by $s/\sqrt{n} = 1.88$. The quantity is used in calculating t and in constructing the confidence limits based on t.

In testing $H_0: \mu = 55$ versus $H_a: \mu > 55$ (ALTERNATIVE _____), the calculated value of t in the TTEST printout agrees with our earlier calculations. However, the value of α is not set in advance; instead, the decision to reject or not reject H_0 is made after examining the p-value for the test. Since the p-value of .017 is (smaller, larger) than the value of $\alpha = .05$ used in Example 9.3, we agree to (reject, not reject) H_0. The lower and upper 95% confidence limits for μ, found in the TINTERVAL printout, are given by 55.63 and 65.31, respectively, the same as the values found in Example 9.4.

60.47; 4.61

1

smaller
reject

Self-Correcting Exercises 9A

1. A school administrator claimed that the average time spent on a school bus by those students in his school district who rode school buses was 35 minutes. A random sample of 20 students who did ride the school buses yielded an average of 42 minutes riding time with a standard deviation of 6.2 minutes. Does this sample of size 20 contain sufficient evidence to indicate that the mean riding time is greater than 35 minutes at the 5% level of significance? What p-value would you report?
2. Find a 95% confidence interval for the mean time spent on a school bus using the data from Exercise 1.
3. The telephone company was interested in measuring the average daily usage in minutes for household telephones in a specific area in order to determine whether this rate is different from a statewide average daily usage for households. Suppose that a random sample of nine households were sampled on random days, producing the following times (in minutes):

   35, 59, 42, 44, 31, 46, 24, 56, 50

   a. Estimate the average daily usage using a 90% confidence interval.
   b. Suppose that the statewide average for households has been found to be 45 minutes. Does this data provide evidence to indicate that the average in this area differs from the statewide average? Use $\alpha = .05$.

9.4 Small-Sample Inferences Concerning the Difference Between Two Means, $\mu_1 - \mu_2$

Inferences concerning $\mu_1 - \mu_2$ based on small samples are founded upon the following assumptions:

198 / Chap. 9: Inference from Small Samples

normal
variances

$\bar{x}_1 - \bar{x}_2$

1. Each population sampled has a _____ distribution.
2. The population _____ are equal; that is, $\sigma_1^2 = \sigma_2^2$.
3. The samples are independently drawn.

An unbiased estimator for $\mu_1 - \mu_2$, regardless of sample size, is _____.
The standard deviation of this estimator is

$$\sqrt{\frac{\sigma_1^2}{n_1} + \frac{\sigma_2^2}{n_2}}.$$

When $\sigma_1^2 = \sigma_2^2$, we can replace σ_1^2 and σ_2^2 by a common variance σ^2. Then the standard deviation of $\bar{x}_1 - \bar{x}_2$ becomes

σ

$$\sqrt{\frac{\sigma^2}{n_1} + \frac{\sigma^2}{n_2}} = (\underline{}) \sqrt{\frac{1}{n_1} + \frac{1}{n_2}}.$$

If σ were known, then in testing an hypothesis concerning $\mu_1 - \mu_2$, we would use the statistic

$$z = \frac{(\bar{x}_1 - \bar{x}_2) - D_0}{\sigma \sqrt{\frac{1}{n_1} + \frac{1}{n_2}}},$$

where $D_0 = \mu_1 - \mu_2$. For small samples with σ unknown, we would use

$$t = \frac{(\bar{x}_1 - \bar{x}_2) - D_0}{s \sqrt{\frac{1}{n_1} + \frac{1}{n_2}}},$$

Student's t

σ^2

where s is the estimate of σ, calculated from the sample values. When the data are normally distributed, this statistic has a _____ _____ distribution with degrees of freedom the same as those available for estimating _____.

variance
variance
$s_1^2; s_2^2$
c

In selecting the best estimate (s^2) for σ^2, we have three immediate choices:
a. s_1^2, the sample _____ from population I.
b. s_2^2, the sample _____ from population II.
c. A combination of _____ and _____.

The best choice is (a, b, c), since it uses the information from both samples. A logical method of combining this information into one estimate, s^2, is

d. $$s^2 = \frac{(n_1 - 1)s_1^2 + (n_2 - 1)s_2^2}{(n_1 - 1) + (n_2 - 1)},$$

a weighted average of the sample variances using the degrees of freedom as weights.

The expression in d can be written in another form by replacing s_1^2 and s_2^2 by their defining formulas. Then, if x_{1j} and x_{2j} represent j^{th} observation in samples 1 and 2, respectively,

$$s^2 = \frac{\sum_{j=1}^{n_1}(x_{1j} - \bar{x}_1)^2 + \sum_{j=1}^{n_2}(x_{2j} - \bar{x}_2)^2}{(n_1 - 1) + (n_2 - 1)}.$$

In this form we see that we have pooled or added the sums of squared deviations from each sample and divided by the pooled degrees of freedom, $n_1 + n_2 - 2$. Hence, s^2 is a *pooled estimate* of the common variance σ^2 and is based on _____ degrees of freedom. Since our samples were drawn from normal populations, the statistic

$$t = \frac{(\bar{x}_1 - \bar{x}_2) - (\mu_1 - \mu_2)}{s\sqrt{\frac{1}{n_1} + \frac{1}{n_2}}}$$

has a _____ _____ distribution with _____ degrees of freedom.

	$n_1 + n_2 - 2$
	Student's t; $n_1 + n_2 - 2$

Example 9.7
A medical student conducted a diet study using two groups of 12 rats each as subjects. Group I received diet I while group II received diet II. After 5 weeks the student calculated the gain in weight for each rat. The data yielded the following information:

Group I	Group II
$\bar{x}_1 = 6.8$ ounces	$\bar{x}_2 = 5.3$ ounces
$s_1 = 1.5$ ounces	$s_2 = .9$ ounce
$n_1 = 12$	$n_2 = 12$

Do these data present sufficient evidence to indicate, at the $\alpha = .05$ level, that rats on diet I will gain more weight than those on diet II? Find a 90% confidence interval for $\mu_1 - \mu_2$, the mean difference in weight gained.

Solution
We will take the gains in weight to be normally distributed with equal variances and calculate a pooled estimate for σ^2.

$$s^2 = \frac{(n_1 - 1)s_1^2 + (n_2 - 1)s_2^2}{n_1 + n_2 - 2}$$

$$= \frac{11(1.5)^2 + 11(.9)^2}{12 + 12 - 2} = \frac{33.66}{22} = \underline{\qquad}$$

1.530

so that

$$s = \sqrt{1.530} = \underline{\qquad}.$$

1.237

The test is as follows.

200 / Chap. 9: Inference from Small Samples

0	1. $H_0: \mu_1 - \mu_2 = $ _____.
0	2. $H_a: \mu_1 - \mu_2 > $ _____.
	3. Test statistic: $$t = \frac{(\bar{x}_1 - \bar{x}_2) - D_0}{s\sqrt{\frac{1}{n_1} + \frac{1}{n_2}}}.$$
	4. Rejection region:
22	With $n_1 + n_2 - 2 = $ _____ degrees of freedom, we would reject H_0 if
1.717	$t > $ _____. Now we calculate the test statistic.
	$$t = \frac{(\bar{x}_1 - \bar{x}_2) - D_0}{s\sqrt{\frac{1}{n_1} + \frac{1}{n_2}}}$$
1.5; 0	$$= \frac{(_____) - (_____)}{1.237\sqrt{.1667}}$$
1.5	$$= \frac{(_____)}{.505}$$
2.97	$= $ _____.
Reject H_0	Decision: _____.
	To find a 90% confidence interval for $\mu_1 - \mu_2$, we need $t_{.05}$ based on 22
1.717	degrees of freedom, $t_{.05} = $ _____. Hence we would use
	$$(\bar{x}_1 - \bar{x}_2) \pm 1.717s\sqrt{\frac{1}{n_1} + \frac{1}{n_2}}$$
1.5; .505	_____ ± 1.717 (_____)
1.5; .867	(_____) $\pm$ (_____).
	Therefore, a 90% confidence interval for $\mu_1 - \mu_2$ would be
.63; 2.37	(_____, _____).

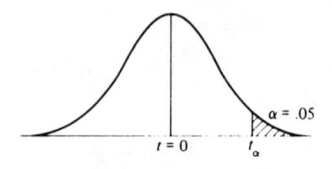

Example 9.8

In an effort to compare the average swimming times for two swimmers, each swimmer was asked to swim freestyle for a distance of 100 yards at randomly selected times. The swimmers were thoroughly rested between laps and did not race against each other, so that each sample of times was an independent random sample. The times for each of 10 trials are shown for the two swimmers.

Swimmer 1	Swimmer 2
59.62	59.81
59.48	59.32
59.65	59.76
59.50	59.64
60.01	59.86
59.74	59.41
59.43	59.63
59.72	59.50
59.63	59.83
59.68	59.51

Suppose that swimmer 2 was last year's winner when the two swimmers raced. Does it appear that the average time for swimmer 2 is still faster than the average time for swimmer 1 in the 100 yard freestyle? Find the approximate level of significance for the test and interpret the results.

Solution

We will assume that the times for the two swimmers are normally distributed with equal variances.

1. The hypothesis to be tested is

$$H_0: \mu_1 - \mu_2 = 0$$

$$H_a: \mu_1 - \mu_2 \underline{} . \qquad > 0$$

The following calculations are necessary:

Swimmer 1	Swimmer 2	
$\Sigma x_{1j} = 596.46$	$\Sigma x_{2j} = \underline{}$	596.27
$\Sigma x_{1j}^2 = 35576.698$	$\Sigma x_{2j}^2 = \underline{}$	35554.109
$\bar{x}_1 = 59.646$	$\bar{x}_2 = \underline{}$	59.627

2. Calculate the pooled estimate for σ^2 using the alternate form of s^2 as

202 / Chap. 9: Inference from Small Samples

$$s^2 = \frac{\sum_{j=1}^{n_1}(x_{1j}-\bar{x}_1)^2 + \sum_{j=1}^{n_2}(x_{2j}-\bar{x}_2)^2}{(n_1-1)+(n_2-1)}$$

$$= \frac{\Sigma x_{1j}^2 - \frac{(\Sigma x_{1j})^2}{n_1} + \Sigma x_{2j}^2 - \frac{(\Sigma x_{2j})^2}{n_2}}{n_1+n_2-2}$$

35554.109; 596.27

$$= \frac{35576.698 - \frac{(596.46)^2}{10} + \underline{\qquad} - \frac{(\underline{\qquad})^2}{10}}{\underline{\qquad}}$$

18

.56255; .0312528

$$= \frac{\underline{\qquad}}{18} = \underline{\qquad}$$

3. The test statistic is

$$t = \frac{\bar{x}_1 - \bar{x}_2 - D_0}{\sqrt{s^2\left(\frac{1}{n_1}+\frac{1}{n_2}\right)}} = \frac{59.646 - 59.627}{\sqrt{.0312528\left(\frac{1}{10}+\frac{1}{10}\right)}}$$

.019; .079; .24

$$= \frac{\underline{\qquad}}{\underline{\qquad}} = \underline{\qquad}$$

one-tailed
.24

4. Level of Significance: The level of significance for this (one-tailed, two-tailed) test is p-value = $P[t > \underline{\qquad}]$. Notice in Table 4 in the text that this value is not tabulated. Rather, for each different value of the degrees of freedom, the table gives t_α such that $P[t > t_\alpha] = \alpha$. This is the value of t that cuts off an area equal to α to its (right, left). However, since the observed value, t = .24 falls to the (left, right) of the smallest tabulated value, $t_{.10}$ = 1.330, the area to the right of t = .24 must be (greater, less) than .10. Hence, the level of significance is p-value > .10. This is too large a value to allow rejection of H_0. Hence, we conclude that there (is, is not) sufficient evidence to detect a difference in the two averages.

right
left

greater

is not

TWOSAMPLE, followed by a confidence coefficient expressed as a decimal fraction and the two column numbers containing the data sets for testing, is one of the two MINITAB commands that can be used to implement a two-independent-sample t-test and confidence interval. The subcommand

ALTERNATIVE can also be used with this command; however, the subcommand POOLED *must* be used so that the sample variances are pooled to estimate the underlying common equal variance. The MINITAB printout using the TWOSAMPLE command on the data for Example 9.8 is given in the following display.

```
MTB > TWOSAMPLE C1 C2;
SUBC> ALTERNATIVE 1;
SUBC> POOLED.

TWOSAMPLE T FOR C1 VS C2
        N    MEAN    STDEV  SE MEAN
C1     10   59.646   0.165   0.0521
C2     10   59.627   0.188   0.0594

95 PCTCI FOR MU C1 − MU C2: (−0.1471, 0.1851)

TTEST MU C1 = MU C2 (VS GT): T = 0.24 P = 0.41 DF = 18

POOLED STDEV =      0.177
```

Notice that the results on the printout agree with our earlier calculations.

At times it is known that the two underlying population variances *are not equal*. (An appropriate test $\sigma_1^2 = \sigma_2^2$ is given in Section 9.7.) In this case the statistic used for testing hypotheses about $D_0 = \mu_1 - \mu_2$ is

$$t^* = \frac{(\bar{x}_1 - \bar{x}_2) - D_0}{\sqrt{\dfrac{s_1^2}{n_1} + \dfrac{s_2^2}{n_2}}}$$

When the sample sizes are small, the appropriate critical values for this test statistic can be read from the critical values of t given in Table 4, using degrees of freedom equal to

$$\frac{\left(\dfrac{s_1^2}{n_1} + \dfrac{s_2^2}{n_2}\right)^2}{\dfrac{\left(\dfrac{s_1^2}{n_1}\right)^2}{n_1 - 1} + \dfrac{\left(\dfrac{s_2^2}{n_2}\right)^2}{n_2 - 1}}$$

This quantity must be rounded to the nearest integer to use Table 4. When n_1 and n_2 are large, the distribution of t^* can be approximated by the standard normal distribution, as we did in Section 8.3.

Self-Correcting Exercises 9B

1. What are the assumptions required for the proper use of the statistic

$$t = \frac{(\bar{x}_1 - \bar{x}_2) - (\mu_1 - \mu_2)}{s\sqrt{\dfrac{1}{n_1} + \dfrac{1}{n_2}}} \ ?$$

2. In the process of making a decision to either continue operating or close a civic health center, a random sample of 25 people who had visited the center at least once was chosen and each person asked whether he or she felt the center should be closed. In addition, the distance between each person's place of residence and the health center was computed and recorded. Of the 25 people responding, 16 were in favor of continued operation. For these 16 people, the average distance from the center was 5.2 miles with a standard deviation of 2.8 miles. The remaining 9 people who were in favor of closing the center lived at an average of 8.7 miles from the center with a standard deviation of 5.3 miles. Do these data indicate that there is a significant difference in mean distance to the health center for these two groups?

3. Estimate the difference in mean distance to the health center for the two groups in Exercise 2 with a 95% confidence interval.

4. In investigating which of two presentations of subject matter to use in a computer-programmed course, an experimenter randomly chose two groups of 18 students each, and assigned one group to receive presentation I and the second to receive presentation II. A short quiz on the presentation was given to each group and their grades recorded. Do the following data indicate that a difference in the mean quiz scores (hence a difference in effectiveness of presentation) exists for the two methods? Find the approximate level of significance and interpret your results.

	$\bar{x}$	s^2
Presentation I	81.7	23.2
Presentation II	77.2	19.8

9.5 A Paired-Difference Test

In many situations an experiment is designed so that a comparison of the effects of two "treatments" is made on the same person, on twin offspring, two animals from the same litter, two pieces of fabric from the same loom, or two plants of the same species grown on adjacent plots. Such experiments are designed so that the pairs of experimental units (people, animals, fabrics, plants) are as much alike as possible. By taking measurements on the two treatments within the relatively homogeneous pairs of experimental units, the difference in the measurements for the two treatments in a pair will primarily reflect the difference between _____ means rather than the difference between experimental units. This experimental design reduces the error of comparison and increases the quantity of information in the experiment.

treatment

To analyze such an experiment using the techniques of the last section would be incorrect. In planning this type of experiment, we *intentionally violate* the assumption that the measurements are *independent* and hope that this violation will work to our advantage by (increasing, reducing) the variability of the differences of the paired observations. Consider the situation in which two sets of identical twin calves are selected for a diet experiment. One of each set of twins is randomly chosen to be fed diet A while the other is given diet B. At the end of a given period of time, the calves are weighed and the data are presented for analysis.

reducing

	Diet		
Set	A	B	Difference
1	A_1	B_1	$A_1 - B_1$
2	A_2	B_2	$A_2 - B_2$

Now A_1 and B_1 are *not* independent since the calves are identical twins and as such have the same growth trend, weight-gain trend, and so on. Although A_1 could be larger or smaller than B_1, if A_1 were large, we would also expect B_1 to be large. A_2 and B_2 are not independent for the same reason. However, by looking at the differences $(A_1 - B_1)$ and $(A_2 - B_2)$, the characteristics of the twin calves no longer cloud the issue, since these differences would represent the difference due to the effects of the two treatments.

In using a paired-difference design, we analyze the differences of the paired measurements and, in so doing, attempt to *reduce* the *variability* that would be present in two *randomly* selected groups without pairing. A test of the hypothesis that the difference in two population means, $\mu_1 - \mu_2$, is equal to a constant, D_0, is equivalent to a test of the hypothesis that the mean of the differences, μ_d, is equal to a constant, D_0. That is, $H_0: \mu_1 - \mu_2 = D_0$ is equivalent to $H_0: \mu_d = D_0$. Usually, we will be interested in the hypothesis that $D_0 = 0$.

Example 9.9

To test the results of a conventional versus a new approach to the teaching of reading, 12 pupils were selected and matched according to IQ, age, present reading ability, and so on. One from each of the pairs was assigned to the conventional reading program and the other to the new reading program. At the end of 6 weeks, their progress was measured by a reading test. Do the following data present sufficient evidence to indicate that the new approach is better than the conventional approach at the $\alpha = .05$ level?

Pair	Conventional	New	$d_i = N - C$
1	78	83	5
2	65	69	4
3	88	87	-1
4	91	93	2
5	72	78	6
6	59	59	0

Find a 95% confidence interval for the difference in mean reading scores.

206 / Chap. 9: Inference from Small Samples

Solution

1. We analyze the set of six differences as we would a single set of six measurements. The change in notation required is straightforward.

$$\sum_{i=1}^{6} d_i = 16 \qquad \sum_{i=1}^{6} d_i^2 = 82.$$

The sample mean is

$$\bar{d} = \frac{1}{6} \sum_{i=1}^{6} d_i = \underline{\qquad}.$$

2.6667

The sample variance of the differences is

$$s_d^2 = \frac{\sum d_i^2 - (\sum d_i)^2/6}{5}$$

$$= \frac{82 - (16)^2/6}{5} = \frac{39.3333}{5} = \underline{\qquad}$$

7.8667

$$s_d = \sqrt{7.8667} = \underline{\qquad}.$$

2.8048

2. The test is conducted as follows. Remember that $\mu_d = \mu_N - \mu_C$.

$$H_0: \mu_d = 0.$$

$$H_a: \mu_d \underline{\qquad}.$$

> 0

Test statistic:

$$t = \frac{\bar{d} - 0}{s_d/\sqrt{n}}.$$

Rejection region: Based on 5 degrees of freedom, we will reject H_0 if the observed value of t is greater than $t_{.05} = \underline{\qquad}$.

2.015

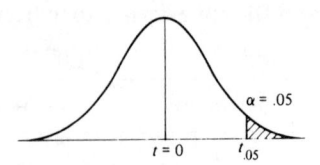

The sample value of t is

$$t = \frac{\bar{d} - 0}{s_d/\sqrt{n}}$$

$$= \frac{\rule{1cm}{0.4pt} - 0}{2.80/\sqrt{6}} = \frac{2.6667}{1.145} = \rule{1cm}{0.4pt}.$$

2.6667; 2.33

Since the value of the test statistic is greater than 2.015, we (reject, do not reject) H_0. This sample indicates that the new method appears to be superior to the conventional method at the $\alpha = .05$ level, if we assume that the reading test is a valid criterion upon which to base our judgment.

reject

3. A 95% confidence interval estimate for μ_d is given by

$$\bar{d} \pm t_{.025}\, s_d/\sqrt{n}.$$

Using sample values and $t_{.025} = \rule{1cm}{0.4pt}$, we have

2.571

$$2.6667 \pm \rule{1cm}{0.4pt} (1.145)$$

2.571

$$2.6667 \pm \rule{1cm}{0.4pt}.$$

2.9439

With 95% confidence, we estimate that μ_d lies within the interval _____ to _____.

-.28
5.61

Notice that in using a paired-difference analysis, the degrees of freedom for the critical value of t drop from $2n - 2$ for an unpaired design to $n - 1$ for the paired, a loss of $(2n - 2) - (n - 1) = \rule{1cm}{0.4pt}$ degrees of freedom. This results in a (larger, smaller) critical value of t. Therefore, a larger value of the test statistic is needed to reject H_0. Fortunately, *proper* pairing will reduce $\sigma_{\bar{d}}$. Hence the paired-difference experiment results in both a loss and a gain of information. However, the *loss* of $(n - 1)$ degrees of freedom is usually far overshadowed by the gain in information when $\sigma_{\bar{d}}$ is substantially reduced.

$n - 1$
larger

The statistical design of the paired-difference test is a simple example of a randomized block design. In such a design, the pairing must occur when the experiment is planned and not after the data are collected. Once the experimenter has used a paired design for an experiment, he no longer has the choice of using the unpaired design for testing the difference between means. This is because we have violated the assumptions needed for the unpaired design, namely, that the samples are random and independent.

A paired-difference experiment can be analyzed using the MINITAB one-sample TTEST or the TINTERVAL commands. If the paired responses are stored in columns one and two, the differences can be stored in column three (or any other column) using the command

LET C3 = C1 − C2.

(Alternately, you may wish to subtract C1 from C2.) In the following display, the data from Example 9.9 has been stored in C1 and C2.

```
MTB > LET C3 = C2 - C1
MTB > PRINT C1 - C3
```

ROW	C1	C2	C3
1	78	83	5
2	65	69	4
3	88	87	-1
4	91	93	2
5	72	78	6
6	59	59	0

```
MTB > TTEST C3;
SUBC> ALTERNATIVE 1.
```

TEST OF MU = 0.00 VS MU G.T. 0.00

	N	MEAN	STDEV	SE MEAN	T	P VALUE
C3	6	2.67	2.80	1.15	2.33	0.034

```
MTB > TINTERVAL C3
```

	N	MEAN	STDEV	SE MEAN	95.0 PERCENT C.I.
C3	6	2.67	2.80	1.15	(-0.28, 5.61)

The command TTEST C3; with the subcommand ALTERNATIVE 1. provides a one-tailed test of the hypothesis $H_0: \mu_d = 0$ versus $H_a: \mu_d > 0$, while the command TINTERVAL C3 provides the 95% confidence limits for the mean difference $\mu_d = \mu_1 - \mu_2$.

Self-Correcting Exercises 9C

1. The owner of a small manufacturing plant is considering a change in salary base by replacing an hourly wage structure with a per-unit rate. She hopes that such a change will increase the output per worker but has reservations about a possible decrease in quality under the per-unit plan. Before arriving at any decision, she forms 10 pairs of workers so that within each pair the two workers have produced about the same number of items per day and their work has been of comparable quality. From each pair, one worker is randomly selected to be paid as usual and the other is to be paid on a per-unit basis. In addition to the number of items produced, a cumulative quality score for the items produced is kept for each worker. The quality scores follow. (A high score is indicative of high quality.)

| | Rate | |
Pair	Per Unit	Hourly
1	86	91
2	75	77
3	87	83
4	81	84
5	65	68
6	77	76
7	88	89
8	91	91
9	68	73
10	79	78

Do these data indicate that the average quality for the per-unit production is significantly lower than that based on an hourly wage? What *p*-value would you report?

2. Refer to Exercise 1. The following data represent the average number of items produced per worker, based on one week's production records:

| | Rate | |
Pair	Per Unit	Hourly
1	35.8	31.2
2	29.4	27.6
3	31.2	32.2
4	28.6	26.4
5	30.0	29.0
6	32.6	31.4
7	36.8	34.2
8	34.4	31.6
9	29.6	27.6
10	32.8	29.8

a. Estimate the mean difference in average daily output for the two pay scales with a 95% confidence interval.
b. Test the hypothesis that a per-unit pay scale increases production at the .05 level of significance.

9.6 Inferences Concerning a Population Variance

In many cases the measure of variability is more important than that of central tendency. For example, an educational test consisting of 100 items has a mean score of 75 with standard deviation of 2.5. The value $\mu = 75$ may sound impressive, but $\sigma = 2.5$ would imply that this test has very poor discriminating ability since approximately 95% of the scores would be between 70 and 80. In like manner, a production line producing bearings with $\mu = .5$ inch and $\sigma = .2$ inch would produce many defective items; the fact that the bearings have a mean diameter of .5 inch would be of little value when the bearings are fitted together. *The precision of an instrument, whether it be an educational test or a machine, is measured by the standard deviation of the error of measurement.* Hence, we proceed to a test of a population variance σ^2.

210 / Chap. 9: Inference from Small Samples

unbiased

σ^2
nonsymmetric
zero
n
σ^2

The sample variance s^2 is an _____ estimator for σ^2. To use s^2 for inference making, we find that in repeated sampling, the distribution of s^2 has the following properties:
1. $E(s^2) =$ _____.
2. The distribution of s^2 is (symmetric, nonsymmetric).
3. s^2 can assume any value greater than or equal to _____.
4. The shape of the distribution changes for different values of _____ and _____.
5. In sampling from a *normal* population, s^2 is independent of the population mean μ and the sample mean $\bar{x}$. As with the z statistic, the distribution for s^2 when sampling from a normal population can be standardized by using

$$\chi^2 = (n-1)s^2/\sigma^2,$$

which is the chi-square random variable having the following properties in repeated sampling:

$n - 1$

1. $E(\chi^2) = $ d.f. = _____.

nonsymmetric

2. The distribution of χ^2 is (symmetric, nonsymmetric).

0

3. $\chi^2 \geq$ _____.

4. The distribution of χ^2 depends on the degrees of freedom, $n - 1$. Since χ^2 does not have a symmetric distribution, critical values of χ^2 have been tabulated for both the upper and lower tails of the distribution in Table 5 of the text. The degrees of freedom are listed along both the right and left margins of the table. Across the top margin are values of χ_α^2, indicating a value of χ^2 having an area equal to α to its right, that is,

$$P(\chi^2 > \chi_\alpha^2) = \alpha.$$

Example 9.10
Use Table 5 to find the following critical values of χ^2:

5.99147
2.55821
37.5662
18.4926
1.734926
27.4884
45.5585
10.0852

	α	d.f.	χ_α^2
a.	.05	2	_____
b.	.99	10	_____
c.	.01	20	_____
d.	.95	30	_____
e.	.995	9	_____
f.	.025	15	_____
g.	.005	24	_____
h.	.90	17	_____

The statistical test of an hypothesis concerning a population variance σ^2 at the α level of significance is given as follows:

1. $H_0: \sigma^2 = \sigma_0^2.$

2. H_a: appropriate one- or two-tailed test.
3. Test statistic:

$$\chi^2 = (n-1)s^2/\sigma_0^2.$$

4. Rejection region:
   a. For $H_a: \sigma^2 > \sigma_0^2$, reject H_0 if $\chi^2 > \chi_\alpha^2$ based on $n-1$ degrees of freedom.
   b. For $H_a: \sigma^2 < \sigma_0^2$, reject H_0 if $\chi^2 < \chi_{(1-\alpha)}^2$ based on $n-1$ degrees of freedom.
   c. For $H_a: \sigma^2 \neq \sigma_0^2$, reject H_0 if $\chi^2 > \chi_{\alpha/2}^2$ or $\chi^2 < \chi_{(1-\alpha/2)}^2$ based on $n-1$ degrees of freedom.

Example 9.11
A producer of machine parts claimed that the diameters of the connector rods produced by his plant had a variance of at most .03 inch². A random sample of 15 connector rods from his plant produced a sample mean and variance of .55 inch and .053 inch², respectively. Is there sufficient evidence to reject his claim at the $\alpha = .05$ level of significance?

Solution
1. Collecting pertinent information:

$$s^2 = .053 \text{ inch}^2$$

$$\text{d.f.} = n - 1 = 14$$

$$\sigma_0^2 = .03 \text{ inch}^2.$$

2. The test of the hypothesis is given as

$$H_0: \sigma^2 = .03.$$

$$H_a: \sigma^2 \underline{\qquad} .03. \qquad >$$

3. Test statistic:

$$\chi^2 = (n-1)s^2/\sigma_0^2.$$

4. Rejection region:

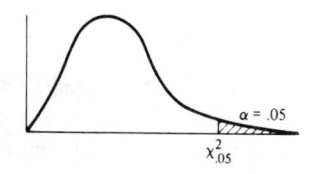

For 14 degrees of freedom, we will reject H_0 if $\chi^2 \geq \underline{\qquad}$. 23.6848

212 / Chap. 9: Inference from Small Samples

Calculate

$$\chi^2 = (n-1)s^2/\sigma_0^2$$

.053; .03

$$= 14(\underline{\hspace{1cm}})/(\underline{\hspace{1cm}})$$

$$= 24.733.$$

Reject

Decision: (Reject, Do not reject) H_0 since

$$24.733 > \chi^2_{.05} = 23.6848.$$

The data produced sufficient evidence to reject H_0. Therefore, we can conclude that the variance of the rod diameters (is, is not) greater than .03 inch^2.

is

Example 9.12
Refer to Example 9.11. Find the approximate level of significance and interpret your results.

Solution
The level of significance for this test is

$$p\text{-value} = P[\chi^2 > 24.733].$$

where χ^2 has a chi-square distribution with $n - 1 = 14$ degrees of freedom. From Table 5 in the text, the observed value falls between $\chi^2_{.05} = \underline{\hspace{1cm}}$

23.6848
26.1190

and $\chi^2_{.025} = \underline{\hspace{1cm}}$. Hence,

$$\underline{\hspace{1cm}} < p\text{-value} < \underline{\hspace{1cm}}.$$

.025; .05

greater than
.05

The null hypothesis can be rejected for any value of α (greater than, less than) or equal to $\alpha = \underline{\hspace{1cm}}$. Since α was .05 in Example 9.11, H_0 was rejected.

The sample variance s^2 is an unbiased point estimator for the population variance σ^2. Utilizing the fact that $(n-1)s^2/\sigma^2$ has a chi-square distribution with $(n-1)$ degrees of freedom, we can show that a $(1-\alpha)100\%$ confidence interval for σ^2 is

$$\frac{(n-1)s^2}{\chi^2_{\alpha/2}} < \sigma^2 < \frac{(n-1)s^2}{\chi^2_{(1-\alpha/2)}},$$

where $\chi^2_{\alpha/2}$ is the tabulated value of the chi-square random variable based on _____ degrees of freedom having an area equal to $\alpha/2$ to its right, while $\chi^2_{(1-\alpha/2)}$ is the tabulated value from the same distribution having an area of $\alpha/2$ to its left or, equivalently, an area of $1 - \alpha/2$ to its right.

$n - 1$

Example 9.13
Find a 95% confidence interval estimate for the variance of the rod diameters from Example 9.11.

Solution

From Example 9.11 the estimate of σ^2 was $s^2 = .053$ with 14 degrees of freedom. For a confidence coefficient of .95, we need

$$\chi^2_{(1-\alpha/2)} = \chi^2_{.975} = \underline{\qquad}$$ 5.62872

and

$$\chi^2_{\alpha/2} = \chi^2_{.025} = \underline{\qquad}.$$ 26.1190

1. Using the confidence interval estimator

$$\frac{(n-1)s^2}{\chi^2_{\alpha/2}} < \sigma^2 < \frac{(n-1)s^2}{\chi^2_{(1-\alpha/2)}}$$

we have

$$\frac{14(.053)}{(\underline{\quad})} < \sigma^2 < \frac{14(.053)}{(\underline{\quad})}$$ 26.1190; 5.62872

$$\underline{\qquad} < \sigma^2 < \underline{\qquad}.$$ .028; .132

2. By taking square roots of the upper and lower confidence limits, we have an equivalent confidence interval for the standard deviation σ. For this problem,

$$\underline{\qquad} < \sigma < \underline{\qquad}.$$ .167; .363

Comment. Although the sample variance is an unbiased point estimator for σ^2, notice that the confidence interval estimator for σ^2 is *not symmetrically located about* $\hat{\sigma}^2$ as was the case with confidence intervals that were based on the z or t distributions. This follows from the fact that a chi-square distribution is not symmetric, while the z and t distributions are symmetric.

Self-Correcting Exercises 9D

1. In an attempt to assess the variability in the time until a pain reliever became effective for a patient, a doctor, on five different occasions, prescribed a controlled dosage of the drug for his patient. The five measurements recorded for the time until effective relief were 20.2, 15.7, 19.8, 19.2, and 22.7 minutes. Would these measurements indicate that the standard deviation of the time until effective relief was less than 3 minutes?
2. An educational testing service, in developing a standardized test, would like the test to have a standard deviation of at least 10. The present form of the test has produced a standard deviation of $s = 8.9$ based on $n = 30$ test scores. Should the present form of the test be revised based on these sample data? What *p*-value would you report?

3. A quick technique for determining the concentration of a chemical solution has been proposed to replace the standard technique, which takes much longer. In testing a standardized solution, 30 determinations using the new technique produced a standard deviation of $s = 7.3$ parts per million.
   a. Does it appear that the new technique is less sensitive (has larger variability) than the standard technique whose standard deviation is $\sigma = 5$ parts per million?
   b. Estimate the true standard deviation for the new technique with a 95% confidence interval.

9.7 Comparing Two Population Variances

An experimenter may wish to compare the variability of two testing procedures or compare the precision of one manufacturing process with another. One may also wish to compare two population variances prior to using a t test.

To test the hypothesis of equality of two population variances,

$$H_0: \sigma_1^2 = \sigma_2^2,$$

we need to make the following assumptions:

1. Each population sampled has a __normal__ distribution.
2. The samples are __independent__.

The statistic s_1^2/s_2^2 is used to test

$$H_0: \sigma_1^2 = \sigma_2^2.$$

A __large__ value of this statistic implies that $\sigma_1^2 > \sigma_2^2$; a __small__ value of this statistic implies that $\sigma_1^2 < \sigma_2^2$; while a value of the statistic close to one (1) implies that $\sigma_1^2 = \sigma_2^2$. In repeated sampling this statistic has an F distribution when $\sigma_1^2 = \sigma_2^2$ with the following properties:

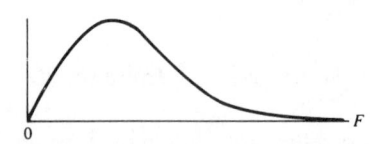

1. The distribution of F is (symmetric, __nonsymmetric__).
2. The shape of the distribution depends on the degrees of freedom associated with __s_1^2__ and __s_2^2__.
3. F is always greater than or equal to __zero__.

The tabulation of critical values of F is complicated by the fact that the distribution is nonsymmetric and must be indexed according to the values of v_1 and v_2, the degrees of freedom associated with the numerator and denominator of the F statistic. As we will see, however, it will be sufficient to have only right-tailed critical values of F for the various combinations of v_1 and v_2.

Tables 6, 7, 8, 9 and 10 in the text have tabulated right-tailed critical values for the F statistic, where F_α is that value of F having an area of α to its right, based on ν_1 and ν_2, the degrees of freedom associated with the *numerator* and *denominator* of F, respectively. F_α satisfies the relationship $P(F > F_\alpha) = \alpha$.

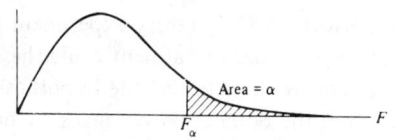

Table 6 has values of F_α for $\alpha = .10$ and various values of ν_1 and ν_2 between 1 and ∞, while Tables 7, 8, 9 and 10 have the same information for $\alpha = .05$, .025, .01 and .005, respectively.

Example 9.14
Find the value of F based on $\nu_1 = 5$ and $\nu_2 = 7$ degrees of freedom such that that
$$P(F > F_{.05}) = .05.$$

Solution
1. We wish to find a critical value of F with an area $\alpha = .05$ to its right based on $\nu_1 = 5$ and $\nu_2 = 7$ degrees of freedom. Therefore, we will use Table 7.
2. Values of ν_1 are found along the *top* margin of the table while values of ν_2 appear on both the right *and* left margins of the table. Find the value of $\nu_1 = 5$ along the top margin and cross-index this value with $\nu_2 = 7$ along the left margin to find $F_{.05(5,7)} = 3.97$.

Example 9.15
Find the critical right-tailed values of F for the following:

	ν_1	ν_2	α	F_α	
a.	5	2	.05	_____	19.30
b.	7	15	.10	_____	2.16
c.	20	10	.025	_____	3.42
d.	30	40	.005	_____	2.40
e.	17	13	.01	_____	3.76

We can always avoid using left-tailed critical values of the F distribution by using the following approach. In testing $H_0: \sigma_1^2 = \sigma_2^2$ against the alternative $H_a: \sigma_1^2 > \sigma_2^2$, we would reject H_0 only if s_1^2/s_2^2 is too large (larger than a right-tailed critical value of F). In testing $H_0: \sigma_1^2 = \sigma_2^2$ against $H_a: \sigma_1^2 < \sigma_2^2$, we would reject H_0 only if s_2^2/s_1^2 were too large. In testing $H_0: \sigma_1^2 = \sigma_2^2$ against the two-tailed alternative $H_a: \sigma_1^2 \neq \sigma_2^2$, we will agree to *designate the population that produced the larger sample variance as population 1 and the larger sample variance as* s_1^2. We then agree to reject H_0 if s_1^2/s_2^2 is *too large*.

When we agree to designate the population with the larger sample variance as population 1, the test of $H_0: \sigma_1^2 = \sigma_2^2$ versus $H_a: \sigma_1^2 \neq \sigma_2^2$ using s_1^2/s_2^2 will be right-tailed. However, in so doing we must remember that the tabulated tail area must be doubled to get the actual significance level of the test. For

.10
.02

example, if the critical right-tailed value of F has been found from Table 7, the actual significance level of the test will be $\alpha = 2(.05) =$ _____. If the critical value comes from Table 9, the actual level will be $\alpha = 2(.01) =$ _____, and so forth.

Example 9.16
An experimenter has performed a laboratory experiment using two groups of rats. One group was given a standard treatment while the second received a newly developed treatment. Wishing to test the hypothesis $H_0: \mu_1 = \mu_2$, the experimenter suspects that the population variances are not equal, an assumption necessary for using the t statistic in testing the equality of the means. Use the following data to test if the experimenter's suspicion is warranted at the $\alpha = .02$ level:

	Old Treatment	New Treatment
	$s = 2.3$	$s = 5.8$
	$n = 10$	$n = 10$

Solution
This problem involves a test of the equality of two population variances. Let population 1 be the population receiving the new treatment.

1. $H_0: \sigma_1^2 = \sigma_2^2$.

2. $H_a: \sigma_1^2 \neq \sigma_2^2$.

3. Test statistic:

$$F = s_1^2/s_2^2.$$

4. Rejection region: With $v_1 = v_2 = 9$ degrees of freedom, reject H_0 if $F > F_{.01} =$ _____.

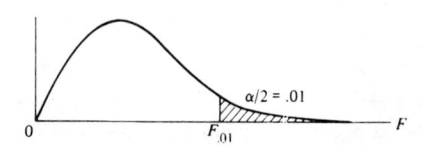

5.35

Now we calculate the test statistic.

$$F = s_1^2/s_2^2$$

$$= (5.8)^2/(2.3)^2$$

33.64; 5.29

$$= (\underline{\hspace{1cm}}) / (\underline{\hspace{1cm}})$$

6.36

$$= \underline{\hspace{1cm}}.$$

Sect. 9.7: Comparing Two Population Variances / 217

Decision: Since F is (greater, less) than $F_{.01} = 5.35$, $H_0: \sigma_1^2 = \sigma_2^2$ (is, is not) rejected. | greater; is

Since the population variances were judged to be different, the experimenter is not justified in using the t statistic to test $H_0: \mu_1 - \mu_2 = 0$. She must resort to other methods, several of which will be discussed in Chapter 14.

Example 9.17
Refer to Example 9.16. Find the level of significance for the test and interpret your results.

Solution
The level of significance for this (one-tailed, two-tailed) test is p-value = $2P[F > \underline{\quad}]$, where F has an F distribution with $v_1 = \underline{\quad}$ and $v_2 = \underline{\quad}$ degrees of freedom. In order to find an approximate p-value, we must first find critical values, F_α, for various values of α with $v_1 = v_2 = 9$. These values are found in Tables 6, 7, 8, 9, and 10. Fill in the missing entries in the table that follows. | two-tailed 6.36; 9 9

α	F_α
.10	2.44
.05	_____
.025	4.03
.01	_____
.005	_____

3.18

5.35
6.54

Since the observed value, $F = 6.36$ falls between $F_{.01} = \underline{\quad}$ and $F_{.005} = 6.54$, the p-value will be between $2(.005) = .01$ and $2(.01) = .02$. That is, $.01 < p$-value $< .02$. Hence, H_0 can be rejected for any value of α (greater, less) than or equal to _____ . | 5.35

greater; .02

Utilizing the fact that $(s_1^2/s_2^2)(\sigma_2^2/\sigma_1^2)$ has an F distribution with $v_1 = (n_1 - 1)$ and $v_2 = (n_2 - 1)$ degrees of freedom, it can be shown that a $(1 - \alpha)100\%$ confidence interval for σ_1^2/σ_2^2 is

$$\frac{s_1^2}{s_2^2} \frac{1}{F_{v_1 v_2}} < \frac{\sigma_1^2}{\sigma_2^2} < \frac{s_1^2}{s_2^2} F_{v_2 v_1}$$

where $F_{v_1 v_2}$ is the tabulated value of F with v_1 and v_2 degrees of freedom and area $\alpha/2$ to its right, and $F_{v_2 v_1}$ is the tabulated value of F with v_2 and v_1 degrees of freedom having area $\alpha/2$ to its right.

Example 9.18
A comparison of the precisions of two machines developed for extracting juice from oranges is to be made using the following data:

Machine A	Machine B
$s^2 = 3.1$ ounces^2	$s^2 = 1.4$ ounces^2
$n = 25$	$n = 25$

218 / Chap. 9: Inference from Small Samples

Is there sufficient evidence to indicate that $\sigma_A^2 > \sigma_B^2$ at the $\alpha = .05$ level? Find a 90% confidence interval for σ_A^2/σ_B^2.

Solution

Let population 1 be the population of measurements on machine A. The test would proceed as follows:

$$H_0: \sigma_1^2 = \sigma_2^2.$$

> $$H_a: \sigma_1^2 \underline{\qquad} \sigma_2^2.$$

Test statistic:

$$F = s_1^2/s_2^2.$$

24
1.98

Rejection region: Based on $\nu_1 = \nu_2 = \underline{\qquad}$ degrees of freedom, we will reject H_0 if $F > F_{.05}$ with $F_{.05} = \underline{\qquad}$.

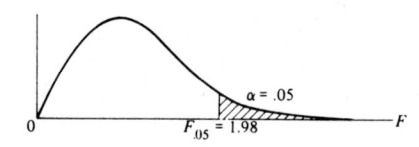

The value of the statistic is

2.21
$$F = s_1^2/s_2^2 = 3.1/1.4 = \underline{\qquad}.$$

is
Decision: We reject H_0 and conclude that the variability of machine A (is, is not) greater than that of machine B.

A 90% confidence interval for $\sigma_A^2/\sigma_B^2 = \sigma_1^2/\sigma_2^2$ is

$$\frac{s_1^2}{s_2^2} \frac{1}{F_{24,24}} < \frac{\sigma_A^2}{\sigma_B^2} < \frac{s_1^2}{s_2^2} F_{24,24}$$

$$\frac{2.21}{1.98} < \frac{\sigma_A^2}{\sigma_B^2} < 2.21 (1.98)$$

1.12; 4.38
$$\underline{\qquad} < \frac{\sigma_A^2}{\sigma_B^2} < \underline{\qquad}.$$

Self-Correcting Exercises 9E

1. Refer to Exercise 2, Self-Correcting Exercises 9B. In using the t statistic in testing an hypothesis concerning $\mu_1 - \mu_2$, one assumes that $\sigma_1^2 = \sigma_2^2$. Based on the sample information, could you conclude that this assumption had been met for this problem? Use $\alpha = .05$.

2. An experiment to explore the pain thresholds to electrical shock for males and females resulted in the following data summary:

	Males	Females
n	10	13
$\bar{x}$	15.1	12.6
s^2	11.3	26.9

Do these data supply sufficient evidence to indicate a significant difference in variability of thresholds for these two groups at the 10% level of significance? What *p*-value would you report?

9.8 Assumptions

The testing and estimation procedures presented in this chapter are based on the t, χ^2, and F statistics. In order that the probability statements associated with these testing and estimation procedures accurately reflect the prescribed probability values, specific assumptions concerning the sampled population(s) and the method of sampling must be satisfied.

The valid use of the t, χ^2, and F statistics requires that all samples be randomly selected from _____ populations. With the exception of the paired-difference experiment, when two samples are drawn, the samples must be drawn _____. In addition, when making inferences about the difference in two population means μ_1 and μ_2 using two independent samples, the population variances σ_1^2 and σ_2^2 must be _____.

It would be unusual to have all these assumptions satisfied in practice. However, if the sampled population were not normal, or $\sigma_1^2 \neq \sigma_2^2$, we would like our procedures to produce error probabilities that are approximately equal to the specified values. A statistical procedure that is insensitive to departures from the assumptions upon which it is based is said to be _____.

Procedures based on the t statistic are fairly robust to departures from normality provided that the sampled population(s) is(are) not strongly skewed. This (is, is not) true for procedures based upon the χ^2 and F statistics. The t statistic used in comparing two means is moderately robust to departures from the assumption $\sigma_1^2 = \sigma_2^2$ when $n_1 = n_2$. However, when $\sigma_1^2 \neq \sigma_2^2$ and one sample size becomes large relative to the other, the procedure fails to be robust.

When the experimenter is aware of possible violations of assumptions, the usual procedure can be used if it is robust with respect to the assumptions violated. Otherwise, the nonparametric procedures presented in Chapter 14 can be used. Nonparametric methods require few or no assumptions concerning the sampled population(s); however, samples must nonetheless be _____ selected, and when appropriate, the samples must also be independently drawn. When the sample sizes are relatively large, techniques such as those presented in Chapter 8 can be used in place of nonparametric procedures.

normal
independently
equal
robust
is not
randomly

EXERCISES

1. Why can we say that the test statistics employed in Chapter 8 are approximately normally distributed?
2. What assumptions are made when Student's t statistic is used to test an hypothesis concerning a population mean μ?
3. How does one determine the degrees of freedom associated with a t statistic?
4. Ten butterfat determinations for brand G milk were carried out yielding $\bar{x} = 3.7\%$ and $s = 1.7\%$. Do these results produce sufficient evidence to indicate that brand G milk contains, on the average, less than 4.0% butterfat? (Use $\alpha = .05$.)
5. Refer to Exercise 4. Estimate the mean percentage of butterfat for brand G milk with a 95% confidence interval.
6. An experimenter has developed a new fertilizing technique that should increase the production of cabbages. Do the following data produce sufficient evidence to indicate that the mean weight of those cabbages grown by using the new technique is greater than the mean weight of those grown by using the standard technique?

	Population I (New Technique)	Population II (Standard Technique)
	$n_1 = 16$	$n_2 = 10$
	$\bar{x}_1 = 33.4$ ounces	$\bar{x}_2 = 31.8$ ounces
	$s_1 = 3$ ounces	$s_2 = 4$ ounces

7. Find a 90% confidence interval for the difference in means $\mu_1 - \mu_2$ for the data given in Exercise 6.
8. To test the comparative brightness of two red dyes, nine samples of cloth were taken from a production line and each sample was divided into two pieces. One of the two pieces in each sample was randomly chosen and red dye 1 applied; red dye 2 was applied to the remaining piece. The following data represent a "brightness score" for each piece. Is there sufficient evidence to indicate a difference in mean brightness scores for the two dyes?

Sample	Dye 1	Dye 2
1	10	8
2	12	11
3	9	10
4	8	6
5	15	12
6	12	13
7	9	9
8	10	8
9	15	13

9. To test the effect of alcohol in increasing the reaction time to respond to a given stimulus, the reaction times of seven persons were measured. After

consuming 3 ounces of 40% alcohol, the reaction time for each of the seven persons was measured again. Do the following data indicate that the mean reaction time after consuming alcohol was greater than the mean reaction time before consuming alcohol? (Use $\alpha = .05$.)

Person	Before (time in seconds)	After (time in seconds)
1	4	7
2	5	8
3	5	3
4	4	5
5	3	4
6	6	5
7	2	5

10. A manufacturer of odometers claimed that mileage measurements indicated on his instruments had a variance of at most .53 mile per 10 miles traveled. An experiment, consisting of eight runs over a measured 10-mile stretch, was performed in order to check the manufacturer's claim. The variance obtained for the eight runs was .62. Does this provide sufficient evidence to indicate that $\sigma^2 > .53$? (Use $\alpha = .05$.)

11. Construct a 99% confidence interval estimate for σ^2 in Exercise 10.

12. In a test of heat resistance involving two types of metal paint, two groups of ten metal strips were randomly formed. Group one was painted with type I paint, while group two was painted with type II paint. The metal strips were placed in an oven in random order, heated, and the temperature at which the paint began to crack and peel recorded for each strip. Do the following data indicate that the variability in the critical temperatures differs for the two types of paint? Use $\alpha = .05$.

	$\bar{x}$	s^2	n
Type I	280.1° F	93.2	10
Type II	269.9° F	51.9	10

13. Construct a 98% confidence interval for σ_1^2/σ_2^2 for the data given in Exercise 12.

14. In an attempt to reduce the variability of machine parts produced by process A, a manufacturer has introduced process B (a modification of A). Do the following data, based on two samples of 25 items, indicate that the manufacturer has achieved his goal? Use $\alpha = .01$.

	n	s^2
Process A	25	6.57
Process B	25	3.19

15. Before contracting to have stereo music piped into each of his suites of offices, an executive had his office manager randomly select seven offices in which to have the system installed. The average time spent outside

these offices per excursion among the employees involved was recorded before and after the music system was installed with the following results.

Office number	Time in minutes	
	No music	Music
1	8	5
2	9	6
3	5	7
4	6	5
5	5	6
6	10	7
7	7	8

Would you suggest that the executive proceed with the installation? Find the approximate level of significance and interpret your results.

16. The weights in grams of 10 male and 10 female juvenile ring-necked pheasants are given below, together with a data summary produced by the MINITAB command DESCRIBE.

Males	Females
1384	1073
1286	1058
1503	1053
1627	1038
1450	1018
1672	1146
1370	1123
1659	1089
1725	1034
1394	1281

MTB > DESCRIBE C1 C2

	N	MEAN	MEDIAN	TREMEAN	STDEV	SEMEAN
C1	10	1507.0	1476.5	1507.4	153.1	48.4
C2	10	1091.3	1065.5	1076.8	77.7	24.6

	MIN	MAX	Q1	Q3
C1	1286.0	1725.0	1380.5	1662.3
C2	1018.0	1281.0	1037.0	1128.8

a. Use a statistical test to determine if the population variance of the weights of the male birds differs from that of the females.

b. Test whether the average weight of juvenile male ring-necked pheasants exceeds that of the females by 300 grams. The procedure that you use should take into account the results of the analysis in part a. (If $\sigma_1^2 \neq \sigma_2^2$, the alternate form of the t statistic must be used and the degrees of freedom estimated.)

The analysis in part b can be implemented using the MINITAB command TWOSAMPLE without the subcommand POOLED. Since the TWOSAMPLE command always tests $H_0: \mu_1 - \mu_2 = 0$, you can subtract 300 from the weights of the males before implementing the analysis using MINITAB.

CHAPTER 10
LINEAR REGRESSION AND CORRELATION

10.1 Introduction

We have investigated the problem of making inferences about population parameters in the case of large and small sample sizes. We will now consider another aspect of this problem. Suppose that $E(y)$, the expected value of a random variable y, depends on the values assigned to other variables, $x_1, x_2, \ldots, x_k$. Then we say that a functional relationship exists between $E(y)$ and $x_1, x_2, \ldots, x_k$. Since the values of $E(y)$ depend on the values assumed by $x_1, x_2, \ldots, x_k$, $E(y)$ is called the *dependent variable* and $x_1, x_2, \ldots, x_k$ are called the *independent variables*. The variables $x_1, x_2, \ldots, x_k$ can also be called *predictor variables,* since they are used for the purpose of predicting the value of y that will be observed for given values of $x_1, x_2, \ldots, x_k$. We restrict our investigation to the case where $E(y)$ is a *linear* function of one variable x. By linear, we mean that the relationship between $E(y)$ and x can be described by a straight line.

Review: The Algebraic Representation of a Straight Line
To understand the development of the following linear models, you must be familiar with the algebraic representation of a straight line and its properties.

The mathematical equation for a straight line is

$$y = \beta_0 + \beta_1 x,$$

where x is the independent variable, y is the dependent variable, and β_0 and β_1 are fixed constants. When values of x are substituted into this equation, pairs of numbers, (x_i, y_i), are generated which, when plotted or graphed on a rectangular coordinate system, form a straight line.

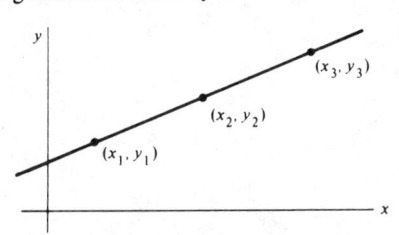

Consider the graph of a linear equation $y = \beta_0 + \beta_1 x$, shown below.

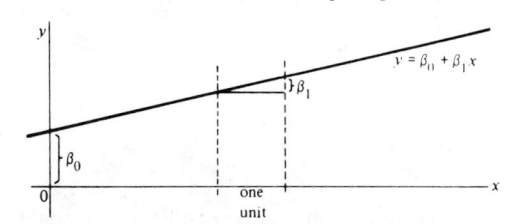

1. By setting $x = 0$, we have $y = \beta_0 + \beta_1(0) = \beta_0$. Because the line intercepts or cuts the y-axis at the value $y = \beta_0$, β_0 is called the y _____ .   intercept
2. The constant β_1 represents the increase in y for a one-unit increase in x and is called the _____ of the line.   slope

Example 10.1
Plot the equation $y = 1 + .5x$ on a rectangular coordinate system.

Solution
Two points are needed to uniquely determine a straight line and therefore a minimum of two points must be found. A third point is usually found as a check on calculations.
1. Using 0, 2, and 4 as values of x, find the corresponding values of y.

When $x = 0, y = 1 + .5(0) =$ _____ .   1

When $x = 2, y = 1 + .5(2) =$ _____ .   2

When $x = 4, y = 1 + .5(4) =$ _____ .   3

2. Plot these points on a rectangular coordinate system and join them by using a straightedge.

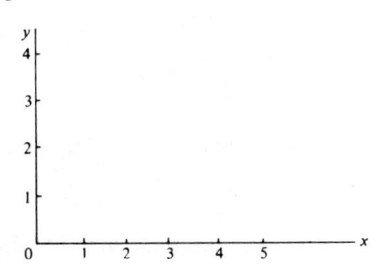

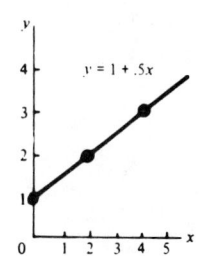

Practice plotting the following linear equations on a rectangular coordinate system:
a. $y = -1 + 3x$.
b. $y = 2 - x$.
c. $y = -.5 - .5x$.
d. $y = x$.
e. $y = .5 + 2x$.

10.2 A Simple Linear Probabilistic Model

Suppose we are given a set consisting of n pairs of values for x and y, each pair representing the value of a response y for a given value of x. Plotting these points might result in the following scatter diagram:

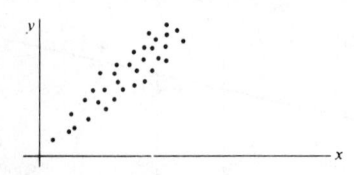

Someone might say that these points appear to lie on a straight line. This person would be hypothesizing that a *model* for the relationship between x and y is of the form

$$y_i = \beta_0 + \beta_1 x_i \qquad i = 1, 2, \ldots, n.$$

According to this model, for a given value of x, the value of y is *uniquely determined*. Therefore, this is called a _____ model.   **deterministic**

Another person might say that these points appear to be *deviations* about a straight line, hypothesizing the model

$$y_i = \beta_0 + \beta_1 x_i + \epsilon_i \qquad i = 1, 2, \ldots, n,$$

where ϵ_i represents the deviation of the ith point (x_i, y_i) from the straight line $y = \beta_0 + \beta_1 x$.

Suppose that we were able to make 4 observations on y at each of the values x_1, x_2, and x_3. We might observe the following 12 pairs of values:

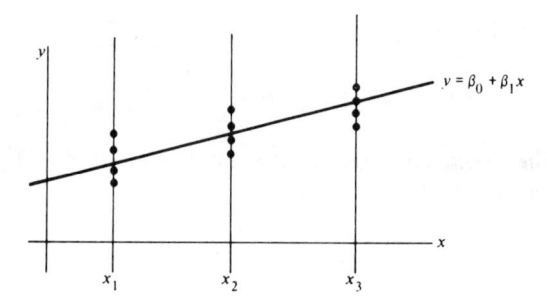

To account for what appear to be random deviations about the deterministic line $y = \beta_0 + \beta_1 x$, we will consider the deviations to be *random errors* with the following properties:

1. For any fixed value of x, in repeated sampling, the random errors have a mean of zero and a variance equal to σ^2.
2. Any two random errors are independent in the probabilistic sense.
3. Regardless of the value of x, the random errors have the same *normal* distribution with mean zero and variance σ^2.

Since this model uses a random error component having a probability distribution, it is referred to as a _____ model.   **probabilistic**

The probabilistic model assumes that the average value of y is linearly related to x and the observed values of y will deviate above and below the line

$$E(y) = \beta_0 + \beta_1 x$$

by a random amount. The random components all have the same normal distribution and are independent of each other. According to the properties given above, repeated observations on y at the values x_1, x_2, and x_3 would result in the following visual representation of the random errors:

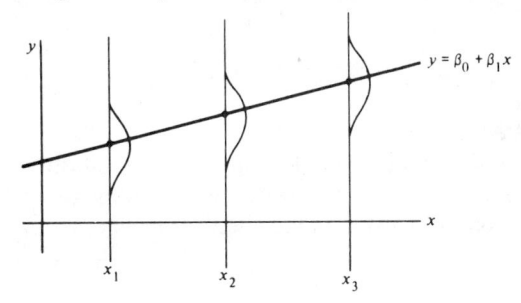

The probabilistic model appears to be the model that best describes the data and we now proceed to find an estimate for this prediction equation, the regression line

$$\hat{y} = \hat{\beta}_0 + \hat{\beta}_1 x.$$

10.3 The Method of Least Squares

The criterion used for estimating β_0 and β_1 in the model

$$y_i = \beta_0 + \beta_1 x_i + \epsilon_i$$

is to find an estimated line

$$\hat{y}_i = \hat{\beta}_0 + \hat{\beta}_1 x_i$$

that in some sense minimizes the deviations of the observed values of y from the fitted line. If the deviation of the ith observed value from the fitted value is $(y_i - \hat{y}_i)$, we define the best estimated line as one that minimizes the sum of squares of the deviations of the observed values of y from the fitted values of y. The quantity

$$\sum_{i=1}^{n} (y_i - \hat{y}_i)^2$$

represents the sum of squares of deviations of the observed values of y from the fitted values and is called the sum of squares for error (SSE):

$$\text{SSE} = \sum_{i=1}^{n} (y_i - \hat{y}_i)^2 = \sum_{i=1}^{n} [y_i - (\hat{\beta}_0 + \hat{\beta}_1 x_i)]^2.$$

The values of $\hat{\beta}_0$ and $\hat{\beta}_1$ are determined mathematically so that SSE will be minimum.

This process of minimization is called the *method of least squares* and produces estimates of β_0 and β_1. If we agree that all summations will be with respect to i as the variable of summation, $i = 1, 2, \ldots, n$, then the least-squares estimates of β_1 and β_0 are

$$\hat{\beta}_1 = S_{xy}/S_{xx} \quad \text{and} \quad \hat{\beta}_0 = \bar{y} - \hat{\beta}_1 \bar{x}$$

where

$$S_{xy} = \Sigma (x_i - \bar{x})(y_i - \bar{y}) = \Sigma x_i y_i - [(\Sigma x_i)(\Sigma y_i)/n]$$

and

$$S_{xx} = \Sigma (x_i - \bar{x})^2 = \Sigma x_i^2 - [(\Sigma x_i)^2/n].$$

Example 10.2

In this chapter we will use the following example to illustrate each type of problem encountered. Be ready to refer to the information tabulated on this page. For the following data, find the best fitting line, $\hat{y} = \hat{\beta}_0 + \hat{\beta}_1 x$:

x_i	y_i	x_i^2	y_i^2	$x_i y_i$
2	1	4	1	2
3	3	9	9	9
5	4	25	16	20
7	7	49	49	49
9	10	81	100	90
Sum ____	____	168	175	170

$\bar{x} = $ _____ $\bar{y} = $ _____

Solution

1. First find all the sums needed in the computations.

$$S_{xy} = \Sigma x_i y_i - [(\Sigma x_i)(\Sigma y_i)/n]$$

$$= 170 - [(\underline{\hspace{1cm}})(\underline{\hspace{1cm}})/5]$$

$$= 170 - \underline{\hspace{1cm}}$$

$$= \underline{\hspace{1cm}}$$

$$S_{xx} = \Sigma x_i^2 - [(\Sigma x_i)^2/n]$$

$$= \underline{\hspace{1cm}} - [(26)^2/5]$$

$$= \underline{\hspace{1cm}} - 135.2$$

$$= \underline{\hspace{1cm}}.$$

2. $\hat{\beta}_1 = S_{xy}/S_{xx} = 4.0/32.8 = \underline{\hspace{1cm}}$.

3. $\hat{\beta}_0 = \bar{y} - \hat{\beta}_1 \bar{x}$

$$= (\underline{\hspace{1cm}}) - 1.22(\underline{\hspace{1cm}})$$

$$= (\underline{\hspace{1cm}}) - (\underline{\hspace{1cm}})$$

$$= -1.3415 \text{ or } -1.34$$

4. The best fitting line is

$$\hat{y} = \underline{\qquad}.$$

$-1.34 + 1.22x$

We can now use the equation

$$\hat{y} = -1.34 + 1.22x$$

to predict values of _____ for values of x in the interval $2 \leqslant x \leqslant 9$. However, we also need to place _____ on the _____ of this prediction. To do this we need σ^2 or its estimator s^2.

y

bounds; error

Self-Correcting Exercises 10A

1. The registrar at a small university noted that the preenrollment figures and the actual enrollment figures for the past 6 years (in hundreds of students) were as shown here:

x, preenrollment	30	35	42	48	50	51
y, actual enrollment	33	41	46	52	59	55

   a. Plot these data. Does it appear that a linear relationship exists between x and y?
   b. Find the least-squares line $\hat{y} = \hat{\beta}_0 + \hat{\beta}_1 x$.
   c. Using the least-squares line, predict the actual number of students enrolled if the preenrollment figure is 5,000 students.

2. An entomologist, interested in predicting cotton harvest using the number of cotton bolls per quadrate counted during the middle of the growing season, collected the following data, where y is the yield in bales of cotton per field quadrate and x is hundreds of cotton bolls per quadrate counted during midseason:

y	21	17	20	19	15	23	20
x	5.5	2.8	4.7	4.3	3.7	6.1	4.5

   a. Fit the least-squares line $\hat{y} = \hat{\beta}_0 + \hat{\beta}_1 x$ using these data.
   b. Plot the least-squares line and the actual data on the same graph. Comment on the adequacy of the least-squares predictor to describe these data.

3. Refer to Exercise 2. The same entomologist also had available a measure of the number of damaging insects present per quadrate during a critical time in the development of the cotton plants. The data follow.

y, yield	21	17	20	19	15	23	20
x, insects	11	20	13	12	18	10	12

   a. Fit the least-squares line to these data.
   b. Plot the least-squares line and the actual data points on the same graph.

10.4 Calculating s^2, an Estimator of σ^2

Before we can proceed with evaluations of the estimates $\hat{\beta}_0$ and $\hat{\beta}_1$, or assess the goodness of any prediction of y based on the estimated regression line, we must first estimate _____ σ^2 , the variance of y for a given value of x.

To estimate σ^2, we use SSE, the sum of squares of deviations about the line, $\hat{y} = \hat{\beta}_0 + \hat{\beta}_1 x$. The n pairs of data points provide n degrees of freedom for estimation. Having estimated β_0 and β_1, we now have _____ $n-2$ remaining degrees of freedom to estimate σ^2.

Therefore, the estimate of σ^2 is

$$s^2 = \frac{SSE}{(n-2)} = \frac{\Sigma(y_i - \hat{y}_i)^2}{(n-2)}$$

The computational form for the quantity SSE is

$$SSE = S_{yy} - \hat{\beta}_1 S_{xy}$$

where

$$S_{yy} = \Sigma(y_i - \bar{y})^2 = \Sigma y_i^2 - (\Sigma y_i)^2/n.$$

S_{xy} is the numerator used in computing $\hat{\beta}_1$ and *has already been found*.

Example 10.3
Calculate s^2 for our data.

Solution

1. Calculate S_{yy}.

$$S_{yy} = \Sigma y_i^2 - \frac{(\Sigma y_i)^2}{n} = \underline{} - \frac{(\underline{})^2}{5}$$

$$= \underline{}$$

2. Using S_{xy} from the calculations for $\hat{\beta}_1$, use the computational formula for SSE.

$$SSE = S_{yy} - \hat{\beta}_1 S_{xy}$$

$$= \underline{} - (\underline{})(\underline{})$$

$$= \underline{}$$

3. Calculate $s^2 = \frac{SSE}{n-2} = \frac{1.2195}{(\underline{})} = \underline{}$

Margin notes:

σ^2

$n-2$

175; 25

50

50; 1.2195; 40

1.2

3; 0.4065

10.5 Inferences Concerning the Slope of the Line, β_1

The slope β_1 is the average increase in _____ for a one-unit increase in _____. The question of the existence of a linear relationship between x and y must be phrased in terms of the slope β_1. If no linear relationship exists between x and y, then $\beta_1 = 0$. Hence a test of the existence of a *linear* relationship between x and y is given as $H_0: \beta_1 =$ _____ versus $H_a: \beta_1 \neq$ _____.

y
x

0
0

When the random error ϵ is *normally* distributed, the estimator $\hat{\beta}_1$ has the following properties:
1. $\hat{\beta}_1$ has a _____ distribution.
2. $\hat{\beta}_1$ is an unbiased estimator for _____ so that $E(\hat{\beta}_1) =$ _____.
3. The variance of $\hat{\beta}_1$ is

normal
$\beta_1; \beta_1$

$$\sigma_{\hat{\beta}_1}^2 = \sigma^2/S_{xx}.$$

The following test statistics can be constructed using the fact that $\hat{\beta}_1$ is a *normally* distributed, *unbiased* estimator of β_1:

1. $z = \dfrac{\hat{\beta}_1 - \beta_1}{\sigma/\sqrt{S_{xx}}}$ if σ^2 is known.

2. $t = \dfrac{\hat{\beta}_1 - \beta_1}{s/\sqrt{S_{xx}}}$ if s^2 is used to estimate σ^2 and hence to estimate $\sigma_{\hat{\beta}_1}^2$.

Since σ^2 is rarely known, we can test for a significant linear relationship using the statistic given in 2, which has a Student's t distribution with _____ degrees of freedom. A test of the hypothesis $H_0: \beta_1 = 0$ versus $H_a: \beta_1 \neq 0$ is given as follows:

$n - 2$

1. $H_0: \beta_1 = 0$.

2. $H_a: \beta_1 \neq 0$.

3. Test statistic:

$$t = \dfrac{\hat{\beta}_1 - (0)}{s\sqrt{S_{xx}}}.$$

4. Rejection region: Reject H_0 if $|t| > t_{\alpha/2}$ based on $n - 2$ degrees freedom.

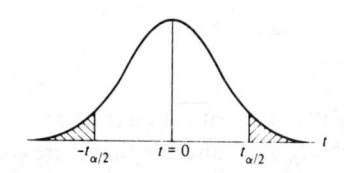

232 / Chap. 10: Linear Regression and Correlation

Example 10.4

For our data, test the hypothesis that there is no linear relationship between x and y at the $\alpha = .05$ level.

Solution

1. $H_0: \beta_1 = $ _____ .   [0]

2. $H_a: \beta_1 \neq $ _____ .   [0]

3. Test statistic:

$$t = \frac{\hat{\beta}_1 - (0)}{s/\sqrt{S_{xx}}}.$$

4. Rejection region: With 3 degrees of freedom, we will reject H_0 if $|t| > t_{.025} = $ _____ .   [3.182]

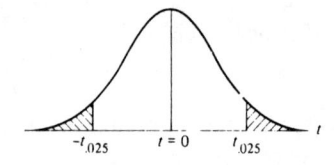

5. To calculate the test statistic, we draw upon earlier calculations for the value of s and S_{xx}.

$$s = \sqrt{s^2} = \sqrt{\rule{1cm}{0.4pt}} = \rule{1cm}{0.4pt}$$ [0.4065; .63758]

and

$$S_{xx} = \rule{1cm}{0.4pt}$$ [32.8]

The test statistic is then

$$t = \frac{\hat{\beta}_1 - 0}{s/\sqrt{S_{xx}}}$$

$$= \frac{(\rule{1cm}{0.4pt})}{.63758/\sqrt{32.8}}$$ [1.2195]

$$= \rule{1cm}{0.4pt}.$$ [10.95]

6. Since 10.95 is larger than the critical value of $t = $ _____ , we (reject, do not reject) H_0 and conclude that there (is, is not) a linear relationship between x and y.   [3.182; reject; is]

Confidence Interval for β_1

If x increases one unit, what is the predicted change in y? Since $\hat{\beta}_1$ is an unbiased estimator for β_1 and has a normal distribution, the t statistic, based on $n - 2$ degrees of freedom, can be used to derive the confidence interval estimator for the slope β_1:

$$\hat{\beta}_1 \pm t_{\alpha/2}\, s/\sqrt{S_{xx}}.$$

Example 10.5

Find a 95% confidence interval for the average change in y for an increase of one unit in x.

Solution

1. $1 - \alpha = .95$; $\alpha = $ _____ ; $\alpha/2 = $ _____ ; $n - 2 = $ _____ ;   .05; .025; 3
   $t_{.025} = $ _____ .   3.182

2. $\quad \hat{\beta}_1 \pm t_{.025}\, s/\sqrt{S_{xx}}$

   $1.22 \pm ($ _____ $)(.63758)/\sqrt{32.8}$ 3.182

   $1.22 \pm ($ _____ $)$.   .35

3. A 95% confidence interval for β_1 is (_____ , _____).   .87; 1.57

Points Concerning Interpretation of Results

If the test $H_0: \beta_1 = 0$ is performed and H_0 is *not rejected*, this (does, does not)   does not
mean that x and y are *not related*, since
1. a type _____ error may have been committed, or   II
2. x and y may be related, but not _____ . For example, the true   linearly
   relationship may be of the form $y = \beta_0 + \beta_1 x + \beta_2 x^2$.

If the test $H_0: \beta_1 = 0$ is performed and H_0 is *rejected*,
1. we (can, cannot) say that x and y are solely linearly related, since there   cannot
   may be other terms (x^2 or x^3) that have not been included in our model;
2. we should not conclude that a *causal* relationship exists between x and y,
   since the related changes we observe in x and y may actually be *caused* by
   an unmeasured third variable, say z.

Consider the problem where the true relationship between x and y is a "curve" rather than a straight line. Suppose we fitted a straight line to the data for values of x between a and b.

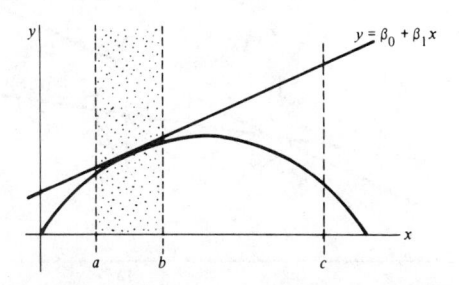

234 / Chap. 10: Linear Regression and Correlation

Using $\hat{y} = \hat{\beta}_0 + \hat{\beta}_1 x$ to predict values of y for $a \leqslant x \leqslant b$ would result in quite an accurate prediction. However, if the prediction line were used to predict y for the value $x = c$, the prediction would be highly _____inaccurate_____. Although the line adequately describes the indicated trend in the region $a \leqslant x \leqslant b$, there is no justification for assuming that the line would fit equally well for values of x outside the region $a \leqslant x \leqslant b$. The process of predicting outside the region of experimentation is called _____extrapolation_____. As our example shows, an experimenter should *not* extrapolate unless he or she is willing to assume the consequences of *gross errors*.

Self-Correcting Exercises 10B

1. Refer to Self-Correcting Exercises 10A, Exercise 1. Calculate SSE, s^2 and s for these data.
   a. Test the hypothesis that there is no linear relationship between actual and pre-enrollment figures at the $\alpha = .05$ level of significance.
   b. Estimate the average increase in actual enrollment for an increase of 100 in pre-enrolled students with a 95% confidence interval.
2. Refer to Self-Correcting Exercises 10A, Exercise 2. Calculate SSE, s^2 and s for these data and test for a significant linear relationship between yield and number of bolls at the $\alpha = .05$ level of significance.
3. Refer to Self-Correcting Exercises 10A, Exercise 3. Test for a significant linear relationship between yield and the number of insects present at the $\alpha = .05$ level of significance.

10.6 Estimating the Expected Value of y for a Given Value of x

Assume that x and y are related according to the model

$$y = \beta_0 + \beta_1 x + \epsilon.$$

We have found an estimator for this line which is

$$\hat{y} = \underline{\hat{\beta}_0 + \hat{\beta}_1 x}.$$

Suppose we are interested in estimating $E(y|x)$ for a given value of x, say x_p.

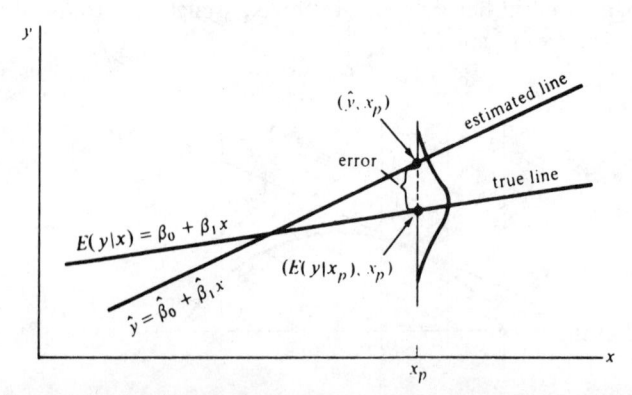

In repeated sampling, the predicted values of y will generate a distribution of estimates, $\hat{y}$, for the value of $x = x_p$, as shown in the diagram. The mean of these estimates is the true value

$$E(y|x = x_p) = \beta_0 + \beta_1 x_p.$$

Therefore, we will use $\hat{y}$ to estimate the expected or average value of y for $x = x_p$, using as our estimator

$$\hat{y} = \underline{\hspace{2cm}}.$$ $\hat{\beta}_0 + \hat{\beta}_1 x_p$

The estimator $\hat{y} = \hat{\beta}_0 + \hat{\beta}_1 x_p$ has the following properties:
1. $E(\hat{y}|x_p) = E(y|x_p)$
   That is, for a fixed value of x, $\hat{y}$ is an unbiased estimator for the average value of y,

$$E(y|x_p) = \underline{\hspace{2cm}}.$$ $\beta_0 + \beta_1 x_p$

2. The variance, $\sigma_{\hat{y}}^2$, of the estimator $\hat{y}|x_p$ is given by

$$\sigma_{\hat{y}}^2 = \sigma^2 \left[\frac{1}{n} + \frac{(x_p - \bar{x})^2}{S_{xx}} \right].$$

3. When the random component ϵ is normally and independently distributed, then $\hat{y}|x$ is normally distributed.

By using these results we can construct a z or t statistic to test an hypothesis concerning the expected value of y when $x = x_p$. Since σ^2 is rarely known, its sample estimate s^2 is used, resulting in a _____ statistic with _____ degrees of freedom. $t; n - 2$

Test of an Hypothesis Concerning $E(y|x_p)$

1. $H_0 : E(y|x_p) = E_0$.
2. H_a: Appropriate one- or two-tailed test.
3. Test statistic:

$$t = \frac{\hat{y} - E_0}{\hat{\sigma}_{\hat{y}}} = \frac{\hat{y} - E_0}{s\sqrt{\frac{1}{n} + \frac{(x_p - \bar{x})^2}{S_{xx}}}}.$$

4. Rejection region: Appropriate one- or two-tailed rejection region based on H_a.

Example 10.6
For our data, test the hypothesis that $\beta_0 = -1$ against the alternative that $\beta_0 < -1$ at the $\alpha = .05$ level.

236 / Chap. 10: Linear Regression and Correlation

Remark: By setting $x_p = 0$, $E(y|x_p = 0) = \beta_0 + \beta_1(0) = \beta_0$. Therefore, the test described above can be used to test an hypothesis about the intercept β_0.

Solution

1. $\quad H_0: E(y|x = 0) = \beta_0 = -1$.

2. $\quad H_a: \beta_0 < -1$.

3. Test statistic:

$$t = \frac{\hat{\beta}_0 - (-1)}{s\sqrt{\dfrac{1}{n} + \dfrac{(0-\bar{x})^2}{S_{xx}}}}$$

.64; -.53

$$= \frac{-1.3415 - (-1)}{.638\sqrt{\dfrac{1}{5} + \dfrac{(0-5.2)^2}{32.8}}} = \frac{-.3415}{\rule{1cm}{0.15mm}} = \rule{2cm}{0.15mm}.$$

3
-2.353

4. Rejection region: With _____ degrees of freedom, we will reject H_0 if $t <$ _____ .

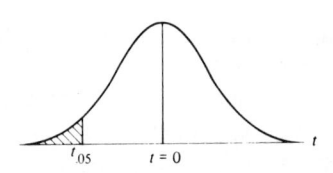

Do not reject; do not

5. Decision: (Reject, Do not reject) H_0. The data (do, do not) present sufficient evidence to indicate that $\beta_0 < -1$.

A $100(1 - \alpha)\%$ confidence interval for $E(y|x_p)$ is given as

$$(\hat{y}|x_p) \pm t_{\alpha/2}\, s\sqrt{\dfrac{1}{n} + \dfrac{(x_p - \bar{x})^2}{S_{xx}}}$$

where $(\hat{y}|x_p)$ is the value of the estimate for $x = x_p$, found by using

$$\hat{y} = \hat{\beta}_0 + \hat{\beta}_1 x_p.$$

Example 10.7
Find a 95% confidence interval for $E(y|x = 6)$.

Solution

5.98

1. $\quad (\hat{y}|x = 6) = -1.3415 + 1.2195(6) =$ _____ .

3.182

2. $\quad t_{.025} =$ _____ .

3. $\hat{\sigma}_{\hat{y}} = .638\sqrt{\dfrac{1}{5} + \dfrac{(6-5.2)^2}{32.8}} =$ _____ .

.2987

4. A 95% confidence interval is constructed as follows:

$(\hat{y}|x=6) \pm 3.182 \hat{\sigma}_{\hat{y}}$

(_____) ± 3.182 (_____)

5.98; .2987

(_____) ± (_____).

5.98; .95

In order to obtain reasonably accurate results when predicting $E(y)$ at a particular x_p, it is desirable that x_p lie within the range of the observed values of x.

10.7 Predicting a Particular Value of y for a Given Value of x

In the last section, we were interested in estimating $E(y|x)$ when $x = x_p$. That is, we estimated a point on the _____ regression line at the value of $x = x_p$. Now we consider the problem of predicting the actual single value of y that occurs (or will occur) when $x = x_p$, rather than the _____ value of all the y_i that would occur at $x = x_p$ in _____ sampling. We have as a predictor for this actual value of y, the quantity

true

expected or average
repeated

$\hat{y} =$ _____

$\hat{\beta}_0 + \hat{\beta}_1 x_p$

By looking at the following graph, we can see that our error in predicting the actual value of y when $x = x_p$ will come from two sources:

1. The difference between the predicted value of y, $\hat{y}$, and the expected value of y, $E(y|x_p)$. This difference is labeled _____ in the diagram and is the source of the variance of $\hat{y}$ as a predictor of $E(y|x_p)$ that was discussed in the last section.

II

2. The difference between the actual value of y and the expected value of y, $E(y|x_p)$. This difference is labeled _____ in the diagram, and is identical to _____, the _____ term in the probabilistic model.

I

ϵ; random error

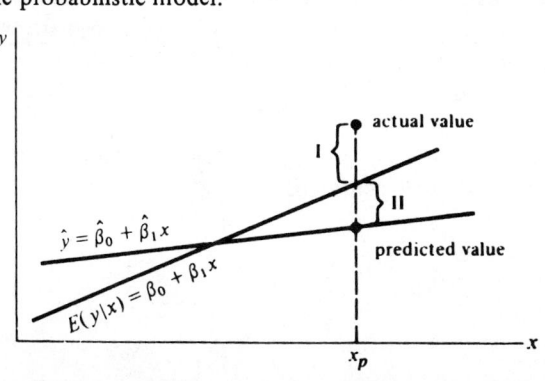

Thus the error associated with using $\hat{y} = \hat{\beta}_0 + \hat{\beta}_1 x_p$ as our prediction for the actual value of y which will occur when $x = x_p$, consists of the two components, I and II.

1. The variance associated with component I (the difference between the actual and expected values of y) is, by assumption, _____ . σ^2

2. The variance associated with component II (the difference between the true and estimated regression lines) is, as shown in the last section,

   _____

   $\sigma^2 \left[\dfrac{1}{n} + \dfrac{(x_p - \bar{x})^2}{S_{xx}} \right]$

Not surprisingly, the variance of the error $(y - \hat{y})$ in predicting a particular value of y with $\hat{y}$ can be shown to be the sum of the variances of the components of that error

$$\sigma^2_{error} = \sigma^2 + \sigma^2 \left[\dfrac{1}{n} + \dfrac{(x_p - \bar{x})^2}{S_{xx}} \right]$$

or more simply,

$$\sigma^2_{error} = \sigma^2 \left[\text{_____} \right]$$

$1 + \dfrac{1}{n} + \dfrac{(x_p - \bar{x})^2}{S_{xx}}$

Notice that this variance is (smaller, larger) than the variance of the error associated with using $\hat{y}$ to predict the expected or average value of y, $E(y|x_p)$, for a given value of $x = x_p$. This is a consequence of the fact the mean of a population of measurements on a random variable has a (smaller, larger) variance than does any individual measurement on the random variable.

larger

smaller

When s^2 is used to estimate σ^2, a prediction interval for the actual value of y when $x = x_p$ can be constructed based on the _____ -statistic, which has an associated confidence coefficient of $(1 - \alpha)$:

t

$(\hat{y}|x_p) \pm$ _____

$t_{\alpha/2} s \sqrt{1 + \dfrac{1}{n} + \dfrac{(x_p - \bar{x})^2}{S_{xx}}}$

Because the variance of the error in predicting the actual value of y when $x = x_p$ is larger than the variance of the error in estimating $E(y|x_p)$, the resulting confidence or prediction interval will be (wider, narrower) when predicting the actual value of y, for a given level of α.

wider

Example 10.8
Continuing the example from previous sections, predict the particular value of y when $x = 6$, with 95% confidence.

Solution
1. $\hat{y} = -1.3415 + 1.2195(6)$

   = _____ .

5.98

2. With 3 degrees of freedom, $t_{.025}$ = _____ .   3.182
3. The 95% prediction interval would be

$$5.98 \pm (\underline{\hspace{1cm}})(\underline{\hspace{1cm}}) \sqrt{1 + \frac{1}{5} + \frac{(6-5.2)^2}{32.8}}$$ 3.182; .638

$$5.98 \pm (\underline{\hspace{1cm}}).$$ 2.24

Recall that the 95% confidence interval for our estimate of $E(y|x = 6)$ was $5.98 \pm .95$. Consequently, the prediction interval is _____ for the actual value of y at $x = 6$.   wider

Self-Correcting Exercises 10C

1. Refer to Self-Correcting Exercises 10A, Exercise 1. Test the hypothesis that the expected enrollment is zero if there are no students pre-enrolled at the $\alpha = .05$ level. Does the line of means pass through the origin? Would you expect it to pass through the origin?
2. For Self-Correcting Exercises 10A, Exercise 2, predict the expected yield in cotton when the mid-season boll count is 450 with a 90% confidence interval. Could you use the prediction line to predict the cotton yield if the mid-season boll count was 250?
3. Refer to Self-Correcting Exercises 10A, Exercise 3. Using the least-squares prediction line for these data, predict the expected cotton yield if the insect count is 12 with a 90% confidence interval. Compare this interval with that found in Exercise 2 and comment on these two predictors of cotton yield.
4. Use the least-squares line from Exercise 1, Self-Correcting Exercises 10A to predict the enrollment with 95% confidence if the pre-enrollment figure is 4,000 students.

10.8 A Coefficient of Correlation

A common measure of the strength of the _____ relationship between two variables is the Pearson product-moment _____ _____ _____, symbolized by _____. This correlation coefficient is (dependent on, independent of) the scales of measurement of the two variables. The Pearson product-moment coefficient of correlation is calculated as

$$r = \underline{\hspace{2cm}}$$

linear
coefficient
of correlation; r
independent of

$$\frac{S_{xy}}{\sqrt{S_{xx}S_{yy}}}$$

where S_{xx}, S_{yy}, and S_{xy} are as defined earlier in this chapter, and where

$$\underline{\hspace{1cm}} \leq r \leq \underline{\hspace{1cm}}$$ $-1; 1$

Examine the formula for r above, and notice the following:

	1. The denominator of r is the square root of the product of two positive quantities and will always be _____ .
positive	2. The numerator of r is identical to the numerator used to calculate _____ , whose denominator is also always positive.
$\hat{\beta}_1$	3. Hence _____ and r will always have the same algebraic sign. When
$\hat{\beta}_1$	
> 0	a. When $\hat{\beta}_1 > 0$, then r _____ .
$= 0$	b. When $\hat{\beta}_1 = 0$, then r _____ .
< 0	c. When $\hat{\beta}_1 < 0$, then r _____ .
positive negative; no	When $r > 0$, there is a _____ linear correlation; when $r < 0$, there is a _____ linear correlation; when $r = 0$, there is _____ linear correlation. See the following examples:
$>$; $<$; $=$	Notice that both r and $\hat{\beta}_1$ measure the linear relationship between x and y. While r is independent of the scale of measurement, $\hat{\beta}_1$ retains the measurement units for both x and y, since $\hat{\beta}_1$ is the number of units increase in
y; x	_____ for a one-unit increase in _____ . When should the investigator use r, and when should a least-squares estimate of β_1 be used? Although the situation is not always clear-cut, r is used when either x or y can be considered as the random variable of interest (i.e., when x and y are both random), while regression estimates and confidence intervals are appropriate when one variable, say x, is not random and the other (y) is random.

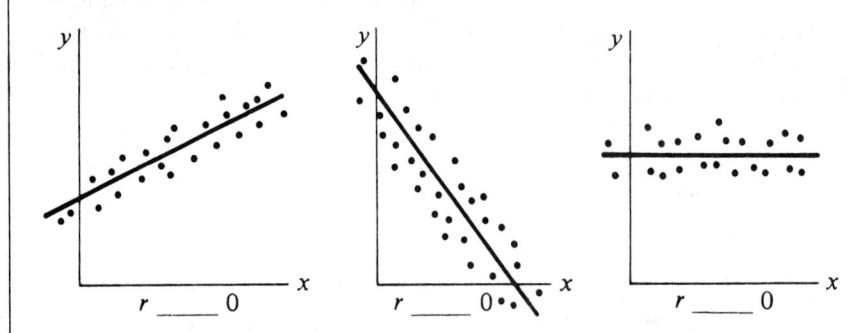

Example 10.9
Two personnel evaluation techniques are available. The first requires a 2-hour test-interview session, while the second can be completed in less than an hour. A high correlation between test scores would indicate that the second test, which is shorter to use, could replace the 2-hour test and hence save time and money. The following data give the scores on test I (x) and test II (y) for $n = 15$ job applicants. Find the coefficient of correlation for the following pairs of scores:

Sect. 10.8: A Coefficient of Correlation / 241

Applicant	Test I (x)	Test II (y)
1	75	38
2	89	56
3	60	35
4	71	45
5	92	59
6	105	70
7	55	31
8	87	52
9	73	48
10	77	41
11	84	51
12	91	58
13	75	45
14	82	49
15	76	47

Solution

1. As with a regression problem, we need all the summations as given in Section 10.4. Using a similar tabulation (or a calculator), find the following summations:

$\Sigma x_i =$ _____ $\Sigma y_i =$ _____   1,192; 725

$\Sigma x_i^2 = 96,990$ $\Sigma y_i^2 = 36,461$

$\Sigma x_i y_i =$ _____   59,324

2. To use the formula for *r*, we need the following:

$S_{xy} = \Sigma x_i y_i - [(\Sigma x_i)(\Sigma y_i)/n]$

$= 59,324 - [(\underline{\hspace{1cm}})(\underline{\hspace{1cm}})/15]$ 1,192; 725

$= \underline{\hspace{1cm}}$,   1,710.6667

$S_{xx} = \Sigma x_i^2 - (\Sigma x_i)^2/n$

$= 96,990 - (\underline{\hspace{1cm}})^2/15$ 1,192

$= 2,265.7333$,

$S_{yy} = \Sigma y_i^2 - (\Sigma y_i)^2/n$

$= 36,461 - (\underline{\hspace{1cm}})^2/15$ 725

$= 1,419.3333$.

Then

$$r = S_{xy}/\sqrt{S_{xx}S_{yy}}$$

$$= (\underline{})/\sqrt{(2,265.7333)(1,419.3333)}$$

1,710.6667

$$= (\underline{})/1,793.1154$$

1,710.6667

$$= \underline{}.$$

.9540

3. Since the correlation coefficient has a maximum value of 1 and a minimum value of −1, it would appear that the correlation between these two test scores is quite strong, with the relationship being a _____ linear one, as indicated by the sign of the correlation coefficient. In other words, a high score on test I would predict a _____ score on test II, or a low score on test I would predict a _____ score on test II.

positive

high
low

It should be noted that a dependent random variable, y, usually depends on _____ predictor variables, rather than just one. Consequently, the correlation between y and a single predictor variable is of doubtful value. It is even more important to bear in mind that r measures only the _____ relationship between two variables, say x and y. So even when $r = 0$, x and y could be *perfectly* related by a _____ function.

several

linear
nonlinear

What more can be gleaned from knowing the value of r? How can one assess the strength of the linear relationship of two variables? If $r = 0$, it is fairly obvious that there appears to be no linear relationship between x and y. To evaluate nonzero values of r, let us consider two possible predictors of y.

If the x variate were not measured, we would be forced to use the model

$$y = \beta_0 + \epsilon$$

with $\hat{\beta}_0 = \bar{y}$, so that we would use

$$\hat{y} = \bar{y}$$

as the predictor of the response y. The sum of squares for error for this predictor would be

$$SSE_1 = \Sigma(y_i - \bar{y})^2 = S_{yy}.$$

Knowing the value of x as well as y, we would use the model

$$y = \beta_0 + \beta_1 x + \epsilon,$$

in which case the resulting sums of squares for error would be

$$SSE_2 = S_{yy} - [(S_{xy})^2/S_{xx}].$$

If a linear relationship between x and y does exist, then SSE_1 would be larger than SSE_2.

A criterion for evaluating this relationship is to look at the ratio SSE_2/SSE_1, which can be simplified in the following manner:

1. Dividing SSE_2 by SSE_1 we obtain

$$SSE_2/SSE_1 = (S_{yy}/S_{yy}) - [(S_{xy})^2/S_{xx}S_{yy}]$$
$$= 1 - r^2.$$

2. Rearranging this equation we find

$$r^2 = 1 - (SSE_2/SSE_1) = (SSE_1 - SSE_2)/SSE_1$$

Since the difference $SSE_1 - SSE_2$ represents a reduction in the sum of squares accomplished by using a linear relationship, then

$r^2 = $ ratio of the reduction in the sum of squares achieved by using the linear model to the total sum of squares about the sample mean which would be used as a predictor of y if x were ignored.

A more understandable way of saying the same thing is to note that r^2 represents the amount of variability in y that is accounted for by knowing x. Thus we see that to evaluate a correlation coefficient r, we should examine r^2 to interpret the strength of the linear relationship between x and y. The quantity r^2 is often called the *coefficient of determination*.

For our example, the value of r was found to be $r = .9540$; therefore, $r^2 = $ _____. Hence we have reduced the variability of our predictor by _____ by knowing the value of x.

.9101
91%

The coefficient r, which is calculated from sample data, is actually an estimator of the _____ coefficient of correlation, symbolized by _____, where _____ $\leq \rho \leq$ _____. Since ρ and _____ both measure the linear relationship between x and y, the test of $H_0: \beta_1 = 0$ is equivalent to testing $H_0: \rho = 0$, and is based on a similar set of assumptions. The test is given below:

population
ρ; -1; 1
β_1

1. $H_0: \rho = 0$
2. Appropriate one- or two-tailed alternative hypothesis.
3. Test statistic:

$$t = \frac{r\sqrt{n-2}}{\sqrt{1-r^2}}$$

4. For a specified value of α and $(n-2)$ degrees of freedom, the appropriate one- or two-tailed rejection region is found using Table 3 of Appendix III.

Example 10.10
For the data given in Example 10.2, calculate r and test for significant correlation at the $\alpha = .05$ level of significance.

Solution
From Example 10.2, $S_{xy} = 40$, $S_{xx} = 32.8$, $S_{yy} = 50$. Hence,

244 / Chap. 10: Linear Regression and Correlation

$$r = \frac{S_{xy}}{\sqrt{S_{xx}S_{yy}}} = \frac{40}{\sqrt{32.8(50)}} = \frac{40}{40.4969} = .9877$$

The test of hypothesis is as follows:
1. $H_0: \rho = 0$
2. $H_a: \rho \neq 0$
3. Test statistic:

.9877

10.94

.9877

$$t = \frac{r\sqrt{n-2}}{\sqrt{1-r^2}} = \frac{\sqrt{3}}{\sqrt{1-(\underline{})^2}} = \underline{}$$

Note that except for rounding error, this is exactly the same value for t as was given in Example 10.4, in which we tested $H_0: \beta_1 = 0$ against $H_a: \beta_1 \neq 0$.
4. Rejection region: With 3 degrees of freedom, we will reject H_0 if

3.182

$$|t| > t_{.025} = \underline{}.$$

falls; reject
is

5. Since $t = 10.94$ (falls, does not fall) in the rejection region, we (reject, do not reject) H_0. There (is, is not) significant correlation between x and y.

Self-Correcting Exercises 10D

1. Refer to Exercise 2, Self-Correcting Exercises 10A.
   a. Find the correlation between the number of bolls and the yield of cotton.
   b. Find the coefficient of determination r^2, and explain its significance in using the number of cotton bolls to predict the yield of cotton.
   c. Test to see if there is a significant positive correlation between x and y. Use $\alpha = .05$.
2. Refer to Exercise 3, Self-Correcting Exercises 10A.
   a. Find the value of r^2 and r for these data and explain the value of using the number of damaging insects present to predict cotton yield.
   b. Compare the values of r^2 using these two predictors of cotton yield. Which predictor would you prefer?
3. The data in Exercises 2 and 3, Self-Correcting Exercises 10A, are related in that for each field quadrate, the yield, the number of bolls, and the number of damaging insects were simultaneously recorded. Using this fact, calculate the correlation between the number of cotton bolls and the number of insects present for the seven field quadrates. Does this value of r explain in any way the similarity of results when using the predictor in Exercise 2 and that in Exercise 3?

10.9 Regression Analysis Using Packaged Computer Programs

The calculations required for a regression analysis can be cumbersome when working with a large set of data points. Since several excellent packaged regression programs are available at most computer facilities, many researchers prefer to rely on these programs for the analysis of a regression problem. Although the available options and the form of the printed results, called output, differ from program to program, they all include the same standard information. We will discuss the salient points of the output resulting from a computer regression analysis of data collected during a calibration study.

Calibration studies are run in order to determine the accuracy of a given instrument in dispensing a fixed amount of material. In the data that follow, the variable x represents the fill setting in ounces on a machine designed to dispense detergent and y represents the actual amount of detergent dispensed.

x	6	8	9	10	12	13	15	16	20	24	28	32
y	6.4	8.2	9.2	10.5	12.9	12.8	14.7	15.9	20.3	24.5	27.6	33.1

An analysis of these data using the command REGRESS in the MINITAB package is given in the following display. The values of y and x were stored in columns 1 and 2 in the MINITAB data format, with columns 1 and 2 named "Y" and "X" using the MINITAB command NAME. The fitted regression equation is given as

$$Y = 0.151 + 1.01X$$

MINITAB Output

MTB > REGRESS C1 1 C2

The regression equation is
Y = 0.151 + 1.01 X

Predictor	Coef	Stdev	t-ratio	p
Constant	0.1509	0.3141	0.48	0.641
X	1.00668	0.01751	57.48	0.000

s = 0.4812   R-sq = 99.7%   R-sq(adj) = 99.7%

Analysis of Variance

SOURCE	DF	SS	MS	F	p
Regression	1	765.03	765.03	3303.98	0.000
Error	10	2.32	0.23		
Total	11	767.35			

Unusual Observations

Obs.	X	Y	Fit	Stdev.Fit	Residual	St.Resid
12	32.0	33.100	32.365	0.311	0.735	2.00R

R denotes an obs. with a large st. resid.

The portion of the printout just below the regression equation contains relevant information about the individual parameters in the model. The column labeled *Predictor* lists *Constant,* the designated name for the intercept, and the coefficient of the independent predictor variable X, listed as either its column number (C2) or the name designated for C2 (X) using the MINI-

Chap. 10: Linear Regression and Correlation

TAB command NAME. Therefore, the intercept is $\hat{\beta}_0 = .1509$ and the slope is $\hat{\beta}_1 = 1.00668$. Hence, using two-decimal accuracy, the fitted line is given by

$$\hat{y} = .15 + 1.01x,$$

where x is the independent variable, fill-setting in ounces.

The estimated standard deviations, $s_{\hat{\beta}_0}$ and $s_{\hat{\beta}_1}$, of the regression coefficients $\hat{\beta}_0$ and $\hat{\beta}_1$ are given in the column headed *Stdev*. These quantities, used in the testing and construction of confidence intervals, are found to be

$$s_{\hat{\beta}_0} = .3141 \text{ and } s_{\hat{\beta}_1} = \underline{\hspace{1cm}}.$$ .01751

The entries in the fourth column, labeled *t-ratio*, are the computed values of the t statistic used in testing hypotheses concerning the regression parameters. In testing $H_0: \beta_1 = 0$ versus $H_a: \beta_1 \neq 0$, the computed value of

$$t = \hat{\beta}_1 / s_{\hat{\beta}_1}$$

is the second entry in this column and is equal to _____. This value of   57.48
t, with a p-value virtually zero, is large enough to conclude that $\beta_1 \neq 0$.
However, in order to find critical values of t, we note that the degrees of freedom with t are $n - 2$, or _____. The degrees of freedom for testing   10
and estimation appear in the *Error* line and *DF* column of the *Analysis of Variance* portion of the printout.

Although tests concerning the intercept β_0 are not usually performed, in this calibration study the setting $x = 0$ should produce a fill of _____.   zero
The value of

$$t = \hat{\beta}_0 / s_{\hat{\beta}_0}$$

for testing $H_0: \beta_0 = 0$ versus $H_a: \beta_0 \neq 0$ is the first entry in the t-ratio column and is equal to _____, which is clearly nonsignificant. If the   .48
instrument being calibrated is working properly, a one-unit increase in the setting x should produce a one-unit increase in fill. We would then be interested in testing $H_0: \beta_1 = $ _____ versus $H_a: \beta_1 \neq $ _____. The   1; 1
appropriate test statistic is

$$t = (\hat{\beta}_1 - 1)/s_{\hat{\beta}_1}$$

$$= (1.00668 - 1)/.01751$$

$$= \underline{\hspace{1cm}}.$$ .38159

which (is, is not) significant. It appears that the instrument is performing as   is not
it should.

A confidence interval estimate of β_1 is easily calculated using the information provided in the printout. A $(1 - \alpha)$ 100% confidence interval estimator of β_1 is given as

$$\hat{\beta}_1 \pm t_{\alpha/2} s_{\hat{\beta}_1},$$

where $t_{\alpha/2}$ is a tabulated value of t from Table 4 of the text with $n - 2$ degrees of freedom. With 10 degrees of freedom, $t_{.025} =$ _____. Therefore, a 95% confidence interval estimate of the slope is | 2.228

$$\hat{\beta}_1 \pm t_{.025} s_{\hat{\beta}_1}$$

1.00668 ± _____ (.01751) | 2.228

1.00668 ± _____ | .03901

or   1.01 ± _____ . | .04

Analysis of Variance

In regression analysis, an analysis of variance is a technique that partitions the total variation in the response y into one portion associated with random error and another portion associated with the variability accounted for by regression. The *Error* line in this portion of the printout provides the degrees of freedom for error in the *DF* column, the value of *SSE* in the *SS* (Sum of Squares) column, and the value of s^2 in the *MS* (Mean Square) column. The value of the standard deviation s, found elsewhere on the printout, is the square root of s^2. For this problem, the degrees of freedom for error is 10, SSE = _____, $s^2 = .232$, and $s = \sqrt{.232} =$ _____. | 2.32; .4812

The value that appears in the *F-ratio* column of the *Regression* line of the *Analysis of Variance* is the calculated value of an F statistic used in testing H_0: $\beta_1 = 0$ versus H_a: $\beta_1 \neq 0$. This value is the square of the *t-ratio* for testing this same hypothesis found in the *t-ratio* column. Note that

$$(\text{t-ratio})^2 = (57.48)^2 = \underline{\qquad} = F$$ | 3,303.99

within rounding errors. This relationship (<u>will, will not</u>) hold when two or more independent variables are included in the regression equation and subsequently in the regression analysis. | will not

Remaining Entries

The coefficient of determination r^2 is labeled *R-sq*. Since $r^2 = 99.7\%$, this means 99.7% of the variation in fills is accounted for by regression. In the case of simple linear regression, the correlation coefficient r is the square root of the *R-sq* entry and takes the same sign as $\hat{\beta}_1$. For this problem, $r = +\sqrt{.997}$ = _____. In a regression analysis involving two or more independent variables, the *R-sq* entry retains its earlier meaning, but its square root does not. This topic will be taken up in the next chapter. | .998

A second generic printout for a regression analysis of these data follows. This printout differs in format from both the MINITAB printout given earlier in this study guide and the SAS printout given in your text. However, the same basic information is provided for almost all regression programs. Take time to examine this printout and to compare the entries with those given in the MINITAB printout.

MULTIPLE R	.99849
R-SQUARE	.99698
STD. ERROR OF EST.	.48120

ANALYSIS OF VARIANCE

SOURCE	DF	SUM OF SQUARES	MEAN SQUARE	F-VALUE
REGRESSION	1	765.03367	765.03367	3303.9729
RESIDUAL	10	2.31549	.23155	
TOTAL	11	767.34917		

INDIVIDUAL ANALYSIS OF VARIABLES

VARIABLE	COEFFICIENT	STD.ERROR	T-VALUE
INTERCEPT	.15092	.31406	.48054
FILL	1.00668	.01751	57.48020

10.10 Assumptions

For the techniques in this chapter to be valid, the pairs of data points must satisfy the following assumptions:
1. The response y can be modeled as

$$y = \beta_0 + \beta_1 x + \epsilon.$$

2. The independent variable x is measured without error.
3. The quantity ϵ is a random variable such that for any fixed value of x,

$$E(\epsilon) = 0, \quad \sigma_\epsilon^2 = \sigma^2,$$

and all pairs ϵ_i, ϵ_j are independent.

4. The random variable ϵ has a _____ distribution.   *normal*

The first assumption requires that the response y be _____   *linearly*
related to the independent variable x. If the response is not just a simple linear function of x, but the fit to the data as evidenced by a high value of r^2 is good, the least-squares equation will produce adequate predictions within the range of the experimental values of x. Extrapolation, however, can cause problems in prediction, since the fitted linear relationship may fail to adequately describe the data outside the range of values used in the analysis. This
is especially true when the independent variable is _____.   *time*

The assumption of constant variance for the ϵ's may not be valid in all situations. For example, the variability in the amount of impurities in a chemical mixture as well as the amount of impurities itself may increase with increasing temperature of the mixture. However, if repeated values of the independent variable x are included within the experiment, a plot of the data points will usually reveal whether the variance of the ϵ's depends on x. When this is the case, weights are assigned to each value of x and a regression analysis is performed on y and the weighted values of x.

It is not unusual to have the error terms correlated when the data are collected over time. For example, a high inflation rate during the first 3 months of the year is likely to be followed by a high inflation rate during the next quarter as well. Ordinary regression techniques applied to such data produce underestimates of the true variance and hence cause distortion in significance levels and confidence coefficients. Time series analysis should be used in the analysis of such data.

Departure from the normality assumption will not distort results too strongly provided the distributions of the ϵ's are not strongly skewed.

EXERCISES

1. For the following equations, (i) give the y intercept, (ii) give the slope, and (iii) graph the line corresponding to the equation:

    a. $y = 3x - 2$.   d. $3x + 2y = 5$.

    b. $2y = 4x$.   e. $y = 2$.

    c. $-y = .5 + x$.

2. a. Find the least-squares line for the following data:

x	-3	-2	-1	0	1	2	3
y	-1	-1	0	1	2	2	3

    b. As a check on your calculations, plot the data points and graph the least-squares line.
    c. Calculate SSE and s^2. Under what conditions could SSE = 0?
    d. Do the data present sufficient evidence to indicate that x and y are linearly related at the $\alpha = .05$ level of significance?
    e. Estimate the average change in y for a one-unit change in x with a 95% confidence interval.
    f. Calculate the coefficient of linear correlation for the data and interpret your results.
    g. Calculate r^2 and state in words the significance of its magnitude.
    h. Construct a 90% confidence interval estimate for a particular value of y when $x = 1$.
    i. Test the hypothesis that $E(y|x = 0) = 0$ at the $\alpha = .05$ level of significance. (This is actually a test of $H_0: \beta_0 = 0$.)

3. For the following data,

x	0	2	4	6	8	10
y	9	7	3	1	-2	-3

    a. Fit the least-squares line, $\hat{y} = \hat{\beta}_0 + \hat{\beta}_1 x$.
    b. Plot the points, and graph the line to check your calculations.
    c. Calculate SSE, s^2, and s.

d. Is there a linear relationship between x and y at the $\alpha = .05$ level of significance?
e. Calculate r^2, and explain its significance in predicting the response, y.
f. Predict the particular value of y when $x = 5$ with 80% confidence.
g. Predict the expected value of y when $x = 5$ with 80% confidence.

4. What happens if the coefficient of linear correlation, r, assumes the value one? The value −1?

5. The following data were obtained in an experiment relating the dependent variable, y (texture of strawberries), with x (coded storage temperature).

x	−2	−2	0	2	2
y	4.0	3.5	2.0	0.5	0.0

a. Find the least-squares line for the data.
b. Plot the data points and graph the least-squares line as a check on your calculations.
c. Calculate SSE, s^2, and s.
d. Do the data indicate that texture and storage temperature are linearly related? ($\alpha = .05$)
e. Predict the expected strawberry texture for a coded storage temperature of $x = -1$ with a 90% confidence interval.
f. Of what value is the *linear* model in increasing the accuracy of prediction as compared to the predictor, $\bar{y}$?
g. Estimate the particular value of y when $x = 1$ with a 98% confidence interval.
h. At what value of x will the width of the confidence interval for a particular value of y be a minimum, assuming n remains fixed?

6. In addition to increasingly large bounds on error, why should an experimenter refrain from predicting y for values of x outside the experimental region?

7. If the experimenter stays within the experimental region, when will the error in predicting a particular value of y be maximum?

8. An agricultural experimenter, investigating the effect of the amount of nitrogen (x) applied in 100 pounds per acre on the yield of oats (y) measured in bushels per acre, collected the following data:

x	1	2	3	4
y	22	38	57	68
	19	41	54	65

a. Fit a least-squares line to the data.
b. Calculate SSE and s^2.
c. Is there sufficient evidence to indicate that the yield of oats is linearly related to the amount of nitrogen applied? ($\alpha = .05$)
d. Predict the expected yield of oats with 95% confidence if 250 pounds of nitrogen per acre are applied.
e. Predict the average increase in yield for an increase of 100 pounds of nitrogen with 90% confidence.
f. Calculate r^2 and explain its significance in terms of predicting y, the yield of oats.

9. In an industrial process, the yield, y, is thought to be linearly related to temperature, x. The following coded data is available:

Temperature	0	0.5	1.5	2.0	2.5
Yield	7.2	8.1	9.8	11.3	12.9
	6.9	8.4	10.1	11.7	13.2

   a. Find the least-squares line for this data.
   b. Plot the points and graph the line. Is your calculated line reasonable?
   c. Calculate SSE and s^2.
   d. Do the data line indicate a linear relationship between yield and temperature at the $\alpha = .01$ level of significance?
   e. Calculate r, the coefficient of linear correlation and interpret your results.
   f. Calculate r^2, and interpret its significance in predicting the yield, y.
   g. Test the hypothesis that $E(y|x = 1.75) = 10.8$ at the $\alpha = .05$ level of significance.
   h. Predict the particular value of y for a coded temperature $x = 1$ with 90% confidence.

10. A horticulturist devised a scale to measure the viability of roses that were packaged and stored for varying periods of time before transplanting. y represents the viability measurement and x represents the length of time in days that the plant is packaged and stored before transplanting.

x	5	10	15	20	25
y	15.3	13.6	9.8	5.5	1.8
	16.8	13.8	8.7	4.7	1.0

   a. Fit a least-squares line to the data.
   b. Calculate SSE and s^2 for the data.
   c. Is there sufficient evidence to indicate that a linear relationship exists between freshness and storage time? (Use $\alpha = .05$.)
   d. Estimate the mean rate of change in freshness for a 1-day increase in storage time by using a 98% confidence interval.
   e. Predict the expected freshness measurement for a storage time of 14 days with 95% confidence.
   f. Of what value is the linear model in preference to $\bar{y}$ in predicting freshness?

CHAPTER 11
MULTIPLE REGRESSION ANALYSIS

11.1 Introduction

We have examined estimation, testing, and prediction techniques for the situation in which y, the response of interest, was linearly related to an independent or predictor variable x in the following way:

$$y = \beta_0 + \beta_1 x + \epsilon.$$

In this chapter, we extend these techniques to the more general situation in which the response y is linearly related to one or more independent or predictor variables. These extended techniques can be used when the response is linearly related to several different independent variables, or when the response is a polynomial function of just one variable. Modeling, testing, and prediction in these cases belong to an area of statistics called multiple regression analysis.

11.2 The Multiple Regression Model and Associated Assumptions

The general results given in the remainder of this chapter are applicable and produce standard solutions for a multiple regression problem when the response y is a _____ function of the unknown regression coefficients. That is, we write

$$y = \beta_0 + \beta_1 x_1 + \beta_2 x_2 + \cdots + \beta_k x_k + \epsilon$$

where

1. y is the response variable we wish to predict,
2. $\beta_0, \beta_1, \cdots, \beta_k$ are (known, unknown) constants,
3. $x_1, x_2, \cdots, x_k$ are independent variables that (are, are not) measured without error,
4. ϵ is a random error having a normal distribution with mean zero and variance σ^2, independent of $x_1, x_2, \cdots, x_k$. Further, the error terms for any two values of y are taken to be _____ .

linear

unknown
are

independent

5. Since $E(\epsilon) = 0$, we can write the mean value of y for a given set $x_1, x_2, \ldots, x_k$ as

$$E(y) = \beta_0 + \beta_1 x_1 + \beta_2 x_2 + \cdots + \beta_k x_k.$$

Although the actual observed values of y will not exactly equal the values generated by the model, they will deviate from $E(y)$ by a random amount ϵ if the model is in fact true.

The methodology that we use requires only that the β_i's occur in a linear fashion. That is, β_i must be the coefficient of a term that does not involve any *unknown* parameters.

Example 11.1
The following are examples of the general linear model:

a. $y = \beta_0 + \beta_1 t + \beta_2 \sin\left(\dfrac{2\pi t}{n}\right) + \epsilon.$

b. $y = \beta_0 + \beta_1 x + \beta_2 x^2 + \epsilon.$

c. $y = \beta_0 + \beta_1 x_1 + \beta_2 x_2 + \beta_3 x_1 x_2 + \epsilon.$

Although the model given in part a involves the sine function, no unknown parameters occur within the function itself. By letting $x_1 = t$ and $x_2 = \sin(2\pi t/n)$, this model could be written as

$$y = \beta_0 + \beta_1 x_1 + \beta_2 x_2 + \epsilon$$

which is a linear function of the unknown regression parameters β_0, _____, and _____. In a similar fashion the model given in part b can be rewritten by letting $x_1 =$ _____ and $x_2 =$ _____. In the third model, we can achieve the same result by letting $x_1 = x_1, x_2 = x_2,$ and $x_3 =$ _____.

$\beta_1; \beta_2$
$x; x^2$

$x_1 x_2$

Terms involving only $x_1, x_2, \ldots x_k$ are called *first order terms*, while terms involving $x_1^2, x_2^2, \ldots, x_k^2$ or the product $x_i x_j$ are called *second order terms*.

Example 11.2
The next three models (do, do not) fit the requirements of the general linear model.

do not

a. $y = \beta_0 e^{\beta_1 x} + \epsilon.$

b. $y = \beta_0 + \beta_1 \sin\left(\dfrac{2\pi t}{n} + \beta_2\right) + \epsilon.$

c. $y = \beta_0 + \beta_1 x^{\beta_2} + \epsilon.$

Notice that models a and c involve an unknown parameter as a power, while model b involves an unknown parameter, β_2, within the sine function itself.

Example 11.3
Determine whether the following models are linear (L) or nonlinear (NL) models:

L   a. $y = \beta_0 + \beta_1 x_1^2 + \beta_2 x_2^2 + \beta_3 x_1 x_2 + \epsilon.$ (L, NL)

NL   b. $y = \beta_0 + \beta_1 \cos\left(\dfrac{2\pi x}{5}\right) + \beta_2 x^{\beta_3} + \epsilon.$ (L, NL)

L   c. $y = \beta_0 e^{-5x} + \epsilon.$ (L, NL)

may   A linear statistical model (may, may not) involve nonlinear terms provided all unknown parameters occur in a linear fashion within the model.

    Formulation of the linear model to be used in the data analysis is perhaps the most difficult aspect of regression analysis since the results we achieve depend strictly on the model we have chosen to fit. For example, if y is related to x in a quadratic fashion and we include only a linear term in x in our model, our analysis would produce a poor estimator of y in general. In like manner, if we include the independent predictor variables x_1 and x_2 in our model, but fail to include x_3, which has high predictive potential, we may indeed end up with a poor estimator of y. Model formulation will be addressed in Section 11.6.

Self-Correcting Exercises 11A

1. Graph the following equations:
   a. $E(y) = 1 + 2x$;
   b. $E(y) = 1 + .5x$;
   c. $E(y) = 2 - 2x$.
2. Graph the following equations, which graph as parabolas:
   a. $E(y) = x^2$;
   b. $E(y) = 1 + x^2$;
   c. $E(y) = -x^2$.
   d. How does the sign of the coefficient of x^2 affect the graph of the parabola?
3. Graph the following equation:
   a. $E(y) = 1 - 2x + x^2$.
   b. Compare the graph in part a with the graph in Exercise 2, part b. What effect does the term $-2x$ have on the graph?
   c. How would the graph change if $-2x$ were replaced by $+2x$?
4. Suppose $E(y)$ is related to two predictor variables x_1 and x_2 by the equation

$$E(y) = 2 + 3x_1 - x_2.$$

a. Graph the relationship between $E(y)$ and x_1 when $x_2 = 0$. Repeat for $x_2 = 1$ and $x_2 = 2$.
b. How are the graphs of the three lines in part a related?
c. Graph the relationship between $E(y)$ and x_2 when $x_1 = 0$. Repeat for $x_1 = 1$ and $x_1 = 2$.
d. How are the graphs of the three lines in part c related?

11.3 A Multiple Regression Analysis

In generalizing the results of simple linear regression to multiple linear regression, we have considered the model

$$y = \beta_0 + \beta_1 x_1 + \beta_2 x_2 + \cdots + \beta_k x_k + \epsilon$$

where ϵ is a normally distributed random error component with a mean of zero and a variance σ^2. In addition, the error terms for any two values of y are taken to be _____. The parameters $\beta_1, \beta_2, \cdots, \beta_k$ are the partial slopes associated with the nonrandom quantities $x_1, x_2, \cdots, x_k$. The slope, β_i, represents the expected increase in the response y corresponding to a one-unit increase in x_i when the values of all other x's are held constant. Estimates of the unknown parameters in the model are found by using the method of least squares. Using this method, the estimates $\hat{\beta}_0, \hat{\beta}_1, \cdots, \hat{\beta}_k$ are chosen so as to _____ the quantity

independent

minimize

$$SSE = \sum_{i=1}^{n} (y_i - \hat{y}_i)^2.$$

This minimization technique leads to a set of $(k + 1)$ simultaneous equations in the unknowns $\hat{\beta}_0, \hat{\beta}_1, \ldots,$ and $\hat{\beta}_k$, which are easily solved using any multiple regression analysis computer program. Such programs are usually available at any computing facility, and provide not only the estimates of the regression parameters, but also additional information required for prediction, estimation and hypothesis testing. These programs require only that the user provide the proper commands to activate the program, and then submit the data in the proper format.

In this section we will analyze two data sets using a multiple regression program and interpret the results of the analyses.

Example 11.4

An agricultural economist interested in predicting cotton harvest using the number of cotton bolls per quadrat counted during the middle of the growing season and the number of damaging insects per quadrat present during a critical time in the development of the plant, collected data on the response y, the yield in bales of cotton, x_1, hundreds of cotton bolls per quadrat counted during midseason, and x_2, the insect count per quadrat. If we can expect a straight line relationship between cotton yield, y, and each of the

256 / Chap. 11: Multiple Regression Analysis

two predictor variables, x_1 and x_2, write a linear model relating y and the predictor variables x_1 and x_2.

Solution

increase

We might expect the yield y to (increase, decrease) linearly as x_1, the number of cotton bolls increases and x_2, the number of damaging insects decreases. The simplest model relating y and the predictors x_1 and x_2 is

$$E(y) = \beta_0 + \beta_1 x_1 + \beta_2 x_2.$$

When x_2 is held constant we can write

$$E(y) = (\beta_0 + \beta_2 x_2) + \beta_1 x_1$$

with β_1 the rate of increase in $E(y)$ for a one unit increase in the boll count, while the intercept, which depends on x_2, is given as $\beta_0(x_2) = \beta_0 + \beta_2 x_2$. For different values of x_2, $E(y)$ would plot as a series of parallel lines, each with slope β_1 as shown in the following graph.

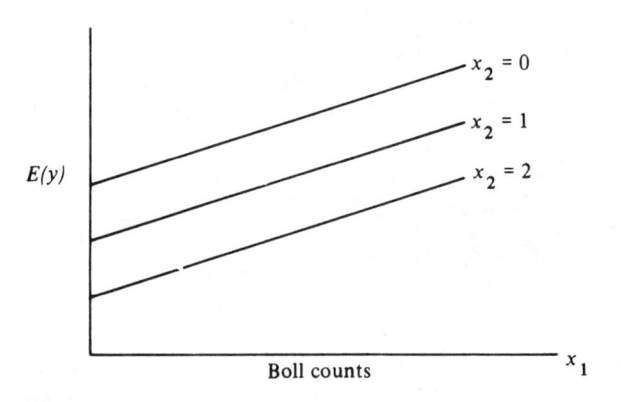

Since yield is expected to decrease as the number of damaging insects increases, the lines in the preceding graph are drawn assuming β_2 is negative.

When x_1 is held constant, we can write the model as

$$E(y) = (\beta_0 + \beta_1 x_1) + \beta_2 x_2,$$

β_2

$\beta_0 + \beta_1 x_1$

which plots as a series of parallel lines with negative slope _____ and varying intercepts that depend upon x_1 and is given by $\beta_0(x_1) =$ _____. Assuming β_2 is negative, a plot of $E(y)$ for three values of x_1 is given in the next graph.

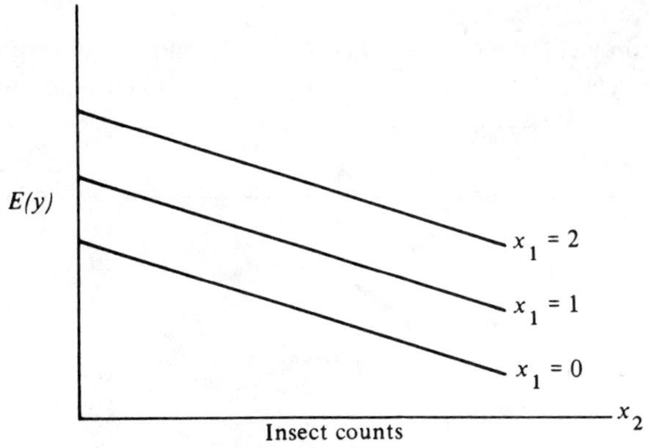
E(y)
x1 = 2
x1 = 1
x1 = 0
x2
Insect counts

An alternative model that is a linear function of x_1 and x_2, but allows for differing slopes as well as differing intercepts is given by

$$E(y) = \beta_0 + \beta_1 x_1 + \beta_2 x_2 + \beta_3 x_1 x_2 \,.$$

In this case, if x_2 is held constant, we can write

$$E(y) = (\beta_0 + \beta_2 x_2) + (\beta_1 + \beta_3 x_2) x_1,$$

where (______________) is the intercept and (______________) is the slope. $E(y)$ would plot as a series of lines with changing slopes and intercepts. If β_3 is negative, the plots would appear as in the following graph. $\beta_0 + \beta_2 x_2; \beta_1 + \beta_3 x_2$

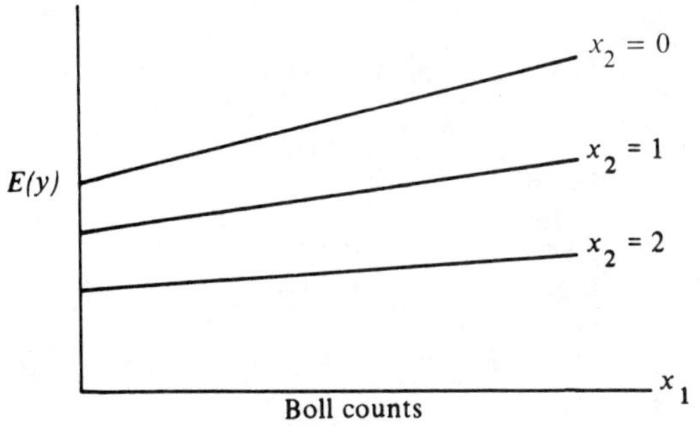
E(y)
x2 = 0
x2 = 1
x2 = 2
x1
Boll counts

If x_2 is held constant, a similar plot would result with varying negative slopes. Hence, both models allow for a straight line relationship between y and each of the predictor variables x_1 and x_2, but the model given by

$$E(y) = \beta_0 + \beta_1 x_1 + \beta_2 x_2 + \beta_3 x_1 x_2$$

is more flexible since it allows for differing slopes and intercepts as either x_1 or x_2 is held constant.

Example 11.5

Data on cotton yield, the number of cotton bolls, and the number of damaging insects as described in Example 11.4 are shown as follows:

y	x_1	x_2
21	5.5	11
17	2.8	20
20	4.7	13
19	4.3	12
15	3.7	18
23	6.1	10
20	4.5	12

These data were analyzed using the MINITAB multiple regression command REGRESS, with the values of the independent variable y stored in column one (C1) as a function of the three variables x_1, x_2, and $x_1 x_2$ stored in columns two, three and four (C2 C3 C4). The regression analysis is based on the model

$$E(y) = \beta_0 + \beta_1 x_1 + \beta_2 x_2 + \beta_3 x_1 x_2.$$

Explain the output of the multiple regression computer printout that follows.

MTB > REGRESS C1 3 C2 C3 C4

The regression equation is
Y = 11.0 + 4.44 X1 + 0.649 X2 − 0.352 X1*X2

Predictor	Coef	Stdev	t-ratio	p
Constant	10.983	7.784	1.41	0.253
X1	4.437	1.619	2.74	0.071
X2	0.6490	0.4707	1.38	0.262
X1*X2	−0.3515	0.1447	−2.43	0.093

s = 0.9364   R-sq = 93.6%   R-sq(adj) = 87.3%

Analysis of Variance

SOURCE	DF	SS	MS	F	p
Regression	3	38.798	12.933	14.75	0.027
Error	3	2.631	0.877		
Total	6	41.429			

SOURCE	DF	SEQ SS
X1	1	32.109
X2	1	1.512
X1*X2	1	5.178

Solution
In explaining the output from the MINITAB regression analysis, we will discuss the items in the order in which they appear on the printout.

1. *Fitted regression equation.* The regression equation that was fitted to these data is given by

$$Y = 11.0 + 4.44\ X1 + 0.649\ X2 - 0.352\ X1*X2$$

2. *Individual coefficients.* The estimates of the model parameters are found in the column labeled *Predictor*. The estimate of the intercept or constant is $\hat{\beta}_0 = $ _____ while $\hat{\beta}_1 = 4.437$, $\hat{\beta}_2 = 0.6490$ and $\hat{\beta}_3 = $ _____. Correct to *three significant digits,* the fitted regression equation was given above. | 10.98
    | -0.3515

The estimated standard deviation of the regression coefficients is given in the column labeled *Stdev*. For example, the standard error of $\hat{\beta}_1$ is

$$s_{\hat{\beta}_1} = \underline{\hspace{2cm}}$$ | 1.619

while the standard error of $\hat{\beta}_3$ is 0.1447. The standard errors of the regression coefficients can be used in testing hypotheses concerning individual coefficients and/or in constructing confidence interval estimates for the coefficients.

For example, in testing the hypothesis $H_0: \beta_1 = 0$ against $H_a: \beta_1 \neq 0$, we use the statistic

$$t = \hat{\beta}_1 / s_{\hat{\beta}_1}$$

which has a Student's *t* distribution with degrees of freedom equal to $(n - 2) = 3$ found in the *Error* line and the *df* column in the *Analysis of Variance* portion of the printout. The calculated values of the *t* statistics are given in the column labeled *t-ratio,* and their *two-tailed* observed significance levels or *p-values* are given under the heading *p*. In testing $H_0: \beta_1 = 0$ against $H_a: \beta_1 \neq 0$, the calculated value of *t* is _____ with a significance level of .071. Hence, we could reject H_0 if we were willing to use a value of α (greater, less) than or equal to .071. | 2.74

    | greater

A $100(1 - \alpha)$ percent confidence interval estimate for β_1 is given by

$$\hat{\beta}_1 \pm t_{\alpha/2} s_{\hat{\beta}_1}$$

with 3 degrees of freedom and $\alpha = .05$, $t_{.025} = $ _____. Therefore, the 95 percent confidence interval estimate for β_1 is | 3.182

$$4.437 \pm 3.182(\underline{\hspace{1.5cm}})$$ | 1.619

or 4.44 ± 5.15, which includes the value of zero. This supports our earlier finding in that we (could, could not) reject the hypothesis $H_0: \beta_1 = 0$ at the $\alpha = .05$ level of significance. | could not

260 / Chap. 11: Multiple Regression Analysis

It is important that tests and confidence interval estimates concerning individual regression coefficients be put in proper perspective. In a multiple regression analysis, the regression coefficients are properly called *partial regression coefficients,* since they are determined in conjunction with other _____ in the model and only partially determine the value of the response y. In fact, the partial regression coefficients in general (would, would not) be the same as those found using several simple linear regression models, each with one predictor variable or just one combination of predictor variables such as $x_1 x_2$. Estimates of the partial regression coefficients are correlated with each other to the extent that the underlying predictor variables share the same predictive information. Therefore, a test of $H_0: \beta_1 = 0$ versus $H_a: \beta_1 \neq 0$ is actually testing whether the term x_1 contributes significant information in predicting y if, in fact, the terms x_2 and $x_1 x_2$ are already in the model. In our example, $\hat{\beta}_1, \hat{\beta}_2$, and $\hat{\beta}_3$ are not significant at the .05 level of significance; but, as we shall see, the model taken as a whole explains a large portion of the variability in the response y.

- variables

- would not

3. *The estimate of σ, the standard deviation of the points about the line.* The values of SSE and s^2 are found in the *Error* line of the *Analysis of Variance* table in the columns labeled SS and MS, respectively. From the printout, we see that SSE = _____ and $s^2 = 0.877$ with _____ degrees of freedom. The estimate of σ is $s = \sqrt{0.877} =$ _____, the entry labeled s following the column of predictors.

- 2.631; 3
- 0.9364

4. *R-sq.* The quantity R^2, called the *coefficient of determination,* measures how well the model fits the data. In a regression analysis, the sum of squared deviations, S_{yy}, can be decomposed as

$$\Sigma (y_i - \bar{y})^2 = \Sigma (y_i - \hat{y}_i)^2 + \Sigma (\hat{y}_i - \bar{y})^2,$$

where $\Sigma(y_i - \bar{y})^2 = S_{yy}$, $\Sigma(y_i - \hat{y}_i)^2 = $ SSE, the sum of squares for error, and $\Sigma(\hat{y}_i - \bar{y})^2 = $ SSR, the sum of squares for regression. SSE measures the discrepancy between observed and predicted values of y so that small values of SSE are indicative of a (poor, good) model fit. SSR measures the difference between the multiple regression predictor $\hat{y}$ and the simple predictor $\bar{y}$, which ignores the predictor variables entirely. Large values of SSR indicate that the predictor $\hat{y}$ is a (poorer, better) predictor than $\bar{y}$. Since $S_{yy} = $ SSE + SSR, if SSE is small, then SSR is (small, large) and visa-versa. The value of R^2 is determined as

- good

- better
- large

$$R^2 = \left(\frac{SSR}{S_{yy}}\right) 100\% = \left(1 - \frac{SSE}{S_{yy}}\right) 100\%$$

and always lies between 0 and 100%. Hence, R^2 represents the proportion of the variation in y that is explained by regression. From the printout, we see that $R^2 = 93.6\%$, so that _____ percent of the variation in y is explained by the regression model.

- 93.6

As independent variables are added one at a time to the predictor list for a fixed set of y-values, the value of R^2 will either stay the same or increase; it will *never decrease* when additional variables are included in the model. Therefore, R^2 can be artificially inflated by including a large number of predictors in the model. For this reason, *R-sq(adj)* is an alternative formulation that adjusts for the number of parameters appearing in the model by using mean squares rather than sums of squares in its calculation. MSE and $S_{yy}/(n-1)$ are used in place of SSE and S_{yy} in the second formula given above, so that

$$R^2_{adj} = \left(1 - \frac{\text{MSE}}{S_{yy}/(n-1)}\right) 100\%$$

The calculated value of the adjusted R^2 is 87.3%. R^2-adjusted will always be (smaller, larger) than unadjusted R^2. | smaller

5. *The Analysis of Variance.* The analysis of variance table shows how the variation in *y* is decomposed into its component parts. The column labeled *SOURCE* shows that the *Total* variation in *y*, given by S_{yy}, can be decomposed into unexplained *Error* variation and the explained variation due to *Regression*. The DF column gives the degrees of freedom with each source of variation, while the Sum of Squares (SS) column gives the calculated sum of squares for each source of variation. In these two columns, the entries for *Regression* and *Error* add to the _____ line. The entries in the Mean Square (MS) column are found by dividing each sum of squares entry by its degrees of freedom. No mean square is calculated for the _____ line. | Total

| Total

If the model contributes information for the prediction of *y*, at least one of the model parameters $\beta_1, \beta_2,$ or β_3 will differ from _____. | zero
In testing the hypothesis $H_0: \beta_1 = \beta_2 = \beta_3 = 0$ against the alternative hypothesis H_a: at least one of $\beta_1, \beta_2,$ or β_3 differs from zero, we use the statistic

$$F = \frac{\text{MSR}}{\text{MSE}} = \frac{R^2/k}{(1-R^2)/[n-(k+1)]}$$

where MSR is the mean square due to regression, and MSE is the mean square due to error. MSR is found in the *Regression* line, and MSE is found in the *Error* line of the Mean Square (MS) column. MSR has degrees of freedom equal to the number of parameters in the model, $v_1 = k$, excluding the intercept, while MSE has degrees of freedom equal to the number of observations minus the number of parameters in the model *including* the intercept, or $v_2 = n - (k+1)$.

Since SSR is the variation explained by the model, and SSE is the variation unexplained by the model, a (small, large) value of MSR when compared to MSE is cause to reject H_0. Therefore, we reject H_0 for (small, large) values of *F*. In this problem $F = 14.75$ with $v_1 = 3$ and $v_2 = 7 - 4 = 3$ degrees of freedom; *F* is found by dividing 38.798 by 2.631. Since this value exceeds the tabulated value of *F* with $v_1 = 3$, $v_2 = 3$ and $\alpha = .05$, given as $F = 9.28$ in Table 8 of your text, we reject H_0 and conclude that the model (does, does not) contribute significant information in predicting *y*. The observed significance level (*p*-value) for this test is _____. | large

| large

| does

| .027

93.6%

.7750

.036
.125

regression
R^2

6. *Sequential Sum of Squares.* Notice that none of the partial regression coefficients is significant at the .05 level of significance, but taken together, they account for _____ of the variation in the values of y. The table below the analysis of variance provides the SEQuential Sum of Squares (SEQ SS) and lists the additional contribution to the sum of squares for regression of each variable (or combination of variables) as they were entered into the model through the MINITAB REGRESS command. In this problem, x_1 was the first variable entered and accounted for (32.109/41.429) = _____, or 77.5% of the variation in y. The variable x_2 was entered next and accounted for an additional (1.512/41.429) = _____, or 3.6% of the variation in y. Entering $x_1 x_2$ accounted for an additional (5.178/41.429) = _____, or 12.5% of the variation in y. Notice that within rounding errors, the sequential sums of squares add up to the sum of squares for _____, and that the three percentages add up to _____.

Example 11.6

An agricultural economist interested in California cotton production gathered the following data concerning the mean number of cotton bolls per plant during the growing season in the San Joaquin Valley of California. Here y is the mean number of bolls per plant and x is the time measured in weeks.

y	110	470	1040	1100	1000	820
x	1	4	7	9	12	15

Use a multiple regression program to fit a second degree polynomial to these data and discuss the computer output from the program.

Solution

1. A second degree polynomial model is given by

$$E(y) = \beta_0 + \beta_1 x + \beta_2 x^2$$

In this case $E(y)$ is a general linear model with $x_1 = x$ and $x_2 = x^2$. If a multiple regression program requires that the data be entered as (y, x_1, x_2) for each of the n data points, the user would enter the triples (y, x, x^2) for each of the n observations. For example, the first data point would be (110,1,1) and the last would be (820, 15, 225). The MINITAB command REGRESS produced the following printout in fitting a second degree polynomial to these data.

```
MTB > PRINT C1-C3

ROW        Y        X       XSQ

  1      110       1         1
  2      470       4        16
  3     1040       7        49
  4     1100       9        81
  5     1000      12       144
  6      820      15       225
```

```
MTB > REGRESS C1 2 C2 C3

The regression equation is
Y = - 176 + 244 X - 11.9 XSQ

Predictor           Coef         Stdev         t-ratio         p
Constant          -175.5         125.4         -1.40           0.256
X                  244.22         35.97         6.79           0.007
XSQ               -11.878          2.172        -5.47          0.012

s = 106.7          R-sq = 95.5%              R-sq(adj) = 92.5%

Analysis of Variance

SOURCE         DF        SS            MS          F         p
Regression      2       727600       363800      31.97     0.009
Error           3        34134        11378
Total           5       761733

SOURCE         DF       SEQ SS
X               1       387292
XSQ             1       340308
```

2. The value of R^2 on the printout is _____, which means that _____ percent of the variation in the mean number of cotton bolls is accounted for by the quadratic model.

    95.5%
    95.5

3. In testing the hypothesis $H_0: \beta_1 = \beta_2 = 0$ against the alternative that at least one of β_1 or β_2 differs from zero, the value of $F = MSR/MSE$ found in the analysis of variance section under F is _____, which is highly significant with a p-value of .009. Therefore, we (can, cannot) conclude that x and x^2 contain significant information in predicting y.

    31.97
    can

4. In the individual analysis of variables section we see that the partial regression coefficients corresponding to x and x^2(X-SQ) are both significant at the .05 level of significance. For example, in testing whether the quadratic term x^2 contributes significant information in predicting y, the test of $H_0: \beta_2 = 0$ produced a value of $t =$ _____, which is significant at the .012 level of significance. The value of t can be verified by calculating

$$t = \hat{\beta}_2/s_{\hat{\beta}_2},$$

    -5.47

where $\hat{\beta}_2 =$ _____ and $s_{\hat{\beta}_2} =$ _____.

    -11.878; 2.172

5. Using the estimated coefficients $\hat{\beta}_0, \hat{\beta}_1$, and $\hat{\beta}_2$ found in the column labeled Predictor, the prediction equation is

$$\hat{y} = -175.5 + 244.22x - 11.878x^2.$$

A plot of the prediction curve and the observed data values are given in the following graph. Notice that the curve appears to fit the observed data points very well.

Time in weeks

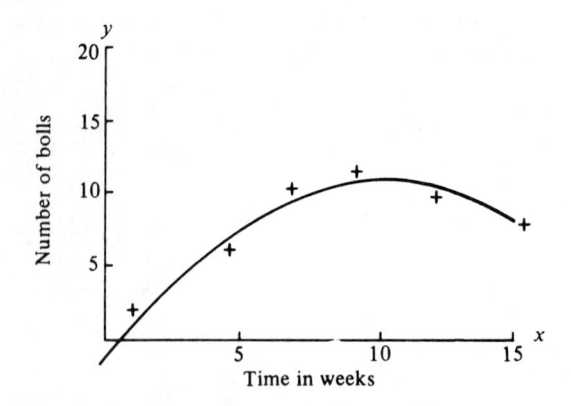

Some computer facilities may have a program written specifically to perform polynomial regression analysis. For such programs it is sufficient to enter the pairs (y, x) since the values of x^2, x^3 and other power terms are generated within the program itself. However, any multiple regression program can be used to fit a polynomial model.

Self-Correcting Exercises 11B

1. In order to study the relationship of advertising and capital investment on corporate profits, the following data, recorded in units of $100,000, was collected for ten medium-sized firms within the same year. The variable y represents profit for the year, x_1 represents capital investment, and x_2 represents advertising expenditures.

y	x_1	x_2
15	25	4
16	1	5
2	6	3
3	30	1
12	29	2
1	20	0
16	12	4
18	15	5
13	6	4
2	16	2

These data were analyzed using the MINITAB REGRESS command based on the mode

$$E(y) = \beta_0 + \beta_1 x_1 + \beta_2 x_2 .$$

The MINITAB printout follows.

```
MTB > REGRESS C1 2 C2 C3

The regression equation is
Y = - 8.18 + 0.292 X1 + 4.43 X2

Predictor        Coef        Stdev      t-ratio         p
Constant       -8.177        4.206       -1.94       0.093
X1              0.2921       0.1357       2.15       0.068
X2              4.4343       0.8002       5.54       0.000

s = 3.303        R-sq = 82.3%        R-sq(adj) = 77.2%

Analysis of Variance

SOURCE        DF         SS          MS         F          p
Regression     2       355.22      177.61     16.28      0.002
Error          7        76.38       10.91
Total          9       431.60

SOURCE        DF      SEQ SS
X1             1       20.16
X2             1      335.05
```

a. Find the values of S_{yy}, SSR, SSE, s^2 and s on the printout.
b. Find R^2 and interpret its value.
c. Verify the value of R^2 using the entries S_{yy}, SSR, and SSE.
d. Do the data provide sufficient evidence to indicate that the model contributes significant information in predicting y? Use $\alpha = .05$.
e. Test the hypothesis $H_0: \beta_2 = 0$ against $H_a: \beta_2 \neq 0$ at the .05 level of significance. Verify the value of t on the printout by using the appropriate entries in the Coef and Stdev columns.
f. Write the prediction equation relating $\hat{y}$ and the predictor variables x_1 and x_2.
g. Estimate the yearly corporate profits for a medium-sized firm whose capital investment was $2,200,000 and whose advertising expenditure was $400,000.

2. A chemical company interested in maximizing the output of a chemical process by selection of the reaction temperature recorded the following data where y is the yield in kilograms and x is the coded temperature:

y	7.5	8.1	8.8	10.9	12.5	11.8	11.1	10.4	9.5
x	-4	-3	-2	-1	0	1	2	3	4

In chemical reactions, the amount of the substance produced may increase until a critical temperature is reached, at which point the amount of substance produced begins to decrease due to its decomposition by the increasing temperature. Anticipating that this would be the case, the model

$$E(y) = \beta_0 + \beta_1 x + \beta_2 x^2$$

was fitted using the MINITAB REGRESS command. The computer printout follows.

```
MTB > REGRESS C1 2 C2 C3

The regression equation is
Y = 11.4 + 0.340 X - 0.205 XSQ

Predictor         Coef         Stdev       t-ratio          p
Constant       11.4346        0.3734        30.62        0.000
X               0.34000       0.09539        3.56        0.012
XSQ            -0.20519       0.04210       -4.87        0.003

s = 0.7389        R-sq = 85.9%         R-sq(adj) = 81.2%

Analysis of Variance

SOURCE         DF            SS            MS           F           p
Regression      2        19.9043        9.9522       18.23       0.003
Error           6         3.2757        0.5459
Total           8        23.1800

SOURCE         DF         SEQ SS
X               1         6.9360
XSQ             1        12.9683
```

a. Find R^2 on the printout and interpret its value.
b. Calculate R^2 directly using the entries S_{yy}, SSR, and SSE.
c. Is there sufficient evidence to indicate that the model contributes significant information in predicting y at the .05 level of significance?
d. Test for significant curvature in the fitted response by testing H_0: $\beta_2 = 0$ against H_a: $\beta_2 \neq 0$ with $\alpha = .05$.
e. Use the fitted prediction equation to predict the yield when the coded temperature is $x = 1$.

11.4 Comparison of Computer Printouts

Multiple regression programs can be found in most statistical program packages. Three commonly used packages that contain multiple regression programs are SAS, MINITAB, and SPSS. The computer printouts for each of these multiple regression programs are given in your text. Although the format (which refers to the use of headings and the placement of the results on the printed page) vary from package to package, essentially the same information appears on all outputs. Rather than reexplain all the entries on these three programs, we will note only the differences as they relate to the information we have presented.

The estimates of the regression coefficients appear in the column labeled ESTIMATE in the SAS package and in the column labeled B in the SPSS package.

The estimated standard deviations of the estimated regression coefficients are labeled STD ERROR OF ESTIMATE in the SAS package and STD

ERROR OF B in the SPSS package. The analysis of variance sections are similar for all three.

The SPSS printout differs from SAS and MINITAB in that individual model parameters are tested using an F-statistic rather than the t-statistic that we have used, given by

$$t = \hat{\beta}_i / s_{\hat{\beta}_i}$$

with error degrees of freedom equal to n minus the number of model parameters. The SPSS package uses the fact that the square of a t-statistic with ν degrees of freedom is the same as an F-statistic with one numerator and ν denominator degrees of freedom. That is

$$(t_\nu)^2 = F_\nu^1$$

or equivalently,

$$t_\nu = \sqrt{F_\nu^1}$$

with the sign of t the same as the sign of the coefficient tested.

Both MINITAB and SPSS printouts give an adjusted value of R^2, which is adjusted for degrees of freedom. The adjusted value of R^2 is calculated as

$$R^2(\text{adjusted}) = 1 - \frac{\text{SSE}/\nu_2}{S_{yy}/(n-1)},$$

where ν_2 represents the degrees of freedom for error and n is the number of observations. Without adjustment,

$$R^2 = 1 - \frac{\text{SSE}}{S_{yy}}.$$

Notice that the adjusted value of R^2 will always be (less, greater) than or equal to the unadjusted value since $\nu_2 \leq n - 1$.

less

11.5 Problems in Using Multiple Linear Regression Analysis

In a multiple regression problem, the regression coefficients are called *partial* regression coefficients, since they are determined in conjunction with other variables in the model and only partially determine the value of y. Further, the values of these partial regression coefficients (would, would not) in general be the same as those found by using several simple linear regression models, each with one independent variable. Estimates of the partial regression coefficients are correlated with each other to the extent that the underlying independent variables share the same predictive information.

would not

When the independent variables included in a regression analysis are correlated among themselves, the values of the estimated β's in the model take into account the amount of shared and independent information available in the x's for estimating the response y. In this situation, individual tests of the regression coefficients are of little value. More information concerning the utility of the independent variables $x_1, x_2, \ldots, x_k$ in predicting y can be obtained by testing the hypothesis

$$H_0: \beta_1 = \beta_2 = \ldots = \beta_k = 0$$

A test of this hypothesis is given in Section 11.4.

When two or more of the independent variables are highly correlated with each other, we are confronted with the problem of *multicollinearity*. Multicollinearity is the technical way of saying, for example, that if one is given pairs of values for two independent variables that are highly correlated with each other, the pairs of values will exhibit a strong linear relationship when plotted on graph paper. When the correlation is very high, the points will almost plot as a _____ _____ [straight line]. Hence, we say that these variables are collinear, and, for all practical purposes, one is working with one independent variable. When this situation is repeated for several pairs of independent variables, we refer to the problem as one of multicollinearity.

Many investigators prefer to use a *stepwise regression program* which at each step adds an independent variable to the regression model only if its inclusion significantly reduces SSE below the value achieved without the variable included. In this way, the investigator can look at the stepwise decrease in SSE and assess the additional contribution of the independent variable just added, above and beyond the contribution of those variables already in the model.

11.6 Some Comments on Model Formulation (Optional)

Variables are classified as being either quantitative or qualitative. A quantitative variable takes values corresponding to the points on the real line. If a variable is not quantitative, then it is said to be qualitative. Variables such as advertising expenditure, number or age of employees, per unit production cost, and number of delivery trucks are examples of quantitative variables while geographic region, plant site, and kind of stock are examples of _____ [qualitative] variables. Although predictor variables can be quantitative or qualitative, a dependent variable must be _____ [quantitative] in order to satisfy the assumptions given in Section 11.2.

The intensity setting of an independent variable is called a _____ [level]. The levels of a quantitative independent variable correspond to the number of distinct values that the variable assumes in an investigation. For example, if an experimenter interested in maximizing the output of a chemical process observed the process when the temperature was set at 100°F, 200°F and 300°F, the independent variable "temperature" was observed at _____ [three] levels. The levels of a quantitative independent variable are defined by describing them. For example, the independent variable "occupational

Sect. 11.6: Some Comments on Model Formulation (Optional) / 269

groups" might be described as white-collar workers, blue-collar workers, service workers, and farm workers. If all four groups were included in an investigation, the qualitative variable "occupational groups" would be taken to have _____ levels. Similarly, if an investigation were to be conducted in three regions, the qualitative variable "regions" would have _____ levels.

| four |
| three |

It is necessary to differentiate between quantitative and qualitative variables to be included in a regression analysis because these variables are entered into a regression model in different ways. Quantitative variables, in general, are entered directly into a regression equation, while qualitative variables are entered through the use of dummy variables, which in effect produce different response curves at each setting of the qualitative independent variable.

When two or more quantiative independent variables appear in a regression model, the resulting response function produces a graph called a response surface in three or more dimensions. These graphs become difficult to produce when three or more independent variables are included in the model. A model involving quantitative variables is said to be a first-order model if each independent variable appears in the model with power _____ . The model

$$E(y) = \beta_0 + \beta_1 x_1 + \beta_2 x_2 + \cdots + \beta_k x_k$$

is a *first-order model* involving _____ independent variables, since the model is linear in each x. The graph of a first-order model is a response plane, which means that the surface is "flat" but has some directional tilt with respect to its axes. Second-order linear models in k quantitative predictor variables include all the terms in a first-order model, all crossproduct terms such as $x_1 x_2, x_1 x_3, x_2 x_3, \ldots, x_{k-1} x_k$, and all pure quadratic terms x_1^2, $x_2^2, \ldots, x_k^2$. A *second order model* with two predictor variables is given as

$$E(y) = \beta_0 + \beta_1 x_1 + \beta_2 x_2 + \beta_3 x_1^2 + \beta_4 x_1 x_2 + \beta_5 x_2^2 .$$

one

k

The quadratic terms x_1^2 and x_2^2 allow for curvature while the crossproduct or interaction term $x_1 x_2$ allows for warping or twisting of the response surface.

Two predictor variables are said to *interact* if the change in $E(y)$ corresponding to a change in one predictor variable depends upon the value of the other variable.

In Examples 11.4 and 11.5, the model included the interaction term $x_1 x_2$. The resulting graphs in Example 11.4 show how the change in $E(y)$ as x_1 changes depends upon the value of x_2 and vice versa.

Qualitative variables are entered into a regression model using dummy variables. For each independent qualitative variable in the model, the number of dummy variables required is one less than the number of _____ associated with that qualitative variable. The following example will demonstrate how this technique is implemented.

levels

Example 11.7
An investigator is interested in predicting the strength of particle board (y) as a function of the size of the particles (x_1) and two types of bonding compounds. If the basic response is expected to be a quadratic function of particle size, write a linear model that incorporates the qualitative variable "bonding compound" into the predictor equation.

Solution
The basic response equation for a specific type of bonding compound would be

$$E(y) = \beta_0 + \beta_1 x_1 + \beta_2 x_1^2$$

Since the qualitative variable "bonding compound" is at two levels, one dummy variable is needed to incorporate this variable into the model. Define the dummy variable x_2 as follows:

$$x_2 = 1 \text{ if bonding compound 2}$$

$$x_2 = \underline{} \text{ if not}$$

0

The expanded model would now be written as

$$E(y) = \beta_0 + \beta_1 x_1 + \beta_2 x_1^2 + \beta_3 x_2 + \beta_4 x_1 x_2 + \beta_5 x_1^2 x_2.$$

1. When $x_2 = 0$, the response has been measured using bonding compound 1 and the resulting equation is

$$E(y) = \beta_0 + \beta_1 \underline{} + \beta_2 \underline{}.$$

$x_1 ; x_1^2$

2. When $x_2 = 1$, the response has been measured using bonding compound 2 and the resulting equation is

$$E(y) = (\beta_0 + \beta_3) + (\underline{}) x_1 + (\beta_2 + \beta_5) x_1^2.$$

$\beta_1 + \beta_4$

3. The use of the dummy variable x_2 has allowed us to simultaneously describe two quadratic response curves for each of the two bonding compounds. Notice that β_3, β_4, and β_5 measure the differences between the intercepts, the linear components, and the quadratic components, respectively, for the two bonding compounds.
4. Had another bonding compound been included in the investigation, x_3, a second dummy variable, woud be defined as

1

$$x_3 = \underline{} \text{ if bonding compound 3}$$

0

$$x_3 = \underline{} \text{ if not}$$

and the model would be expanded to include the terms $x_3, x_1 x_3$, and

$x_1^2 x_3$

$\underline{}$ to produce in effect a third quadratic response curve for compound 3.

The formulation of the model is perhaps the most important aspect of a regression analysis since the fit of the model will depend not only upon the independent variables included in the model, but also upon the way in which the variables are introduced into the model. If, for example, the response increases with some variable x, achieves a maximum, and then begins to decrease, both linear and quadratic terms in x should be included in the model. Failure to include a term in x^2 may cause the model to fit poorly and/or fail in predicting the response y for all values of x. Accurate formulation of a model requires experience and a knowledge of the mechanism underlying the response of interest. The latter is sometimes achieved by running several exploratory investigations, and combining this information within a more elaborate model.

Self-Correcting Exercises 11C

1. Suppose the response y is related to two predictor variables x_1 and x_2.
   a. Write a first-order model relating $E(y)$ and the variables x_1 and x_2.
   b. Describe the graph relating $E(y)$ and x_1 when x_2 is held constant.
   c. Describe the graph relating $E(y)$ and x_2 when x_1 is held constant.
   d. How can the first-order model be extended in order to allow both the slope and the intercept relating $E(y)$ and x_1 to vary with the value of x_2?
2. Suppose y is related to three predictor variables $x_1, x_2,$ and x_3.
   a. Write a first-order model relating $E(y)$ and $x_1, x_2,$ and x_3.
   b. Write a second-order model relating $E(y)$ and $x_1, x_2,$ and x_3.
   c. What is the effect of including $x_1^2, x_2^2,$ and x_3^2 in the model in part b?
   d. Which terms in the model in part b are interaction terms? What effect do they have on the response surface?
3. Consider a situation in which the output (y) of an industrial plant is linearly related to the number of individuals employed (x_1) and the area in which the plant is located. In describing two areas we define the dummy variable x_2 as:

   $x_2 = 1$ if area 2

   $x_2 = 0$ if not.

   Write a linear model relating output to x_1 and x_2 if we assume that the relationship between y and x_1 is linear for both areas.
4. Refer to Exercise 3. Suppose that three areas were involved in the experiment. Define a second dummy variable x_3 as

   $x_3 = 1$ if area 3

   $x_3 = 0$ if not.

   Write a linear model relating output to $x_1, x_2,$ and x_3 if we assume that the relationship between y and x_1 is linear for all areas.

5. Suppose that an experiment as described in Exercise 4 is conducted and the following prediction equation is obtained:

$$\hat{y} = 2 + x_1 + x_2 + 3x_1x_2 + 2x_3 + x_1x_3.$$

Graph the three prediction lines for each of areas 1, 2, and 3.

11.7 Summary

Multiple regression analysis is an extension of simple linear regression analysis to accommodate situations in which the response y is a function of a number of independent predictor variables $x_1, x_2, \ldots, x_k$. The procedures associated with the simple linear model have analogues in the case of a multiple regression model. Hence, any simple linear regression problem (can, cannot) be analyzed using multiple regression techniques.

    Although identical in concept, simple and multiple regression analysis differ in two important aspects. Simple linear regression analysis can be done (with, without) the use of a computer; for multiple regression analysis, this is generally not the case. However, multiple regression analysis programs are available at most computing facilities. Secondly, very few real life situations (can, cannot) be adequately described by a simple linear regression model. Multiple regression analysis provides greater utility and latitude in data analysis by allowing the user to include k independent predictor variables in the regression model.

margin notes: can; without; can

EXERCISES

1. A manufacturer, concerned about the number of defective items being produced within his plant, recorded the number of defective items produced on a given day (y) by each of 10 machine operators, recording also the average output per hour (x_1) for each operator and the time from the last machine servicing (x_2) in weeks The data were

y	x_1	x_2
13	20	3
1	15	2
11	23	1.5
2	10	4
20	30	1
15	21	3.5
27	38	0
5	18	2
26	24	5
1	16	1.5

The following computer output resulted when these data were analyzed using the MINITAB REGRESS command based on the model

$$E(y) = \beta_0 + \beta_1 x_1 + \beta_2 x_2.$$

```
MTB > REGRESS C1 2 C2 C3

The regression equation is
Y = - 28.4 + 1.46 X1 + 3.84 X2

Predictor         Coef        Stdev       t-ratio        p
Constant       -28.3906      0.8273       -34.32      0.000
X1              1.46306      0.02699       54.20      0.000
X2              3.8446       0.1426        26.97      0.000

s = 0.5484     R-sq = 99.8%          R-sq(adj) = 99.7%

Analysis of Variance

SOURCE        DF        SS          MS          F         p
Regression     2      884.79      442.40     1470.84    0.000
Error          7        2.11        0.30
Total          9      886.90

SOURCE        DF       SEQ SS
X1             1       666.04
X2             1       218.76
```

a. Interpret R^2 and comment on the fit of the model.
b. Is there sufficient evidence to indicate that the model contributes significant information in predicting y at the .01 level of significance?
c. What is the prediction equation relating $\hat{y}$ and x_1 when $x_2 = 4$?
d. Use the fitted prediction equation to predict the number of defective items produced for an operator whose average output per hour is 25 and whose machine was serviced three weeks ago.

2. An experiment was conducted to investigate the relationship between the degree of metal corrosion and the length of time the metal is exposed to the action of soil acids. In the data which follow, y is the percentage corrosion and x is the exposure time measured in weeks.

y	0.1	0.3	0.5	0.8	1.2	1.8	2.5	3.4
x	1	2	3	4	5	6	7	8

The following computer output resulted in fitting the model:

$$E(y) = \beta_0 + \beta_1 x + \beta_2 x^2$$

```
MTB > REGRESS C1 2 C2 C3

The regression equation is
Y = 0.196 - 0.100 X + 0.0619 XSQ

Predictor         Coef        Stdev       t-ratio        p
Constant        0.19643     0.07395        2.66       0.045
X              -0.10000     0.03770       -2.65       0.045
XSQ             0.061905    0.004089      15.14       0.000

s = 0.05300    R-sq = 99.9%          R-sq(adj) = 99.8%
```

Analysis of Variance

SOURCE	DF	SS	MS	F	p
Regression	2	9.4210	4.7105	1676.61	0.000
Error	5	0.0140	0.0028		
Total	7	9.4350			

SOURCE	DF	SEQ SS
X	1	8.7771
XSQ	1	0.6438

a. What percent of the total variation is explained by the quadratic regression of y on x?

b. Is the regression of y on x and x^2 significant at the $\alpha = .05$ level of significance?

c. Is the linear regression coefficient significant at the .05 level of significance?

d. Is the quadratic regression coefficient significant at the .05 level of significance?

e. Fitting the model with the linear term omitted given by $E(y) = \beta_0 + \beta_2 x^2$ resulted in the output that follows. What could you say about the contribution of the linear term in x in explaining the total variation in y?

MTB > REGRESS C1 1 C3

The regression equation is
Y = 0.0164 + 0.0513 XSQ

Predictor	Coef	Stdev	t-ratio	p
Constant	0.01643	0.04160	0.39	0.707
XSQ	0.051317	0.001256	40.84	0.000

s = 0.07507   R-sq = 99.6%   R-sq(adj) = 99.6%

Analysis of Variance

SOURCE	DF	SS	MS	F	p
Regression	1	9.4012	9.4012	1668.24	0.000
Error	6	0.0338	0.0056		
Total	7	9.4350			

3. In a study to examine the relationship between the time required to complete a construction project and several pertinent independent variables, an analyst compiled a list of four variables that might be useful in predicting the time to completion. These four variables were size of the contract (in $1,000 units) (x_1), number of workdays adversely affected by the weather (x_2), number of subcontractors involved in the project (x_4), and a variable (x_3) that measured the presence or absence of a workers' strike during the construction. In particular,

$x_3 = 0$ if no strike

$x_3 = 1$ if strike

Fifteen construction projects were randomly chosen, and each of the four variables as well as the time to completion were measured. The data are given in the following table:

y	x_1	x_2	x_3	x_4
29	60	7	0	7
15	80	10	0	8
60	100	8	1	10
10	50	14	0	5
70	200	12	1	11
15	50	4	0	3
75	500	15	1	12
30	75	5	0	6
45	750	10	0	10
90	1200	20	1	12
7	70	5	0	3
21	80	3	0	6
28	300	8	0	8
50	2600	14	1	13
30	110	7	0	4

An analysis of these data using a first-order model in x_1, x_2, x_3, and x_4 produced the following computer printout. Notice that the t-statistic has been replaced by its equivalent F-statistic, which appears in the column labeled F VALUE and the intercept is referred to as CONSTANT.

```
R-SQUARE             .8471
STD ERROR OF EST    11.8450

ANALYSIS OF VARIANCE

                        SUM OF        MEAN
             DF         SQUARES       SQUARE      F RATIO    PR>F
REGRESSION    4         7770.2972    1942.5743    13.8455    0.0004
RESIDUAL     10         1403.0362     140.3036

INDIVIDUAL ANALYSIS OF VARIABLES

VARIABLE    COEFFICIENT    STD ERROR    F VALUE    PR>F
(CONSTANT    -1.5887)
X1           - .00784        .00623      1.5846    0.2367
X2             .67533        .99978       .4563    0.5147
X3           28.01342       11.37143     6.0688    0.0335
X4            3.4889         1.93516     3.2504    0.1016
```

Give a complete analysis of the printout and interpret your results. What can you say about the apparent contribution of x_1 and x_2 in predicting y?

4. A particular savings and loan corporation is interested in determining how well the amount of money in family savings accounts can be predicted using the three independent variables, annual income, number in the family unit, and area in which the family lives. Suppose that there are two specific areas of interest to the corporation. The following data were collected, where

y = amount in all savings accounts

x_1 = annual income

x_2 = number in family unit

x_3 = 0 if area 1; 1 if not

Both y and x_1 were recorded in units of $1,000.

y	x_1	x_2	x_3
0.5	19.2	3	0
0.3	23.8	6	0
1.3	28.6	5	0
0.2	15.4	4	0
5.4	30.5	3	1
1.3	20.3	2	1
12.8	34.7	2	1
1.5	25.2	4	1
0.5	18.6	3	1
15.2	45.8	2	1

The following computer printout resulted when the data was analyzed using MINITAB.

MTB > REGRESS C1 3 C2 C3 C4

The regression equation is
Y = − 3.11 + 0.503 X1 − 1.61 X2 − 1.15 X3

Predictor	Coef	Stdev	t-ratio	p
Constant	−3.112	3.600	−0.86	0.421
X1	0.50314	0.07670	6.56	0.001
X2	−1.6126	0.6579	−2.45	0.050
X3	−1.155	1.791	−0.64	0.543

s = 1.896   R-sq = 92.2%   R-sq(adj) = 88.4%

Analysis of Variance

SOURCE	DF	SS	MS	F	p
Regression	3	256.621	85.540	23.78	0.001
Error	6	21.579	3.597		
Total	9	278.200			

SOURCE	DF	SEQ SS
X1	1	229.113
X2	1	26.012
X3	1	1.496

a. Interpret R^2 and comment on the fit of the model.
b. Test for a significant regression of y on $x_1, x_2,$ and x_3 at the .05 level of significance.
c. Test the hypothesis $H_0: \beta_3 = 0$ against $H_a: \beta_3 \neq 0$ using a significance level of $\alpha = .05$. Comment on the results of your test.
d. What can be said about the utility of x_3 as a predictor variable in this problem?

CHAPTER 12
ANALYSIS OF ENUMERATIVE DATA

12.1 The Multinomial Experiment

Examine the following experimental situations for any general similarities:
1. Two hundred people are classifed according to their blood type and the number of people in each blood type group is recorded.
2. A sample of 100 items is randomly selected from a production line. Each item is classified as belonging to one of three groups: acceptables, seconds, or rejects. The number in each group is recorded.
3. A random sample of 50 books is taken from the local library. Each book is assigned to one of three categories: science, art, or fiction. The number of books in each category is recorded.

Each of these situations is similar to the others in that classes or categories are defined and the number of items falling into each category is recorded. Hence, these experiments result in enumerative or _____ (count) data and have the following general characteristics, which define the _____ (multinomial) experiment:
1. The experiment consists of n identical trials.
2. The outcome of each trial falls into one of k classes or cells.
3. The probability that the outcome of a single trial falls into cell i is p_i, $i = 1, 2, \ldots, k$, where p_i is _____ (constant) from trial to trial and

$$\sum_{i=1}^{k} p_i = \underline{\quad} .$$ (1)

4. The trials are _____ (independent).
5. We are interested in $n_1, n_2, n_3, \ldots, n_k$, where n_i is the number of trials in which the outcome falls into cell i, and

$$\sum_{i=1}^{k} n_i = \underline{\quad} .$$ (n)

The binomial experiment is a special case of the multinomial experiment. This can be seen by letting $k = \underline{\quad}$ (2), and noting the following correspondences:

	Binomial	Multinomial (k = 2)	
a.	n	n	
b.	p	p_1	
c.	q	_____	p_2
d.	x	n_1	
e.	$n - x$	_____	n_2
f.	$E(x) = np$	$E(n_1) = np_1$	
g.	$E(n - x) = nq$	$E(n_2) =$ _____	np_2

For the multinomial experiment, we wish to make inferences about the associated population parameters $p_1, p_2, \ldots, p_k$. A statistic that allows us to make inferences of this sort was developed by the British statistician Karl Pearson around 1900.

12.2 The Chi-Square Test

For a multinomial experiment consisting of n trials with known (or hypothesized) cell probabilities p_i, $i = 1, 2, \ldots, k$, we can find the expected number of items falling into the ith cell by using

$$E(n_i) = np_i, \quad i = 1, 2, \ldots, k.$$

The cell probabilities are rarely known in practical situations. Consequently, we wish to estimate or test hypotheses concerning their values. If the hypothesized cell probabilities given are the correct values, then the *observed* number of items falling in each of the cells, n_i, should differ but slightly from the expected number $E(n_i) = np_i$. Pearson's statistic (given below) utilizes the squares of the deviations of the observed from the expected number in each cell:

Pearson's Chi-Square Test Statistic

$$X^2 = \sum_{i=1}^{k} \frac{[n_i - E(n_i)]^2}{E(n_i)}$$

$$= \sum_{i=1}^{k} \frac{(n_i - np_i)^2}{np_i}.$$

Note that the deviations are divided by the expected number so that the deviations are weighted according to whether the expected number is large or small. A deviation of 5 from an expected number of 20 contributes $(5)^2/20 =$ _____ to X^2, while a deviation of 5 from an expected number of 10 contributes $(5)^2/10 =$ _____, or *twice* as much, to X^2.

1.25
2.50

280 / Chap. 12: Analysis of Enumerative Data

When n, the number of trials, is large, this statistic has an approximate χ^2 distribution, provided the expected numbers in each cell are not too small. We will require as a rule of thumb that $E(n_i) \geq$ __5__. This requirement can be satisfied by combining those cells with small expected numbers until every cell has an expected number of at least __5__. For small deviations from the expected cell counts, the value of the statistic would be (large, __small__), supporting the hypothesized cell probabilities. However, for large deviations from the expected counts, the value of the statistic would be (__large__, small), and the hypothesized values of the cell probabilities would be __rejected__. Hence a one-tailed test is used, rejecting H_0 when X^2 is __large__.

To find the critical value of χ^2 used for testing, the degrees of freedom must be known. Since the degrees of freedom change as Pearson's chi-square statistic is applied to different situations, the degrees of freedom will be specified for each application that follows. In general, the degrees of freedom are equal to the number of cells less one degree of freedom for each independent linear restriction placed upon the cell counts. One linear restriction that will always be present is that

$$n_1 + n_2 + n_3 + \ldots + n_k = \underline{n}.$$

Other restrictions may be imposed by the necessity to estimate certain unknown cell parameters or by the method of sampling employed in the collection of the data.

12.3 A Test of an Hypothesis Concerning Specified Cell Probabilities

Let us consider the following problems concerning cell probabilities in a multinomial experiment:

Example 12.1
Previous enrollment records at a large university indicate that of the total number of persons that apply for admission, 60% are admitted unconditionally, 5% are admitted on a trial basis, and the remainder are refused admission. Of 500 applications to date for the coming year, 329 applicants have been admitted unconditionally, 43 have been admitted on a trial basis, and the remainder have been refused admission. Do these data indicate a departure from previous admission rates?

Solution
This experiment consists of classifying 500 applicants into one of three cells: cell 1 (unconditional admission), cell 2 (conditional admission), and cell 3 (admission refused), where, under previous observation, $p_1 = .60$, $p_2 = .05$, and $p_3 = .35$. The expected cell numbers are found to be as follows:

$$E(n_1) = np_1 = 500(.60) = \underline{300}.$$

$$E(n_2) = np_2 = 500(\underline{.05}) = \underline{25}.$$

$$E(n_3) = np_3 = \underline{500}(\underline{.35}) = \underline{175}.$$

Sect. 12.3: A Test of an Hypothesis Concerning Specified Cell Probabilities

Tabulating the results, we have the following:

	Admissions			
	Unconditional	Conditional	Refused	Total
Observed	329	43	128	500
Expected	300	25	175	500

Using Pearson's chi-square statistic we can test the hypothesis that the cell probabilities remain as before against the alternative hypothesis that at least one cell probability is different from those specified. The value of X^2 will be compared with a critical value of $\chi^2_{.05}$ based on the degrees of freedom found as follows. The degrees of freedom are equal to the number of cells ($k = 3$) less one degree of freedom for the linear restriction $n_1 + n_2 + n_3 = n = 500$. Therefore, the degrees of freedom are $k - 1 = 3 - 1 = 2$ and $\chi^2_{.05} =$ _____.   5.991

Formalizing this discussion we have the following statistical test of the enrollment data:

1. $H_0: p_1 = .60, p_2 = .05, p_3 = .35$.
2. H_a: at least one value of p_i is different from that specified by H_0.
3. Test statistic:

$$X^2 = \sum_{i=1}^{3} \frac{(n_i - np_i)^2}{np_i}.$$

4. Rejection region: Reject H_0 if $X^2 \geq \chi^2_{.05} =$ _____. Now we calculate the test statistic.   5.991

$$X^2 = \frac{(329 - 300)^2}{300} + \frac{(43 - 25)^2}{25} + \frac{(128 - 175)^2}{175}$$

$$= 2.803 + 12.96 + 12.62$$

$$= 28.383.$$

5. Decision: $X^2 = 28.383 > 5.991$; hence we (reject; do not reject) H_0. It appears that the admissions to date are not following the previously stated rates.   reject

Some light can be shed on the situation by looking at the sample estimates of the cell probabilities. $\hat{p}_1 = 329/500 = .658, \hat{p}_2 = 43/500 = .086$, and $\hat{p}_3 =$ _____. Notice that the percentage of unconditional admissions has risen slightly, the number of conditional admissions has increased, and the percentage refused admission has decreased at the expense of the first two categories. A final judgment would have to be made when admissions are closed and final figures in.   .256

Example 12.2

A botanist performs a secondary cross of petunias involving independent factors controlling leaf shape and flower color where the factor A represents red color, a represents white color, B represents round leaves, and b represents long leaves. According to the Mendelian model, the plants should exhibit the characteristics *AB*, *Ab*, *aB*, and *ab* in the ratio 9:3:3:1. Of 160 experimental plants, the following numbers were observed: *AB*, 95; *Ab*, 30; *aB*, 28; *ab*, 7. Is there sufficient evidence to refute the Mendelian model at the $\alpha = .01$ level?

Solution
Translating the ratios into proportions, we have the following:

$$P(AB) = p_1 = 9/16$$

$$P(Ab) = p_2 = 3/16$$

$$P(aB) = p_3 = 3/16$$

$$P(ab) = p_4 = 1/16$$

The data are tabulated as follows:

Cell	AB	Ab	aB	ab
Expected	90	30	30	10
Observed	95	30	28	7

Perform a statistical test of the Mendelian model by using the chi-square statistic.
1. $H_0: p_1 = 9/16, p_2 = 3/16, p_3 = 3/16, p_4 = 1/16$.
2. $H_a: p_i \neq p_{i0}$ for at least one value of $i = 1, 2, 3, 4$.
3. Test statistic:

$$X^2 = \sum_{i=1}^{4} \frac{[n_i - E(n_i)]^2}{E(n_i)}.$$

4. Rejection region: With 3 degrees of freedom, we will reject H_0 if

$$X^2 > \chi^2_{.01} = \underline{}.$$

11.3449

Now we calculate

$$X^2 = \frac{(95-90)^2}{90} + \frac{(30-30)^2}{30} + \frac{(28-30)^2}{30} + \frac{(7-10)^2}{10}$$

$$= .2778 + .0000 + .1333 + .9000$$

1.3111

$$= \underline{}.$$

do not reject
is not

5. Decision: Since $X^2 = 1.3111 < 11.3449$, (reject, do not reject) H_0. There (is, is not) sufficient evidence to refute the Mendelian model.

Self-Correcting Exercises 12A

1. A company specializing in kitchen products has produced a mixer in five different colors. A random sample of $n = 250$ sales has produced the following data:

Color	White	Almond	Mauve	Blue	Yellow
Number sold	62	48	56	39	45

   Test the hypothesis that there is no preference for color at the $\alpha = .05$ level of significance. (Hint: If there is no color preference, then $p_1 = p_2 = p_3 = p_4 = p_5 = 1/5$.)

2. The number of Caucasians possessing the four blood types A, B, AB, and O are said to be in the proportions .41, .12, .03, and .44, respectively. Would the observed frequencies of 90, 16, 10, and 84, respectively, furnish sufficient evidence to refute the given proportions at the $\alpha = .05$ level of significance?

12.4 Contingency Tables

We now examine the problem of determining whether independence exists between two methods for classifying observed data. If we were to classify people first according to their hair color and second according to their complexion, would these methods of classification be independent of each other? We might classify students first according to the college in which they are enrolled and second according to their grade-point average. Would these two methods of classification be independent? In each problem we are asking if one method of classification is *contingent* on another. We investigate this problem by displaying our data according to the two methods of classification in an array called a _____ table.

    contingency

Example 12.3

A criminologist studying criminal offenders who have a record of one or more arrests is interested in knowing if the educational achievement level of the offender influences the frequency of arrests. He has classified his data using four educational achievement level classifications:

    A: completed 6th grade or less.
    B: completed 7th, 8th or 9th grade.
    C: completed 10th, 11th, or 12th grade.
    D: education beyond 12th grade.

Number of Arrests	A	B	C	D	Total
1	55 (45.39)	40 (43.03)	43 (43.03)	30 (36.55)	168
2	15 (21.61)	25 (20.49)	18 (20.49)	22 (17.40)	80
3 or more	7 (10.00)	8 (9.48)	12 (9.48)	10 (8.05)	37
Total	77	73	73	62	285

Educational Achievement

284 / Chap. 12: Analysis of Enumerative Data

The contingency table shows the number of offenders in each cell together with the expected cell frequency (in parentheses). The expected frequencies are obtained as follows:

1. Define p_A as the unconditional probability that a criminal offender will have completed grade 6 or less. Define p_B, p_C, and p_D in a similar manner.
2. Define p_1, p_2, and p_3 to be the unconditional probability that the offender has 1, 2, or 3 or more arrests, respectively.

If two events A and B are independent, then $P(AB) = $ _____ .   [$P(A) \cdot P(B)$]

Hence, if the two classifications are independent, a cell probability will equal the product of the two respective unconditional row and column probabilities. For example, the probability that an offender who has completed grade 6 is arrested 3 or more times is

$$p_{A3} = p_A \cdot p_3,$$

whereas the probability that a person with a 10th grade education is arrested twice is

$$p_{C2} = \underline{\qquad}.$$ [$p_C \cdot p_2$]

Since the row and column probabilities are unknown, they must be estimated from the sample data. The estimators for these probabilities are defined in terms of r_i, the row totals, c_j, the column totals, and n.

$$\hat{p}_A = c_1/n = 77/285.$$
$$\hat{p}_B = c_2/n = 73/285.$$
$$\hat{p}_C = c_3/n = 73/285.$$
$$\hat{p}_D = c_4/n = 62/285.$$
$$\hat{p}_1 = r_1/n = 168/285.$$
$$\hat{p}_2 = r_2/n = 80/285.$$
$$\hat{p}_3 = r_3/n = 37/285.$$

If the observed cell frequency for the cell in row i and column j is denoted by n_{ij}, then an estimate for the expected cell number in the ijth cell under the hypothesis of independence can be calculated by using the estimated cell probabilities.

$$E(n_{ij}) = n(p_{ij}) = n(p_i)(p_j)$$

and

$$\hat{E}(n_{ij}) = n(r_i/n)(c_j/n) = r_i c_j/n.$$

The expected cell numbers enclosed in parentheses for the contingency table are found in this way. For example,

$$\hat{E}(n_{11}) = (168)(77)/285 = 45.39$$

$$\hat{E}(n_{12}) = (168)(73)/285 = 43.03$$

$$\hat{E}(n_{34}) = (37)(62)/285 = \underline{\qquad}$$ [8.05]

$$\hat{E}(n_{24}) = (\underline{\qquad})(\underline{\qquad})/285 = \underline{\qquad}.$$ [80; 62; 17.40]

The chi-square statistic can now be calculated as

$$X^2 = \sum_{i=1}^{3} \sum_{j=1}^{4} \frac{[n_{ij} - \hat{E}(n_{ij})]^2}{\hat{E}(n_{ij})}$$

$$= \frac{(55 - 45.39)^2}{45.39} + \frac{(40 - 43.03)^2}{43.03} + \ldots + \frac{(12 - 9.48)^2}{9.48}$$

$$+ \frac{(10 - 8.05)^2}{8.05}$$

$$= 10.23.$$

Recall that the number of degrees of freedom associated with the X^2 statistic equals the number of cells less one degree of freedom for each independent linear restriction on the cell counts. The first restriction is that $\Sigma\, n_i =$ _____ ; hence, _____ degree of freedom is lost here. Then $(r - 1)$ independent linear restrictions have been placed on the cell counts due to the estimation of $(r - 1)$ row probabilities. Note that we need only estimate $(r - 1)$ independent row probabilities since their sum must equal _____ . In like manner, $(c - 1)$ independent linear restrictions have been placed on the cell counts due to the estimation of the column probabilities.

n

one

one (1)

Since there are rc cells, the number of degrees of freedom for testing X^2 in an $r \times c$ contingency table is

$$rc - (1) - (r - 1) - (c - 1),$$

which can be factored algebraically as

$$(r - 1)(c - 1).$$

In short, the number of degrees of freedom for an $r \times c$ contingency table, where all expected cell frequencies must be estimated from sample data (that is, from estimated row and column probabilities), is the number of rows minus one times the number of columns minus one.

For the problem concerning criminal offenders, the degrees of freedom are

$$(r - 1)(c - 1) = (\text{\_\_\_\_})(\text{\_\_\_\_}) = \text{\_\_\_\_}.$$

2; 3; 6

We can now formalize the test of the hypothesis of independence of the two methods of classification at the $\alpha = .05$ level.
1. H_0: the two classifications are independent.
2. H_a: the two classifications are not independent.
3. Test statistic:

$$X^2 = \sum_{i=1}^{3} \sum_{j=1}^{4} \frac{[n_{ij} - \hat{E}(n_{ij})]^2}{\hat{E}(n_{ij})}.$$

12.5916

do not
do not

4. Rejection region: With 6 degrees of freedom, we will reject H_0 if $X^2 > \chi^2_{.05} =$ _____. The calculation of X^2 results in the value $X^2 = 10.23$.
5. Decision: Since $X^2 < 12.5916$, (do, do not) reject H_0. The data (do, do not) present sufficient evidence to indicate that educational achievement and the number of arrests are dependent.

Example 12.4
A sociologist wishes to test the hypothesis that the number of children in a family is independent of the family income. A random sample of 385 families resulted in the following contingency table:

Number of Children	Income Brackets (in thousands of dollars)				Total
	0–$10	$10–20	$20–30	Above $30	
0	10 (14.26)	9 (15.05)	18 (16.48)	24 (15.21)	61
1	8 (17.77)	12 (18.75)	25 (20.53)	31 (18.95)	76
2	14 (21.74)	28 (22.95)	23 (25.12)	28 (23.19)	93
3	26 (17.77)	24 (18.75)	20 (20.53)	6 (18.95)	76
4 or more	32 (18.47)	22 (19.49)	18 (21.34)	7 (19.70)	79
Total	90	95	104	96	385

If the number in parentheses is the estimated expected cell number, do these data present sufficient evidence at the $\alpha = .01$ level to indicate an independence of family size and family income?

Solution
The estimated cell counts have been found by using

$$\hat{E}(n_{ij}) = \frac{r_i c_j}{n}$$

12

and are given in the parentheses within each cell. The degrees of freedom are $(r-1)(c-1) =$ _____.
1. H_0: the two classifications are independent.
2. H_a: the classifications are not independent.
3. Test statistic:

$$X^2 = \sum_{i=1}^{5} \sum_{j=1}^{4} \frac{[n_{ij} - \hat{E}(n_{ij})]^2}{\hat{E}(n_{ij})}.$$

4. Rejection region: With 12 degrees of freedom, we will reject H_0 if

$$X^2 > \chi^2_{.01} = \underline{\hspace{2em}}.$$

26.2170

Calculate X^2:

$$X^2 = \frac{(10 - 14.26)^2}{14.26} + \frac{(9 - 15.05)^2}{15.05} + \ldots + \frac{(18 - 21.34)^2}{21.34}$$

$$+ \frac{(7 - 19.70)^2}{19.70}$$

$$= 63.4783.$$

5. Decision: (Reject, Do not reject) H_0. Therefore, we can conclude that family size and family income (are, are not) independent classifications.

    Reject
    are not

6. From Table 5, with _____ degrees of freedom, the observed value $X^2 = 63.4783$ exceeds $X^2_{.005} = $ _____ . Hence, the approximate level of significance is

    12
    28.2995

    p-value _____

    < .005

The null hypothesis could be rejected for any α (greater than, less than) or equal to _____ .

greater than
.005

Self-Correcting Exercises 12B

1. On the basis of the following data, is there a significant relationship between levels of income and political party affiliation at the $\alpha = .05$ level of significance?

Party Affiliation	Income		
	Low	Average	High
Republican	33	85	27
Democrat	19	71	56
Other	22	25	13

2. Three hundred people were interviewed to determine their opinions regarding a uniform driving code for all states:

Sex	Opinion	
	For	Against
Male	114	60
Female	87	39

Is there sufficient evidence to indicate that the opinion expressed is dependent on the sex of the person interviewed?

12.5 $r \times c$ Tables with Fixed Row or Column Totals: Tests of Homogeneity

To avoid having rows or columns that are absolutely empty, it is sometimes desirable to fix the row or column totals of a contingency table in the design of the experiment. In Example 12.4, the plan could have been to randomly sample 100 families in each of the four income brackets, thereby insuring that each of the income brackets would be represented in the sample. On the other hand, a random sample of 80 families in each of the family size categories could have been taken so that all family size categories would appear in the overall sample.

When using fixed row or column totals, the number of independent linear restrictions on the cell counts is the same as for an $r \times c$ contingency table. Therefore, the data are analyzed in the same way that an $r \times c$ contingency table is analyzed, using the X^2 statistic based on $(r-1)(c-1)$ degrees of freedom. In the following example we examine a case in which the column totals are fixed in advance.

Example 12.5
Fifty 5th grade students from each of four city schools were given a standardized 5th grade reading test. After grading, each student was rated as satisfactory or not satisfactory in reading ability, with the following results:

	School			
	1	2	3	4
Not satisfactory	7	10	13	6

Is there sufficient evidence to indicate that the percentage of 5th grade students with an unsatisfactory reading ability varies from school to school?

Solution
The preceding table displays only half the pertinent information. Extend the table to include the satisfactory category, allowing space to write in the expected cell frequencies.

	School				Total
	1	2	3	4	
Satisfactory	7 (9)	10 (____)	13 (____)	6 (____)	36
Not satisfactory	43 (____)	40 (41)	37 (____)	44 (____)	164
Total	50	50	50	50	200

9; 9; 9

41; 41; 41

By fixing the column total at 50, we have made certain that the unconditional probability of observing a student from each of the schools is constant and equal to _____ for each school.

1/4

If the percentage of unsatisfactory tests does not vary from school to school, then the probability of observing an unsatisfactory reading grade is the same for each school and equal to a common value p. Therefore, the unconditional probability of observing an unsatisfactory grade is p and, in like manner, the probability of observing a satisfactory grade is $1 - p$ = _____. If the percentage of unsatisfactory grades is the same for the four schools, then the probability of observing an unsatisfactory grade for a student in the jth school will be

q

$$p_{1j} = (1/4)p \quad \text{for} \quad j = 1, 2, 3, 4,$$

while the probability of observing a satisfactory grade in the jth school will be

$$p_{2j} = (1/4)q \quad \text{for} \quad j = 1, 2, 3, 4.$$

However, if the probability of an unsatisfactory grade varies from school to school, then

$$p_{1j} \neq (1/4)p \quad \text{and} \quad p_{2j} \neq (1/4)q$$

for at least one value of j, where $j = 1, 2, 3, 4$. But this is the same as asking if the row and column classifications are independent; hence the test is equivalent to a test of the independence of two classifications based on $(r-1)(c-1)$ degrees of freedom.

Proceeding with the required test, we have the following:

$$H_0: p_1 = p_2 = p_3 = p_4 = p.$$

H_a: at least one proportion differs from at least one other.

Test statistic:

$$X^2 = \sum_i \sum_j \frac{[n_{ij} - \hat{E}(n_{ij})]^2}{\hat{E}(n_{ij})}.$$

Rejection region: For $(2-1)(4-1) =$ _____ degrees of freedom, we will reject H_0 if $X^2 > x^2_{.05} =$ _____.

	3
	7.81473

To calculate the value of the test statistic, we must first find the estimated expected cell counts.

$$\hat{E}(n_{11}) = \hat{E}(n_{12}) = \hat{E}(n_{13}) = (36)(50)/200 = \underline{\quad}.$$

9

$$\hat{E}(n_{21}) = \hat{E}(n_{22}) = \hat{E}(n_{23}) = (164)(50)/200 = \underline{\quad}.$$

41

Then

$$X^2 = \frac{(7-9)^2}{9} + \frac{(10-9)^2}{9} + \frac{(\underline{\quad})^2}{9} + \frac{(6-9)^2}{9}$$

13 - 9

$$+ \frac{(43-41)^2}{41} + \frac{(40-41)^2}{41} + \frac{(\underline{\quad})^2}{41} + \frac{(44-41)^2}{41}$$

37 - 41

$$= \frac{\underline{\quad}}{9} + \frac{\underline{\quad}}{41} = 3.3333 + .7317 = \underline{\quad}.$$

30; 30; 4.0650

Decision: Since $X^2 = 4.0650 < x^2_{.05} = 7.8147$, we (can, cannot) reject the hypothesis that reading ability for 5th graders as measured by this test does not vary from school to school.

cannot

When we fixed the column totals in this example, we found that testing for independence of the row and column categories was, in fact, equivalent to a test concerning the equality of four binomial parameters, $p_1, p_2, p_3,$ and p_4. Such tests are called *tests of homogeneity* of several binomial populations. With three or more row categories with fixed column totals, the test becomes a test of homogeneity of several multinomial populations.

12.6 Analysis of an $r \times c$ Contingency Table Using Computer Packages

Packaged computer programs for performing a contingency table analysis are available at most computer facilities, or can be done on a home computer with the proper statistical software. The SAS, MINITAB, and SPSS computer program packages are referenced in the text and are commonly used for contingency table analysis. All of these programs, and others which may be available at a specific computer facility, produce the same basic information. The reader need only become familiar with the different printouts in order to be able to interpret the results.

Example 12.6
Refer to Example 12.3. The following computer printout resulted when the data in this example were analyzed using the SAS program.

STATISTICAL ANALYSIS SYSTEM

TABLE OF ARRESTS BY EDUCATION

ARRESTS FREQUENCY EXPECTED	A	B	C	D	TOTAL
1	55 45.4	40 43.0	43 43.0	30 36.6	168
2	15 21.6	25 20.5	18 20.5	22 17.4	80
3	7 10.0	8 9.5	12 9.5	10 8.0	37
TOTAL	77	73	73	62	285

STATISTICS FOR 2-WAY TABLES

```
CHI-SQUARE                       10.227     DF = 6     PROB = .1154
PHI                                 .189
CONTINGENCY COEFFICIENT             .186
CRAMER'S V                          .109
LIKELIHOOD RATIO CHI-SQUARE                 DF =       PROB =
```

Discussion

The TABLE OF ARRESTS BY EDUCATION, which appears first in the printout, gives observed and estimated expected cell counts for each cell (rounded to one decimal place) as well as row and column totals. The program allows the user to enter specific names for the designations "ROW" and "COLUMN" in its printout. In particular, we have asked that the word "ROW" be replaced by "_____" and that the word "COLUMN" be replaced by "_____."

ARRESTS
EDUCATION

Directly below the table, several statistics are given. The first is CHI-SQUARE, which is the value of the test statistic calculated in Section 12.4. In this example, $X^2 = $ _____ with DF = _____ degrees of freedom. In the same row, the value PROB = .1154 is the observed significance level for the test. Using this portion of the printout, the student may test the hypothesis of independence of the two classifications. The quantities listed below the line labeled CHI-SQUARE are descriptive statistics used in measuring the degree of association or dependence between the two qualitative variables. They often appear on computer printouts, but will not be discussed in this text. The interested student may check the references at the end of Chapter 12 in the text.

10.227; 6

Example 12.7

Refer to Example 12.3. The following computer printouts resulted when the data were analyzed using the MINITAB and SPSS programs.

1. *MINITAB:*

```
MTB > PRINT C1-C4

ROW      A        B        C        D

 1       55       40       43       30
 2       15       25       18       22
 3        7        8       12       10
```

MTB > CHISQUARE C1-C4

Expected counts are printed below observed counts

	A	B	C	D	Total
1	55	40	43	30	168
	45.39	43.03	43.03	36.55	
2	15	25	18	22	80
	21.61	20.49	20.49	17.40	
3	7	8	12	10	37
	10.00	9.48	9.48	8.05	
Total	77	73	73	62	285

ChiSq = 2.035 + 0.214 + 0.000 + 1.173 +
        2.024 + 0.992 + 0.303 + 1.214 +
        0.898 + 0.230 + 0.672 + 0.473 = 10.227

df = 6

2. SPSS:

		EDUCATION				
COUNT ROW PCT COL PCT TOT PCT		A	B	C	D	ROW TOTAL
ARRESTS	1	55 32.7 71.4 19.3	40 23.8 54.8 14.0	43 25.6 58.9 15.1	30 17.9 48.4 10.5	168 58.9
	2	15 18.8 19.5 5.3	25 31.2 34.2 8.8	18 22.5 24.7 6.3	22 27.5 35.5 7.7	80 28.1
	3	7 18.9 9.1 2.5	8 21.6 11.0 2.8	12 32.4 16.4 4.2	10 27.0 16.1 3.5	37 13.0
COLUMN TOTAL		77 27.0	73 25.6	73 25.6	62 21.8	285 100.0

CHI-SQUARE   D.F.   SIGNIFICANCE   MIN E.F.   CELLS WITH E.F. <5
  10.227   6   0.115412   8.05   NONE

Discussion

The MINITAB printout gives the observed and estimated expected cell counts in the body of the table, as did the SAS printout. The chi-square statistic is calculated below the table, with the individual elements in the sum listed as $(n_{ij} - \hat{E}(n_j))^2/\hat{E}(n_{ij})$. For example, the contribution to X^2 made by row 1, column 1 is

55 — 45.39; 2.035

$$\frac{(n_{11} - \hat{E}(n_{11}))^2}{\hat{E}(n_{11})} = \frac{(\underline{\quad})^2}{45.39} = \underline{\quad}$$

ChiSq

is not

Then $X^2 = 10.227$ is given by the label _____. Notice that the level of significance (is, is not) given on the MINITAB printout. The experimenter must consult Table 5, Appendix III, to obtain an appropriate rejection region for the test.

does not
does

The SPSS printout (does, does not) display the estimated expected cell counts. It (does, does not) display in each cell the observed cell count, the cell count as a percentage of the row total, as a percentage of the column total, and as a percentage of the total observations, n. The value of the test statistic X^2 is

CHI-SQUARE

labeled _____ and is shown directly below the table along with the appropriate degrees of freedom and significance level.

The minimum expected frequency (MIN E.F.) in the table is given by the entry 8.05; hence, the number of CELLS WITH E.F. less than 5 is reported as NONE. Other statistics, such as CRAMER's V LAMBDA and KENDALL'S

TAU may be requested by the user and will appear below the line labeled CHI-SQUARE.

Self-Correcting Exercises 12C

1. A survey of voter sentiment was conducted in four mid-city political wards to compare the fraction of voters favoring a "city manager" form of government. Random samples of 200 voters were polled in each of the four wards with results as follows:

	\multicolumn{4}{c}{Ward}
	1	2	3	4
Favor	75	63	69	58
Against	125	137	131	142

Can you conclude that the fractions favoring the city manager form of government differ in the four wards?

2. A personnel manager of a large company investigating employee satisfaction with their assigned jobs collected the following data for 200 employees in each of four job categories:

Satisfaction	I	II	III	IV	Totals
High	40	60	52	48	200
Medium	103	87	82	88	360
Low	57	53	66	64	240
Totals	200	200	200	200	800

Use the MINITAB printout given below. Do these data indicate that the satisfaction scores are dependent on the job categories? (Use $\alpha = .05$.)

```
MTB > CHISQUARE C1-C4

Expected counts are printed below observed counts

             C1        C2        C3        C4     Total
   1         40        60        52        48       200
            50.00     50.00     50.00     50.00

   2        103        87        82        88       360
            90.00     90.00     90.00     90.00

   3         57        53        66        64       240
            60.00     60.00     60.00     60.00

Total       200       200       200       200       800

ChiSq  =   2.000  +   2.000  +   0.080  +   0.080  +
           1.878  +   0.100  +   0.711  +   0.044  +
           0.150  +   0.817  +   0.600  +   0.267  =  8.727

df = 6
```

12.7 Other Applications

The specific uses of the chi-square test that we have dealt with in this chapter can be divided into two categories:

1. The first category is called "goodness-of-fit tests," whereby observed frequencies are compared with hypothesized frequencies which depend upon the hypothesized cell probabilities for a multinomial probability distribution. A decision is made as to whether the data fit the hypothesized model.
2. The second category is called "tests of independence," whereby a decision is made as to whether two methods of classifying the observations are statistically independent. If it is decided that the classifications are independent, then the probability that an observation would be classified as belonging to a specific row classification would be constant across the columns, or vice versa. If it is decided that the classifications are not independent, then the implication is that the probability that an observation would be classified as belonging to a specific row classification varies from column to column.

To illustrate the general nature of the goodness-of-fit test, we could test whether a set of data comes from any specified distribution such as the normal distribution with mean μ and variance σ^2, or a binomial distribution based on n trials with probability of success p, or perhaps a Poisson distribution with mean λ. Binomial data produce their own natural grouping corresponding to the cells of a multinomial experiment if one counts the number of zeros, ones, twos, and so on occurring in the data. If the expected cell frequencies are less than the required number, cells can be combined before using the chi-square statistic. Data from a normal distribution, on the other hand, do not produce an inherent natural grouping and must be grouped as in a frequency histogram. In conjunction with a table of normal curve areas and the hypothesized normal distribution, the boundary points for the histogram should be chosen so that each "cell" has approximately the same probability and an expected frequency greater than _____ [5]. Grouping the sample data accordingly, one can compare the "observed" group frequencies against the theoretical ones using Pearson's chi-square statistic. If population parameters need to be estimated, the point estimates given in earlier chapters are used and _____ [one] degree of freedom subtracted for each independent estimate.

Tests of independence of two methods of classification are easily extended to three or more classifications by first estimating the expected cell frequencies and applying the chi-square statistic with the proper degrees of freedom. For example, in testing the independence of three classifications with c_1, c_2, and c_3 categories in the respective classifications, the test statistic would be

$$\chi^2 = \sum_{i=1}^{c_1} \sum_{j=1}^{c_2} \sum_{k=1}^{c_3} \frac{[n_{ijk} - \hat{E}(n_{ijk})]^2}{\hat{E}(n_{ijk})},$$

which has an approximate χ^2 distribution with $(c_1 - 1)(c_2 - 1)(c_3 - 1)$ degrees of freedom.

Further applications involving the X^2 test are usually specifically tailored solutions to special problems. An example would be the test of a linear trend in a binomial proportion observed over time, as discussed in the text. Modifications such as this usually require the use of calculus and are beyond the scope of this text.

An alternate approach to the analysis of contingency tables is based on *log-linear models,* in which the logarithm of the cell probabilities $ln\, p_{ij}$ is assumed to be a linear function of row and column parameters. The analysis is somewhat more complicated than that presented here, and almost always requires the use of a computer and appropriate computer programs. However, the analysis of log-linear models is very flexible, and hence can accommodate a wide variety of restrictions and sampling situations. An excellent presentation of log-linear modelling can be found in *Discrete Multivariate Analysis: Theory and Practice,* by Y. M. M. Bishop, S. E. Fienberg, and P. W. Holland (Massachusetts Institute of Technology Press, 1975).

Self-Correcting Exercises 12D

1. A company producing wire rope has recorded the number of "breaks" occurring for a given type of wire rope within a 4-hour period. These records were kept for fifty 4-hour periods. If x is the number of "breaks" recorded for each 4-hour period and μ is the mean number of "breaks" for a 4-hour period, does the following Poisson model adequately describe these data when $\mu = 2$?

$$p(x) = \frac{\mu^x e^{-\mu}}{x!}, \quad x = 0, 1, 2, \ldots.$$

x	0	1	2	3 or more
Number observed	4	15	16	15

Hint: Find $p(0)$, $p(1)$, and $p(2)$. Use the fact that

$$P(x \geq 3) = 1 - p(0) - p(1) - p(2).$$

After finding the expected cell numbers, you can test the model by applying Pearson's chi-square test.

2. In standardizing a score, the mean is subtracted and the result divided by the standard deviation. If 100 scores are so standardized and then grouped, test, at the $\alpha = .05$ level of significance, whether these scores were drawn from the standard normal distribution.

Interval	Frequency
less than −1.5	8
−1.5 to −.5	20
−.5 to .5	40
.5 to 1.5	29
greater than 1.5	3

12.8 Assumptions

In order that the statistic

$$X^2 = \sum_{i=1}^{k} \frac{[n_i - E(n_i)]^2}{E(n_i)}$$

have an approximate χ^2 distribution, the following assumptions are made:
1. The cell counts, $n_1, n_2, \ldots, n_k$, must satisfy the conditions of a _____ experiment (or several multinomial experiments).
2. All expected cell counts should be at least _____.

Although valid multinomial data arise under various sampling plans, in order to be confident in the use of the X^2 statistic, the sample size should be large enough to ensure that all the expected cell counts are 5 or more. This is a conservative figure; some authors have stated that some expected cell counts can be as small as one. By asking for expected cell counts of 5 or more, we automatically satisfy experimental situations in which these counts can in fact be allowed to be less than 5. In so doing, we should realize that sensitivity may be sacrificed for the sake of safety and simplicity.

multinomial
5

EXERCISES

1. What are the characteristics of a multinomial experiment?
2. Do the following situations possess the properties of a multinomial experiment?
    a. A large number of red, white, and blue flower seeds are thoroughly mixed and a sample of $n = 30$ seeds is taken. The numbers of red, white, and blue flower seeds are recorded.
    b. A game of chance consists of picking three balls at random from an urn containing one white, three red, and six black balls. The game "pays" according to the number of white, red, and black balls chosen.
    c. Four production lines are checked for defectives during an 8-hour period and the number of defectives for each production line recorded.
3. The probability of receiving grades of A, B, C, D, and E are .07, .15, .63, .10, and .05, respectively, in a certain humanities course. In a class of 120 students:
    a. What is the expected number of A's?
    b. What is the expected number of B's?
    c. What is the expected number of C's?
4. A department store manager claims that her store has twice as many customers on Fridays and Saturdays than on any other day of the week (the store is closed on Sundays). That is, the probability that a customer visits the store Friday is 2/8, the probability that a customer visits the store Saturday is 2/8, while the probability that a customer visits the store on each of the remaining weekdays is 1/8. During an average week, the following numbers of customers visited the store:

        Monday: 95   Thursday: 75
        Tuesday: 110   Friday: 181
        Wednesday: 125 Saturday: 214

Can the manager's claim be refuted at the $\alpha = .05$ level of significance?

5. If the probability of a female birth is 1/2, according to the binomial model, in a family containing four children, the probability of 0, 1, 2, 3, or 4 female births is 1/16, 4/16, 6/16, 4/16, and 1/16, respectively. A sample of 80 families each containing four children resulted in the following data:

Female births	0	1	2	3	4
Number of families	7	18	33	16	6

Do the data contradict the binomial model with $p = 1/2$ at the $\alpha = .05$ level of significance?

6. A serum thought to be effective in preventing colds was administered to 500 individuals. Their records for 1 year were compared to those of 500 untreated individuals, with the following results:

	No Colds	One Cold	More Than One Cold
Treated	252	146	102
Untreated	224	136	140

Test the hypothesis that the two classifications are independent, at the $\alpha = .05$ level of significance.

7. A manufacturer wished to know whether the number of defectives produced varied for four different production lines. A random sample of 100 items was selected from each line and the number of defectives recorded:

Production lines	1	2	3	4
Defectives	8	12	7	9

Do these data produce sufficient evidence to indicate that the percentage of defectives varies from line to line?

8. In a random sample of 50 male and 50 female undergraduates, each member was asked if he or she was for, against, or indifferent to the practice of having unannounced in-class quizzes. Do the following data indicate that attitude toward this practice is dependent on the sex of the student interviewed?

	Male	Female
For	20	10
Against	15	30
Indifferent	15	10

9. In an experiment performed in a laboratory, a ball is bounced with a container whose bottom (or floor) has holes just large enough for the ball to pass through. The ball is allowed to bounce until it passes through one of the holes. For each of 100 trials, the number of bounces until the ball

falls through one of the holes is recorded. If x is the number of bounces until the ball does fall through a hole, does the model

$$p(x) = (.6)(.4)^x, \quad x = 0, 1, 2, 3, \ldots$$

adequately describe the following data?

x	0	1	2	3 or more
Number observed	65	28	4	3

Hint: First find $p(0), p(1), p(2)$, and $P(x \geq 3)$ from which the expected numbers for the cells can be calculated using np_0, np_1, np_2, and so on. Then a goodness-of-fit test will adequately answer the question posed.

CHAPTER 13
EXPERIMENTAL DESIGN AND THE ANALYSIS OF VARIANCE

13.1 The Design of an Experiment

We have stated that the objective of statistics is to make inferences about a population from information contained in a sample drawn from that population. However, we wish to make these inferences using the maximum amount of information obtained in an experiment at a fixed cost, or possibly at the *least* cost.

An *experiment* is the process by which an observation or measurement is obtained.

The design of an experiment involves deciding how many observations will be taken, and the manner in which these observations will be taken to maximize the sample information at the least cost. The amount of information in a sample is determined by two factors.

1. The size of the sample or samples used in the experiment,
2. The amount of variation in the data.

For example, a $(1 - \alpha)100\%$ confidence interval estimate of the mean of a normal population is given by

$$\bar{x} \pm \underline{\qquad}.\qquad\qquad z_{\alpha/2}\sigma/\sqrt{n}.$$

The width of this interval is equal to

$$w = 2(\underline{\qquad})\qquad\qquad (z_{\alpha/2}\sigma/\sqrt{n})$$

As the width of this interval decreases, the (less, more) precise is the estimate   more
of μ; hence, the length of a confidence interval estimator is a measure of the amount of information in the sample. Notice that, for a fixed level of confidence, the length of the interval can be decreased by either

1. Increasing the sample size n, or
2. Decreasing the variability σ^2

300 / Chap. 13: Experimental Design and the Analysis of Variance

increases
increases

or both. In general, the length of a confidence interval increases as the population standard deviation σ (decreases, increases), and decreases as the sample size n (decreases, increases).

In addition to deciding on the sample sizes to be used in the experiment, the design of an experiment focuses on ways to control the variation in the data. In the design of an experiment, the individual or object on which an observation or measurement is made is called an _____.

experimental unit

Treatments are procedures or techniques applied to the experimental units; their effects on these units are to be estimated and compared with other procedures.

Treatments might be surgical techniques, drug protocols, pharmaceutical suppliers, engine designs, or irrigation methods. Treatments involve one or more *independent experimental variables* called *factors,* whose values are controlled and varied by the experimenter. The amount or intensity setting of a factor is called a _____. If pressure were a factor in an experiment and the pressure settings were 25, 50, and 100 pounds per square inch, then the factor "pressure" had _____ levels. If an experiment involved the northeast, southeast, and midwest geographic regions of the United States, then the factor "regions" had _____ levels.

level

three

three

A *treatment* is a specific combination of factor levels.

Designing an experiment consists of four steps.

1. Specifying the factors to be included in the experiment and the number of levels of each factor. In short, we must decide on which treatments are to be included in the experiment.
2. Determining the accuracy required in testing and estimation.
3. Determining the number of experimental units to be included for each treatment, based on the accuracy in estimation required.
4. Specifying the manner in which the treatments will be allocated to the experimental units.

Designs can be classified as either *volume increasing* or *noise reducing* designs. Volume increasing designs are those that minimize the length of a confidence interval for a fixed number of observations. Noise reducing designs reduce the variation in the data by isolating sources of variation that can be controlled by the experimenter. The paired difference experiment introduced in Chapter 9 is an example of a _____ design called a *randomized block design;* this design will be presented in Section 13.6. The next several examples demonstrate how the amount of information can be increased for a fixed number of observations; hence, these are _____ designs.

noise reducing

volume increasing

Example 13.1
In an experiment to compare two population means, the variances in the population are known to be $\sigma^2 = 100$ and $\sigma^2 = 25$. Compare the standard deviation of $(\bar{x}_1 - \bar{x}_2)$ in estimating $(\mu_1 - \mu_2)$ when

a. $n_1 = 15, n_2 = 15$
b. $n_1 = 20, n_2 = 10$.

Solution
The standard deviation of $(\bar{x}_1 - \bar{x}_2)$ is

$$\sigma_{(\bar{x}_1 - \bar{x}_2)} = \sqrt{\frac{\sigma_1^2}{n_1} + \frac{\sigma_2^2}{n_2}}$$

a. When $n_1 = n_2 = 15$,

$$\sigma_{(\bar{x}_1 - \bar{x}_2)} = \sqrt{\frac{100}{15} + \frac{25}{15}} = \sqrt{125/15} = \sqrt{8.3}$$

b. When $n_1 = 20$ and $n_2 = 10$,

$$\sigma_{(\bar{x}_1 - \bar{x}_2)} = \sqrt{\frac{100}{20} + \frac{25}{10}} = \sqrt{150/20} = \sqrt{7.5}$$

The allocation in part b produced the (smaller, larger) standard deviation when $n_1 + n_2 = 30$. | smaller

In general, $\sigma_{(\bar{x}_1 - \bar{x}_2)}$ is a *minimum* when $n_1/n_2 = \sigma_1/\sigma_2$, so that the ratio of the sample sizes is the same as the ratio of the standard deviations. Notice that this implies that equal allocation is best only if σ_1 (is, is not) equal to σ_2. | is

Example 13.2
Sixty experimental units are available to compare the means of two populations, A and B. What is the optimal allocation of the units if

a. $\sigma_A^2 = \sigma_B^2$?

b. $\sigma_A^2 = 100, \sigma_B^2 = 25$?

c. $\sigma_A^2 = 64, \sigma_B^2 = 4$?

Solution
a. For the case of equal variances, $n_A = n_B = n$ implies that $n_A = n_B$ = _____. | 30

b. Since $\sigma_A = 10$ and $\sigma_B = 5$, twice as many units should be allocated to population A as to population B. Hence, $n_A = $ _____ and n_B = _____. | 40; 20

c. Since $\sigma_A = 8$ and $\sigma_B = 2$, population A should receive four times as many

302 / Chap. 13: Experimental Design and the Analysis of Variance

48
12

units. That is, one-fifth of the total will be allocated to population B and four-fifths to population A. Hence, $n_A = $ _____ and $n_B = $ _____.

Suppose we are interested in estimating the slope of the line, β_1, in the linear model

$$y = \beta_0 + \beta_1 x + \epsilon$$

by the method of least squares. When we can control the n values of x for which y will be observed, we should choose those values of x so that the width of the confidence interval estimate will be a minimum. We seek to choose x to minimize

$$\sigma_{\hat{\beta}_1} = \sqrt{\frac{\sigma^2}{\Sigma(x_i - \bar{x})^2}}.$$

decreases

Note that as $\Sigma(x_i - \bar{x})^2$ increases, $\sigma_{\hat{\beta}_1}$ (increases, <u>decreases</u>). The experimenter will specify the largest and smallest values of x that are of interest, and it will remain for us to allocate the experimental units over this range.

Example 13.3
In fitting a least-squares line, $n = 5$ measurements are to be taken with the objective being the estimation of the slope β_1 with a maximum amount of information. The range of interest for x lies between 1 and 5. Compute $\sigma_{\hat{\beta}_1}$ for the following allocations:
1. Equal spacing of the five values of x.
2. Two points at each end and one in the center.

Solution

15
55

10

1. The five values of x are 1, 2, 3, 4, and 5. Hence, $\Sigma x_i = $ _____ and $\Sigma x_i^2 = $ _____, so that

$$\Sigma(x_i - \bar{x})^2 = \Sigma x_i^2 - (\Sigma x_i)^2/5 = 55 - 45 = \underline{\qquad}.$$

Then

$$\sigma_{\hat{\beta}_1} = \sigma/\sqrt{10}.$$

15
61

16

2. The five values of x are 1, 1, 3, 5, and 5 so that $\Sigma x_i = $ _____ and $\Sigma x_i^2 = $ _____. Then

$$\Sigma(x_i - \bar{x})^2 = 61 - (225/5) = \underline{\qquad}$$

and

$$\sigma_{\hat{\beta}_1} = \sigma/\sqrt{16}.$$

2; second

Note that $\sigma_{\hat{\beta}_1}$ is smaller for part (1, <u>2</u>) so that the (first, <u>second</u>) allocation represents a gain in information.

In general, the smallest value of $\sigma_{\hat{\beta}_1}$ occurs when the n data points are equally divided with half at the lower boundary of x and the other half at the upper boundary of x. A few points should be selected near the center of the experimental region to detect curvature if it is present.

Self-Correcting Exercises 13A

1. In comparing two population means with $\sigma_1^2 = 36$ and $\sigma_2^2 = 16$, what is the optimal allocation of $n = 50$ units between the two populations?
2. A chemist wishes to study the effect of temperatures between 120°F and 300°F on the reaction rate for a given chemical. She suspects that a linear relationship exists between temperature and rate of reaction and plans to run the experiment at ten temperature settings.
   a. For equal spacings of temperature (120°, 140°, ..., 300°), what is the standard deviation of $\hat{\beta}_1$ if $\sigma = 10$?
   b. If five settings are taken at 120° and five at 300°, calculate $\sigma_{\hat{\beta}_1}$ if $\sigma = 10$.
   c. How much larger is $\sigma_{\hat{\beta}_1}$ in a than in b?
   d. Which is the optimal allocation if the estimation of β_1 is of primary interest?

13.2 The Analysis of Variance

To perform an analysis of variance is to partition the total variation in a set of measurements, given by

$$\sum_{i=1}^{n} (x_i - \bar{x})^2,$$

into portions associated with each independent variable in the experiment as well as a remainder attributable to random error. Let us investigate the partitioning of the total variation into two components with the following example.

Example 13.4
The impurities in parts per million were recorded for five batches of chemicals supplied by two different suppliers.

	Supplier 1		Supplier 2
	25		32
	33		43
	42		38
	27		47
	36		30
Sum	163	Sum	190
$\bar{x}_1 =$ ____		$\bar{x}_2 =$ ____	

32.6; 38

It will save confusion if we use two subscripts to identify each observation rather than just one. Let x_{ij} designate the jth observation recorded in the ith

304 / Chap. 13: Experimental Design and the Analysis of Variance

sample. When i is either 1 or 2, j can take the values 1, 2, 3, 4, or 5. We could then write

Supplier 1	Supplier 2
$x_{11} = 25$	$x_{21} = 32$
$x_{12} = 33$	$x_{22} = 43$
$x_{13} = 42$	$x_{23} = 38$
$x_{14} = 27$	$x_{24} = 47$
$x_{15} = 36$	$x_{25} = 30$

1. Total variation: Let us consider all measurements as one large sample of size 10. Then the total of the 10 measurements is

$$\sum_{i=1}^{2} \sum_{j=1}^{5} x_{ij} = 353$$

and the grand mean is

35.3

$$\bar{x} = 353/10 = \underline{\qquad}.$$

The total variation then is given by

$$\text{Total SS} = \sum_i \sum_j (x_{ij} - \bar{x})^2$$

$$= \sum_i \sum_j x_{ij}^2 - (\sum_i \sum_j x_{ij})^2/10$$

$$= 12{,}929 - (353)^2/10$$

$$= 12{,}929 - 12{,}460.9$$

468.1

$$= \underline{\qquad}.$$

This total sum of squares will be partitioned into two sources of variation: treatments and error.

2. Treatment variation: Recall that the variance of a sample mean is given to be σ^2/n, where n is the number of observations used to calculate the mean and σ^2 is the variance of the population of measurements sampled. Suppose we had two samples of size n from the same population. Then if $\bar{x}$ is the grand mean, the sample variance of the means, calculated as

$$s_{\bar{x}}^2 = \sum_{i=1}^{2} (\bar{x}_i - \bar{x})^2/(2 - 1),$$

estimates σ^2/n with one degree of freedom. If we multiply the sum of squares,

$$\sum_{i=1}^{2}(\bar{x}_i - \bar{x})^2$$

by n, we return this sum of squares to a "per measurement" basis. Then, the sum of squares due to variation of the treatment means will be

$$n\sum_{i=1}^{2}(\bar{x}_i - \bar{x})^2.$$

If the sample sizes are not equal, then the sum of squares for treatments is modified to be

$$SST = \sum_{i=1}^{2} n_i(\bar{x}_i - \bar{x})^2.$$

As the difference between the sample means increases, this sum of squares also _____. For the problem at hand, $p = 2$ and

| | increases |

$$\bar{x}_1 = 32.6, \bar{x}_2 = 38, \bar{x} = 35.3$$

$$n_1 = n_2 = 5.$$

Therefore the treatment sum of squares is

$$SST = n_1(\bar{x}_1 - \bar{x})^2 + n_2(\bar{x}_2 - \bar{x})^2$$

$$= 5(32.6 - 35.3)^2 + 5(38 - 35.3)^2$$

$$= 5(-2.7)^2 + 5(2.7)^2$$

$$= 5(7.29) + 5(7.29)$$

$$= 2(36.45)$$

$$= \underline{\qquad}.$$

| | 72.9 |

3. Error variation: If the two samples have come from the same population we can use a pooled estimate of error given by

$$SSE = \sum_{j=1}^{5}(x_{1j} - \bar{x}_1)^2 + \sum_{j=1}^{5}(x_{2j} - \bar{x}_2)^2$$

For sample 1,

$$\Sigma (x_{1j} - \bar{x}_1)^2 = \Sigma x_{1j}^2 - (\Sigma x_{1j})^2/5$$
$$= 5{,}503 - (163)^2/5$$
$$= 5{,}503 - 5{,}313.8$$
$$= \underline{}.$$

189.2

For sample 2,

$$\Sigma (x_{2j} - \bar{x}_2)^2 = \Sigma x_{2j}^2 - (\Sigma x_{2j})^2/5$$
$$= 7{,}426 - (190)^2/5$$
$$= 7{,}426 - 7{,}220$$
$$= \underline{}.$$

206

It follows that

$$\text{SSE} = 189.2 + 206 = \underline{}.$$

395.2

4. Therefore we see directly that

$$\text{SST} = 72.9$$
$$\text{SSE} = 395.2$$
$$\text{Total SS} = 468.1$$

and that

$$\text{Total SS} = \text{SST} + \text{SSE}.$$

Since simpler calculational forms will be given presently, we defer further calculations until then.

The F Test and the Analysis of Variance

For the two-sample problem discussed in Example 13.4, the t statistic is readily available for testing the hypothesis

$$H_0: \mu_1 = \mu_2$$

versus

$$H_a: \mu_1 \neq \mu_2$$

The two-sample unpaired t test requires that both samples can be drawn randomly and independently from two normal populations with equal variances. Using these same assumptions, we can construct an F statistic, which is the ratio of two variances, to test these same hypotheses. The advantage to using the F statistic is that the procedure can be easily extended for testing the equality of several population means.

If $H_0: \mu_1 = \mu_2$ is true, then the partitioning of the total sum of squares provides us with two estimators of the common variance σ^2.

1. $\quad\quad\quad \text{MSE} = \text{SSE}/(n_1 + n_2 - 2)$

   where

   $$\text{SSE} = \sum_{j=1}^{n_1} (x_{1j} - \bar{x}_1)^2 + \sum_{j=1}^{n_2} (x_{2j} - \bar{x}_2)^2$$

   with $n_1 + n_2 - 2$ degrees of freedom.

2. $\quad\quad\quad \text{MST} = \text{SST}/(2 - 1)$

   where

   $$\text{SST} = n_1 (\bar{x}_1 - \bar{x})^2 + n_2 (\bar{x}_2 - \bar{x})^2$$

   with $2 - 1$ degrees of freedom. Therefore, when H_0 is true,

   $$F = \text{MST}/\text{MSE}$$

   has an F distribution with $\nu_1 =$ _____ and $\nu_2 =$ _____ degrees of freedom. $\quad 1; n_1 + n_2 - 2$

   If H_0 is false and $H_a: \mu_1 \neq \mu_2$ is true, this fact should be reflected in MST, and MST should in probability be larger than _____. This $\quad$ MSE
   implies that if H_0 is false, the F ratio,

   $$F = \text{MST}/\text{MSE},$$

   will be too large. Hence, the rejection region for this test will consist of all values of F satisfying

   $$F > F_\alpha$$

   where F_α is the right-tailed critical value of F based on $\nu_1 = 1$ and $\nu_2 = n_1 + n_2 - 2$ degrees of freedom having an area of α to its right.

Example 13.5
For the data in Example 13.4, test the null hypothesis $H_0: \mu_1 = \mu_2$ versus $H_a: \mu_1 \neq \mu_2$ at the $\alpha = .05$ level of significance.

308 / Chap. 13: Experimental Design and the Analysis of Variance

Solution
Let us first gather the information that we have compiled so far.

SST = _____ 72.9; 72.9	MST = 72.9/1 = _____
SSE = _____ 395.2; 49.4	MSE = 395.2/8 = _____

1. $H_0: \mu_1 = \mu_2$ versus $H_a: \mu_1 \neq \mu_2$.
2. The test statistic will be

$$F = MST/MSE$$

with $\nu_1 = $ _____ and $\nu_2 = $ _____ degrees of freedom. [1; 8]

3. Rejection region: For $\alpha = .05$, a right-tailed value of F with $\nu_1 = 1$ and $\nu_2 = 8$ degrees of freedom is $F_{.05} = $ _____ [5.32]. Therefore, we will reject H_0 if $F > $ _____ [5.32].

4. Using the sample values,

$$F = 72.9/49.4 = \underline{\quad\quad} \text{ [1.48]}$$

Since this value is (less, greater) than 5.32, we (reject, do not reject) the null hypothesis. There is not sufficient evidence to indicate that $\mu_1 \neq \mu_2$. [less; do not reject]

Had we tested using the t statistic,

$$t = \frac{(\bar{x}_1 - \bar{x}_2) - 0}{\sqrt{s^2 \left(\frac{1}{n_1} + \frac{1}{n_2}\right)}}$$

with $n_1 + n_2 - 2 = 8$ degrees of freedom, the calculated value would have been

$$t = \frac{32.6 - 38.0}{\sqrt{49.4 \left(\frac{1}{5} + \frac{1}{5}\right)}}$$

$$= -5.4/\sqrt{19.76}$$

$$= -5.4/4.445 = \underline{\quad\quad} \text{ [−1.215]}$$

The rejection region would have consisted of values of t such that $|t| > t_{.025} = $ _____ [2.306]. Hence, we would not have rejected H_0: $\mu_1 = \mu_2$ even had we used the t statistic. Noting that

$$t^2 = (-1.215)^2 = 1.48 = \underline{\quad\quad} \text{ [}F\text{]}$$

and

$$(t_{.025})^2 = (2.306)^2 = 5.32 = \underline{\quad\quad} \text{ [}F_{.05}\text{]}$$

the results should be identical. This verifies the fact that an F with $v_1 = 1$ and v_2 degrees of freedom is the same as t^2 with v_2 degrees of freedom; further, $F_\alpha = t^2_{\alpha/2}$ only if $v_1 =$ _____ .

1

13.3 The Completely Randomized Design

In extending the problem of testing for a significant difference between two population means to one of testing for significant differences among several population means, let us consider an experiment in which independent random samples have been drawn from each of k populations.

A *completely randomized design* involves the selection of randomly and independently drawn samples from each of k populations.

Consider an experiment designed to compare the growth of rats subjected to three specific diets, I, II, and III, for a given length of time. Fifteen rats have been randomly divided into three groups of five and each group randomly assigned to receive one of the diets. Since each rat is "treated" or subjected to one of the three diets, in statistical terminology the diets are called _____ . This type of randomization procedure is called a _____ _____ design.

treatments
completely randomized

The completely randomized design involves one independent variable, treatments. In this design the total sum of squares of deviations of the measurements about their overall mean can be partitioned into two parts.

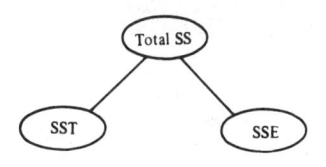

```
        Total SS
         /    \
       SST    SSE
```

Assumptions for the Analysis of Variance of a Completely Randomized Design:

1. Independent random samples have been drawn from k normal populations with means $\mu_1, \mu_2, \ldots, \mu_k$.
2. The variability of the measurements in each of the sampled populations is equal to σ^2.

Let T_i and $\overline{T}_i$ be the sum and the mean of the n_i observations in the sample from the ith population, with $n = n_1 + n_2 + \ldots + n_k$ being the total number of observations. Then we have the following:

1. $$\text{Total SS} = \sum_{i=1}^{k} \sum_{j=1}^{n_i} (x_{ij} - \overline{x})^2$$

with _____ degrees of freedom. $n - 1$

2. $$\text{SST} = \sum_{i=1}^{k} n_i(\overline{T}_i - \overline{x})^2$$

with _____ degrees of freedom. $k - 1$

3. $$\text{SSE} = \sum_{i=1}^{k} \sum_{j=1}^{n_i} (x_{ij} - \overline{T}_i)^2$$

with $\sum_{i=1}^{k} (n_i - 1) = n_1 + n_2 + \ldots + n_k - k =$ _____ degrees of $n - k$

freedom. Not only does

$$\text{Total SS} = \text{SST} + \text{SSE}$$

but the same relationship holds for the degrees of freedom associated with each sum of squares.

$$\text{d.f.}_{\text{Total}} = \text{d.f.}_{\text{treatments}} + \text{d.f.}_{\text{error}}$$

since

$$n - 1 = (k - 1) + (n - k).$$

The formulas actually used for computing these sums of squares are given below. Let

$$\text{CM} = \left(\sum_{i=1}^{k} \sum_{j=1}^{n_i} x_{ij} \right)^2 \bigg/ n = (\text{grand total})^2/n.$$

Then

1. $$\text{Total SS} = \sum_{i=1}^{k} \sum_{j=1}^{n_i} x_{ij}^2 - \text{CM}.$$

2. $$SST = \sum_{i=1}^{k} \frac{T_i^2}{n_i} - CM.$$

3. SSE = Total SS − SST.

Notice that in SST the square of each treatment total is divided by the number of _____ in that total. Although SSE can be computed directly as a pooled sum of squared deviations within each sample, it is computationally easier to use the additivity property, SST + SSE = _____.

	observations
	Total SS

The mean squares for treatments and error are calculated by dividing each sum of squares by its degrees of freedom. Therefore

$$s^2 = MSE = SSE/(_____)$$ $n - k$

and

$$MST = SST/(_____).$$ $k - 1$

If all samples are from the same normal population, then MST and MSE are each estimators of the population variance σ^2. In this case the statistic

$$F = MST/MSE$$

has an _____ distribution with $v_1 = k - 1$ and $v_2 = n - k$ degrees of freedom. F

Consider testing the hypothesis $H_0: \mu_1 = \mu_2 = \ldots = \mu_k$ against the alternative that at least one mean is different from at least one other. *If H_0 is true, then all samples have come from the same normal population* and the statistic

$$F = MST/MSE$$

has the F distribution specified above. However, if H_a is true (at least one of the equalities does not hold), then

$$MST = \frac{1}{k-1}[n_1(\overline{T}_1 - \overline{x})^2 + n_2(\overline{T}_2 - \overline{x})^2 + \ldots + n_k(\overline{T}_k - \overline{x})^2]$$

will in all probability be _____ than MSE and F will tend to be _____ than expected. Hence, H_0 will be rejected for _____ values of F; that is, we will reject H_0 if

	larger
	larger; large

$$F > F_\alpha$$

with $v_1 = k - 1$ and $v_2 = n - k$ degrees of freedom.

Example 13.6
Do the following data provide sufficient evidence to indicate a difference in the means of the three underlying treatment populations?

312 / Chap. 13: Experimental Design and the Analysis of Variance

	Treatment			
	1	2	3	
	3	7	5	
	4	9	4	
	2	8	5	
		7		
T_i	_____	_____	_____	Total = 54
n_i	3	4	3	$n = 10$
$\overline{T}_i$	3	7.75	4.67	

9; 31; 14

Solution

1. We must first partition the total variation into SST and SSE.

$$CM = (54)^2/10 = 2{,}916/10 = \underline{\hspace{1cm}}$$

291.6

a. Total SS = $3^2 + 4^2 + 2^2 + \ldots + 5^2 + 4^2 + 5^2 - 291.6$

$$= 338 - 291.6$$

46.4

$$= \underline{\hspace{1cm}}.$$

b. $$SST = \frac{9^2}{3} + \frac{31^2}{4} + \frac{14^2}{3} - 291.6$$

$$= \frac{81}{3} + \frac{961}{4} + \frac{196}{3} - 291.6$$

$$= 27 + 240.25 + 65.33 - 291.6$$

332.58

$$= \underline{\hspace{1cm}} - 291.6$$

40.98

$$= \underline{\hspace{1cm}}.$$

c. SSE = 46.4 - 40.98

5.42

$$= \underline{\hspace{1cm}}.$$

d. To compute the degrees of freedom, we need the values $n = 10$ and $k = 3$. Hence SST has $k - 1 = \underline{\hspace{1cm}}$ degrees of freedom while SSE has $n - k = \underline{\hspace{1cm}}$ degrees of freedom. The resulting mean squares are

2
7

20.49

MST = 40.98/2 = \underline{\hspace{1cm}}

0.77

MSE = 5.42/7 = \underline{\hspace{1cm}}

2. We are now in a position to test $H_0: \mu_1 = \mu_2 = \mu_3$ versus H_a: at least one equality does not hold.
   a. The test statistic is

   $$F = \text{MST}/\text{MSE}$$

   with $v_1 = $ _____ and $v_2 = $ _____ degrees of freedom. | 2; 7
   b. Rejection region: Using $\alpha = .05$, we will reject H_0 if $F > F_{.05}$
   = _____. | 4.74
   c. Calculate

   $$F = 20.49/.77 = \underline{\quad\quad},$$ | 26.61

   which is (greater, less) than $F_{.05} = 4.74$. Hence we _____ H_0 and conclude that there is evidence to indicate a difference in means for the three treatment populations at the $\alpha = .05$ level of significance. | greater; reject

Example 13.7
The length of time required for kindergarten-age children to assemble a device was compared for four different lengths of preexperiment instructional times. Four students were randomly assigned to each group but two were eliminated during the experiment due to sickness. The data (assembly times in minutes) are shown below:

	.5	1.0	1.5	2.0
	8	9	4	4
	14	7	6	7
	9	5	7	5
	12		8	
T_i	43	21	25	16
$\bar{T}_i$	10.75	7.00	6.25	5.33
n_i	4	3	4	3

Preexperiment Instructional Time (hours)

Do the data present sufficient evidence to indicate a difference in mean time to assemble the device for the four different lengths of instructional time? (Use $\alpha = .01$.)

Solution
1.   CM = $(105)^2/14 = 787.50$.

   a.   Total SS = 895.0 - 787.5 = _____. | 107.5

   b.   SST = $\dfrac{43^2}{4} + \dfrac{21^2}{3} + \dfrac{25^2}{4} + \dfrac{16^2}{3}$ - CM = 63.33.

   c.   SSE = 107.5 - 63.33 = _____. | 44.17

314 / Chap. 13: Experimental Design and the Analysis of Variance

    d. With $n = 14$ and $k = 4$,

21.11
$$MST = SST/(k - 1) = 63.33/3 = \underline{\hspace{1cm}}$$

4.42
$$MSE = SSE/(n - k) = 44.17/10 = \underline{\hspace{1cm}}$$

2. Test of the null hypothesis.
  a. $H_0: \mu_1 = \mu_2 = \mu_3 = \mu_4$.
  b. H_a: at least one equality does not hold.
  c. Test statistic:

$$F = MST/MSE$$

3; 10   with $v_1 = \underline{\hspace{1cm}}$ and $v_2 = \underline{\hspace{1cm}}$ degrees of freedom.
6.55   d. Rejection region: Reject H_0 if $F > F_{.01} = \underline{\hspace{1cm}}$.
  e. For these data,

$$F = MST/MSE = 4.78.$$

do not reject; cannot   Hence, we (reject, do not reject) H_0; we (can, cannot) conclude that sufficient evidence exists to indicate a difference in the mean time to assemble for the four levels of preexperiment instructional time.

13.4 The Analysis of Variance Table for a Completely Randomized Design

The results of an analysis of variance are usually displayed in an analysis of variance (ANOVA or AOV) summary table. The table displays the sources of variation together with the degrees of freedom, sums of squares, and mean squares for each source listed in the table. The results of the F test appear as a final entry in the table. For a completely randomized design, the ANOVA table is as follows:

ANOVA

Source	d.f.	SS	MS	F
Treatments	$k - 1$	SST	MST	MST/MSE
Error	$n - k$	SSE	MSE	
Total	$n - 1$	Total SS		

MST/MSE appears in the F column for Treatments.

This display gives all the pertinent information leading to the F test and further emphasizes the fact that the degrees of freedom and the sums of squares are both additive.

Example 13.8
Display the results of the analysis of the data in Example 13.6.

Solution
We need but collect the results that we have for this example.

ANOVA

Source	d.f.	SS	MS	F	
Treatments	____	40.98	20.49	____	2; 26.61
Error	____	5.42	____		7; 0.77
Total	____	____			9; 46.40

Example 13.9
Display the results of the analysis of the data in Example 13.7.

Solution
Since the term "treatments" is a general way of describing the differences in the sampled populations, we can replace the term "treatments" in this problem by the more descriptive word "times."

ANOVA

Source	d.f.	SS	MS	F	
Times	3	63.33	____	____	21.11; 4.78
Error	10	44.17	____		4.42
Total	____	107.50			13

Computer packages that can be used to perform an analysis of variance are available through your computer center; smaller versions of these packages are also available for personal computers. Your text presents an analysis of variance using the SAS and the MINITAB packages. The MINITAB package includes several commands that will implement an analysis of variance for a completely randomized design. The MINITAB command AOVONEWAY can be used when each sample is stored in a separate column. The following printout using this command contains the analysis of variance table together with the sample means and sample standard deviations as well as the pooled standard deviation.

```
MTB > PRINT C1-C4

  ROW   C1   C2   C3   C4

   1     8    9    4    4
   2    14    7    6    7
   3     9    5    7    5
   4    12         8

MTB > AOVONEWAY C1-C4

ANALYSIS OF VARIANCE
SOURCE   DF     SS      MS     F       p
FACTOR    3   63.33   21.11   4.78   0.026
ERROR    10   44.17    4.42
TOTAL    13  107.50
```

316 / Chap. 13: Experimental Design and the Analysis of Variance

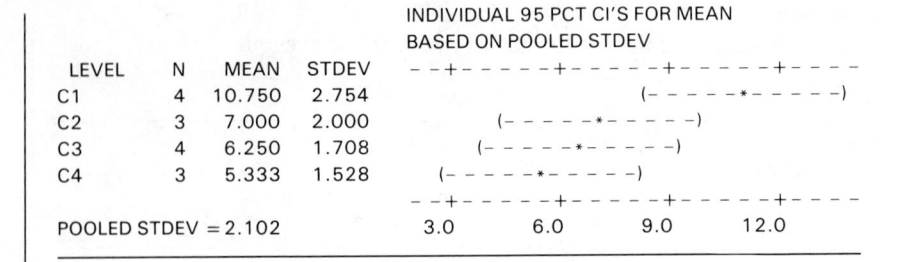

```
                                        INDIVIDUAL 95 PCT CI'S FOR MEAN
                                        BASED ON POOLED STDEV
LEVEL    N     MEAN    STDEV     --+------+------+------+----
C1       4    10.750   2.754                          (------*-----)
C2       3     7.000   2.000             (------*-----)
C3       4     6.250   1.708         (------*-----)
C4       3     5.333   1.528    (------*-----)
                                 --+------+------+------+----
POOLED STDEV = 2.102             3.0    6.0    9.0    12.0
```

The MINITAB command ONEWAY is used when the data are stored in one column and the treatment subscripts (numbers) are stored in another column. The printout using this command is identical to that using the AOVONEWAY command.

```
MTB > PRINT C1 C2

ROW    Y    TIME

 1     8     1
 2    14     1
 3     9     1
 4    12     1
 5     9     2
 6     7     2
 7     5     2
 8     4     3
 9     6     3
10     7     3
11     8     3
12     4     4
13     7     4
14     5     4

MTB > ONEWAY C1 C2

ANALYSIS OF VARIANCE ON Y
SOURCE    DF      SS       MS      F       p
TIME       3    63.33    21.11   4.78   0.026
ERROR     10    44.17     4.42
TOTAL     13   107.50
```

```
                                        INDIVIDUAL 95 PCT CI'S FOR MEAN
                                        BASED ON POOLED STDEV
LEVEL    N     MEAN    STDEV     --+------+------+------+----
  1      4    10.750   2.754                          (------*-----)
  2      3     7.000   2.000             (------*-----)
  3      4     6.250   1.708         (------*-----)
  4      3     5.333   1.528    (------*-----)
                                 --+------+------+------+----
POOLED STDEV = 2.102             3.0    6.0    9.0    12.0
```

13.5 Estimation for the Completely Randomized Design

In using the analysis of variance F test to test for significant differences among a group of population means, an experimenter can conclude that either (a) there is no difference among the means or (b) at least one mean is different from at least one other. In the second case, (b), an experimenter may wish to proceed with estimating the value of a treatment mean or with estimating the difference between two treatment means.

Since the analysis of variance requires that all samples be drawn from _____ populations with a common variance, confidence intervals can be constructed using the t statistic with error degrees of freedom. Hence for estimating the ith treatment mean with a $(1 - \alpha)$ 100% confidence interval, use

normal

$$\bar{T}_i \pm t_{\alpha/2} \sqrt{MSE/n_i},$$

where $\bar{T}_i$ is the ith sample mean, n_i is the number of observations in the ith sample, and MSE is the pooled estimate of σ^2 from the analysis of variance with _____ degrees of freedom.

$n - k$

To estimate the difference between two population means with a $(1 - \alpha)$ 100% confidence interval, use

$$(\bar{T}_i - \bar{T}_j) \pm t_{\alpha/2} \sqrt{MSE\left(\frac{1}{n_i} + \frac{1}{n_j}\right)}.$$

Example 13.10
Refer to Example 13.7. Estimate the mean time to assemble for treatment 1 with a 95% confidence interval.

Solution
The required information can be obtained from Example 13.7 and Example 13.9.

$\bar{T}_1 = 10.75$ d.f. = 10

$n_1 = 4$ $t_{.025} = $ _____ .   2.228

MSE = 4.42

Therefore, the estimate is given by

$10.75 \pm 2.228 \sqrt{4.42/4}$

$10.75 \pm 2.228 ($ _____ $)$ 1.05

$10.75 \pm$ _____   2.34

or

_____ $< \mu_1 <$ _____ .   8.41; 13.09

Example 13.11
Refer to Example 13.7. Construct a 95% confidence interval for $\mu_1 - \mu_2$.

Solution
Collect pertinent information.

$$\bar{T}_1 = 10.75 \qquad \bar{T}_2 = 7.0 \qquad MSE = 4.42 \qquad t_{.025} = 2.228$$

$$n_1 = 4 \qquad n_2 = 3 \qquad d.f. = 10.$$

The confidence interval is found by using

$$(\bar{T}_1 - \bar{T}_2) \pm t_{.025} \sqrt{MSE\left(\frac{1}{n_1} + \frac{1}{n_2}\right)},$$

which when evaluated becomes

$$(10.75 - 7.0) \pm 2.228 \sqrt{4.42\left(\frac{1}{4} + \frac{1}{3}\right)}$$

3.75; 3.58

_____ ± _____ .

Hence, with 95% confidence, we estimate that $\mu_1 - \mu_2$ lies between

.17; 7.33

_____ and _____ minutes.

Self-Correcting Exercises 13B

1. In the evaluation of three rations fed to chickens grown for market, the dressed weights of five chickens fed from birth on one of the three rations were recorded.

	Rations		
	1	2	3
	7.1	4.9	6.7
	6.2	6.6	6.0
	7.0	6.8	7.3
	5.6	4.6	6.2
	6.4	5.3	7.1
Total	32.3	28.2	33.3
Average	6.46	5.64	6.66

   a. Do the data present sufficient evidence to indicate a difference in the mean growth for the three rations as measured by the dressed weights?
   b. Estimate the difference in mean weight for rations 2 and 3 with a 95% confidence interval.

2. In the investigation of a citizens' committee complaint about the availability of fire protection within the county, the distance in miles to the nearest fire station was measured for each of five randomly selected residences in each of four areas.

	Areas				
	1	2	3	4	
	7	1	7	4	
	5	4	9	6	
	5	3	8	3	
	6	4	7	7	
	8	5	8	5	
T_i	31	17	39	25	Total = 112
n_i	5	5	5	5	$n = 20$
$\bar{T}_i$	6.2	3.4	7.8	5.0	

a. Do these data provide sufficient evidence to indicate a difference in mean distance for the four areas at the $\alpha = .01$ level of significance?
b. Estimate the mean distance to the nearest fire station for those residents in area 1 with a 95% confidence interval.
c. Construct a 95% confidence interval for $\mu_1 - \mu_3$.

13.6 The Randomized Block Design

The randomized block design is a natural extension of the _____ _____ experiment. The purpose is to increase the _____ in the design by making comparisons between treatments within relatively homogeneous blocks of experimental material. The randomized block design for k treatments and b blocks assumes blocks of relatively homogeneous material with each block containing _____ experimental units. Each treatment is applied to one experimental unit in each block. Consequently, the number of observations for a given treatment for the entire experiment will equal _____. Thus for the randomized block design, $n_1 = n_2 = \ldots = n_k = b$. A randomized block design for $k = 3$ treatments and $b = 4$ blocks is shown below. Denote the treatments as T_1, T_2, and T_3.

paired-difference; information

k

b

Blocks

1	2	3	4
T_3	T_2	T_1	T_3
T_1	T_1	T_3	T_2
T_2	T_3	T_2	T_1

The total number of observations for a randomized block design with b blocks and k treatments is $n =$ _____.

bk

A randomized block design containing k treatments consists of b blocks of _____ experimental units each. The treatments are randomly assigned to the units in each _____, with each treatment appearing exactly _____ in each block.

k
block
once

320 / Chap. 13: Experimental Design and the Analysis of Variance

Let us look at some situations where a block design can be used to reduce uncontrolled variation.

1. The potencies of several drugs are to be compared by three analysts. If each analyst makes one determination for each drug, the variability of these determinations should be homogeneous for a given analyst. Hence we can consider the _____ as blocks. [analysts]

2. An experiment is to be conducted to assess the relative merits of five different gasolines. Since vehicle-to-vehicle variation is inevitable in such experiments, four vehicles are chosen and each of the five gasolines are used in each vehicle. In this case, we would take _____ as blocks. [vehicles]

3. An experiment to assess the effects of three raw material suppliers and four different mixtures on the crushing strength of concrete blocks is to be run. To eliminate the variability from supplier to supplier, each of the four mixtures is prepared using the material from each of the three suppliers. Thus we have reduced the variability by measuring the crushing strength of the four mixtures in each of three relatively homogeneous blocks, which are the _____. [suppliers]

The word "randomized" means that the _____ are randomly [treatments] distributed over the experimental units within each block. The randomized block design involves two independent variables: _____ and [blocks] _____. For an experiment run in a randomized block design, the [treatments] total variation can now be partitioned into three sources of variation; blocks (B), treatments (T), and error (E).

Total SS
 ├── SSB
 ├── SST
 └── SSE

Randomized block designs prove to be very useful since many investigations involve human (or animal) subjects that exhibit a large subject-to-subject variability.

1. By using a subject as a "_____" and having each subject [block] receive all the treatments in a random order, treatment comparisons made within subjects would exhibit less variation than treatment comparisons made between subjects.

2. Since every subject receives each treatment in some random order, a _____ number of subjects would be required in a randomized [smaller] block design than in a completely randomized design.

13.7 The Analysis of Variance for a Randomized Block Design

In partitioning the sums of squares for an experiment run in a randomized block design with b blocks and k treatments, the calculational formulas for

Total SS and SST remain the same except that now every treatment total will contain exactly b measurements and the total number of observations will be $n = kb$. Hence,

$$CM = \frac{\left(\sum_{i=1}^{k}\sum_{j=1}^{b} x_{ij}\right)^2}{kb} = \frac{(\text{grand total})^2}{kb}$$

$$\text{Total SS} = \sum_{i=1}^{k}\sum_{j=1}^{b} x_{ij}^2 - CM$$

and

$$SST = \frac{\sum_{i=1}^{k} T_i^2}{b} - CM.$$

The calculation of SSB follows the same pattern as the calculation of SST—namely, square each block total; sum the squares of each total; divide by k, the number of observations per total; subtract the correction for the mean. If $B_1, B_2, \ldots, B_b$ represent the block totals, then

$$SSB = \frac{(B_1^2 + B_2^2 + \ldots + B_b^2)}{k} - CM.$$

Using the additivity of the sums of squares, we can find SSE by subtraction:

$$SSE = \text{Total SS} - \underline{\hspace{1cm}} - \underline{\hspace{1cm}}. \qquad \text{SSB; SST}$$

The analysis of variance table for a randomized block design with b blocks and k treatments follows.

ANOVA

Source	d.f.	SS	MS	
Blocks	_____	SSB	SSB/(b - 1)	(b - 1)
Treatments	_____	SST	SST/(k - 1)	(k - 1)
Error	_____	SSE	SSE/(b - 1)(k - 1)	(b - 1)(k - 1)
Total	bk - 1	Total SS		

To test the null hypothesis "there is no difference in treatment means," we use

$$F = MST/MSE,$$

which has an F distribution with $v_1 = k - 1$ and $v_2 = (k - 1)(b - 1)$ degrees of freedom when H_0 is true. If H_0 is false, the statistic will tend to be larger than expected; hence, we would reject H_0 if

$$F > F_\alpha.$$

Although a test of H_0: "there is no difference in block means" is not always required, we can test this hypothesis by using

$$F = \text{MSB/MSE}.$$

When H_0 is true, this statistic has an F distribution with $v_1 = b - 1$ and $v_2 = (b - 1)(k - 1)$ degrees of freedom. H_0 is rejected if $F > F_\alpha$, where F_α is an α level critical value of F with $(b - 1)$ and $(b - 1)(k - 1)$ degrees of freedom.

A significant test of block means provides a method of assessing the efficiency of the experimenter's blocking procedure, since if

$$F = \text{MSB/MSE}$$

increased

is significant, the experimenter has _____ the available information in the experiment by blocking and would certainly use this same technique in subsequent experiments. In situations where subjects are often used as blocks, a nonsignificant test of block means should not be taken as a license to discontinue blocking in subsequent experiments involving different subjects, since the next group of subjects selected for participation could exhibit strong subject-to-subject variability and provide a highly significant test of blocks.

Example 13.12
The readability of four different styles of textbook types was compared using a speed-reading test. The amount of reading material was identical for all four type styles. The sample material for each of the four type styles was read in random order by each of five readers in order to eliminate the natural variation in reading speed between readers. The length of time to completion of reading was measured. Thus, each reader corresponds to a

block

_____ and comparisons of the four styles were made within readers. Do the data present evidence of a difference in mean reading times for the four type styles?

Type Style	Reader 1	2	3	4	5	Total	Mean
1	15	18	13	21	15	82	16.40
2	19	19	16	22	15	91	18.20
3	13	20	14	21	16	84	16.80
4	11	18	12	17	12	70	14.00
Total	58	75	55	81	58	327	
Mean	14.50	18.75	13.75	20.25	14.50		

Solution
Before analyzing these data, it should be pointed out that the order in which the type style was presented was randomized for each reader. The data layout presented *does not* represent the order of presentation for each reader.

1. *Partitioning the Sum of Squares.*

$$CM = (Total)^2/bk = (327)^2/20 = 5{,}346.45.$$

$$\text{Total SS} = \sum_{i=1}^{k}\sum_{j=1}^{b} x_{ij}^2 - CM$$

$$= 5{,}555 - \underline{} \qquad\qquad 5{,}346.45$$

$$= \underline{}. \qquad\qquad 208.55$$

For SST and SSB remember that the respective totals are each squared and divided by the number of measurements per total. Block totals contain $k = 4$ measurements and treatment totals contain $b = 5$ measurements.

$$SSB = \sum_{j=1}^{b} \frac{B_j^2}{k} - CM$$

$$= \frac{58^2 + 75^2 + \ldots + 58^2}{4} - \underline{} \qquad\qquad 5{,}346.45$$

$$= 5{,}484.75 - 5{,}346.45$$

$$= \underline{}. \qquad\qquad 138.30$$

$$SST = \sum_{i=1}^{k} \frac{T_i^2}{b} - CM$$

$$= \frac{82^2 + 91^2 + 84^2 + 70^2}{5} - \underline{} \qquad\qquad 5{,}346.45$$

$$= \underline{}\,\underline{} - 5{,}346.45 \qquad\qquad 5{,}392.20$$

$$= \underline{}. \qquad\qquad 45.75$$

SSE = Total SS - SSB - SST

$$= 208.55 - 138.30 - 45.75$$

$$= \underline{}. \qquad\qquad 24.50$$

2. *The Analysis of Variance Table.*
Complete the following ANOVA table.

ANOVA

Source	d.f.	SS	MS	F
Blocks	4	138.30	_____	_____
Treatments	3	45.75	_____	_____
Error	12	24.50	_____	
Total	19	208.55		

[margin: 34.58; 16.95 / 15.25; 7.48 / 2.04]

In testing H_0: "no difference in treatment means," we use

$$F = MST/MSE = 15.25/2.04 = _____.$$

[margin: 7.48]

With $v_1 = 3$ and $v_2 = 12$ degrees of freedom, $F_{.05} = _____$. Hence we reject H_0 and conclude that there is a significant difference among the mean reading times for the four type styles.

[margin: 3.49]

Because we expected significant differences in mean reading times for the five readers, we used a randomized block design with the readers as blocks. Let us test whether these five readers have significantly different mean reading times. To test H_0: "no difference in block means," we use

$$F = MSB/MSE = 34.58/2.04 = _____.$$

[margin: 16.95]

The 5% critical value of F with $v_1 = 4$ and $v_2 = 12$ degrees of freedom is _____; hence we _____ H_0 and conclude that the mean reading times for the five readers are significantly different. A significant test of blocks indicates that our experiment has been made (more, less) precise by using the randomized block design with readers as blocks.

[margin: 3.26; reject]

[margin: more]

The MINITAB package features several commands that will produce an analysis of variance for a randomized block design. However, these commands require that the data, the treatment subscripts, and the block subscripts be stored in separate columns. The data from Example 13.12 were stored in the following way. Column 1 contained the data (X), column 2 contained the treatment (TRTS) subscripts, and column 3 contained the block (BLOCKS) subscripts.

```
MTB > PRINT C1 C2 C3

ROW    X    TRTS   BLOCKS

 1    15      1      1
 2    19      2      1
 3    13      3      1
 4    11      4      1
 5    18      1      2
 6    19      2      2
 7    20      3      2
 8    18      4      2
```

9	13	1	3
10	16	2	3
11	14	3	3
12	12	4	3
13	21	1	4
14	22	2	4
15	21	3	4
16	17	4	4
17	15	1	5
18	15	2	5
19	16	3	5
20	12	4	5

A randomized block experiment can be thought of as a two-way classification, since every observation is classified by treatment and block. The MINITAB command TWOWAY, followed by the data, treatment and block column numbers, will implement an analysis of variance for a randomized block experiment. As you can see from the following printout, adding the subcommand MEANS followed by C2 and C3 provides an output similar to that for ONEWAY.

```
MTB > TWOWAY C1 C2 C3;
SUBC> MEANS C2 C3.

ANALYSIS OF VARIANCE Y

SOURCE     DF      SS        MS
TRTS        3     45.75    15.25
BLOCKS      4    138.30    34.58
ERROR      12     24.50     2.04
TOTAL      19    208.55
```

```
                   Individual 95% CI
TRTS   Mean     - - - - -+- - - - - - -+- - - - - -+- - - - - -+- - - -
  1    16.40                        (- - -*- - -)
  2    18.20                                  (- - -*- - -)
  3    16.80                         (- - -*- - -)
  4    14.00         (- - -*- - -)
                - - - - -+- - - - - - -+- - - - - -+- - - - - -+- - - -
                       14.00        16.00       18.00       20.00
```

```
                   Individual 95% CI
BLOCKS Mean     - -+- - - - - -+- - - - - -+- - - - - -+- - - - - - - -
  1    14.50        (- - -*- - -)
  2    18.75                         (- - -*- - -)
  3    13.75     (- - -*- - -)
  4    20.25                                  (- - -*- - -)
  5    14.50        (- - - -*- - -)
                - -+- - - - - -+- - - - - -+- - - - - -+- - - - - - - -
                  12.50       15.00       17.50       20.00
```

Another command called ANOVA is more general than the TWOWAY command, and will implement an analysis of variance for various designs. The design factors are entered in the main command as ANOVA DATA =

TREATMENTS BLOCKS;. Aside from the plots of the confidence limits, this printout contains the same information given by the TWOWAY command.

```
MTB > ANOVA C1 = C2 C3;
SUBC> MEANS C2 C3.
```

Factor	Type	Levels	Values				
TRTS	fixed	4	1	2	3	4	
BLOCKS	fixed	5	1	2	3	4	5

Analysis of Variance for Y

Source	DF	SS	MS	F	P
TRTS	3	45.750	15.250	7.47	0.004
BLOCKS	4	138.300	34.575	16.93	0.000
Error	12	24.500	2.042		
Total	19	208.550			

MEANS

TRTS	N	Y
1	5	16.400
2	5	18.200
3	5	16.800
4	5	14.000

BLOCKS	N	Y
1	4	14.500
2	4	18.750
3	4	13.750
4	4	20.250
5	4	14.500

13.8 Estimation for the Randomized Block Design

Since a randomized block design involves two classifications, not only can we estimate differences in treatment means, but we can also estimate the differences in block means. In either situation we can construct a confidence interval estimate based on the Student's t distribution with $(b-1)(k-1)$ degrees of freedom. Hence, $(1 - \alpha)$ 100% confidence intervals would be found by using

$$(\bar{T}_i - \bar{T}_j) \pm t_{\alpha/2} \sqrt{2\text{MSE}/b}$$

and

$$(\bar{B}_i - \bar{B}_j) \pm t_{\alpha/2} \sqrt{2\text{MSE}/k}.$$

Example 13.13
Refer to Example 13.12. Estimate the difference in mean reading time for type styles 1 and 2 with a 95% confidence interval.

Solution
Pertinent information:

$\bar{T}_1 = 16.40$ MSE = _____   2.04

$\bar{T}_2 = 18.20$ d.f. = _____   12

$b = 5$ $t_{.025} = $ _____.   2.179

Using

$$(\bar{T}_1 - \bar{T}_2) \pm t_{.025} \sqrt{2\text{MSE}/b},$$

we have

$$(16.40 - 18.20) \pm 2.179 \sqrt{2(2.04)/5}$$

or

_____ ± _____ minutes.   −1.80; 1.97

Example 13.14
For the same problem, find a 95% confidence interval for the difference in mean reading time for readers 2 and 3.

Solution
To use the estimator

$$(\bar{B}_2 - \bar{B}_3) \pm t_{.025} \sqrt{2\text{MSE}/k}$$

we need the following additional information: $\bar{B}_2 = $ _____,   18.75
$\bar{B}_3 = $ _____, $k = 4$. Then   13.75

$$(18.75 - 13.75) \pm 2.179 \sqrt{2(2.04)/4}$$

simplifies to

_____ ± _____ minutes.   5.00; 2.20

Confidence interval estimates for block means or the difference between two block means are not always required nor are they always useful. Often, however, blocks constitute an important factor in many experiments. For example, if a marketing research survey investigating potential sales for four products is carried out in six areas, the areas would constitute blocks for this experiment and the differences in mean sales between areas would be of strong interest to the experimenter.

Self-Correcting Exercises 13C

1. An experiment was conducted to compare four feed additives on the growth of pigs. To eliminate genetic variability, pig litters were used as

blocks. Five litters were employed, with four pigs selected randomly from each litter. Each group of four pigs, tending to be more homogeneous than those between litters, was considered a block. The data (growth in pounds) are shown below.

	Additive			
Litter	1	2	3	4
1	78	69	78	85
2	66	64	70	70
3	81	78	72	83
4	76	66	77	74
5	61	66	69	70

a. Do the data present sufficient evidence to indicate a difference in mean growth for the four additives?
b. Do the data present sufficient evidence to indicate a difference in mean growth between litters? Was blocking desirable?
c. Find a 95% confidence interval for the difference in mean growth for additives 1 and 2.
d. Find a 95% confidence interval for the mean growth for additive 3.

2. Example 9.9 was analyzed in Chapter 9 as a paired-difference experiment. The data have been reproduced below.

Pair	Conventional	New
1	78	83
2	65	69
3	88	87
4	91	93
5	72	78
6	59	59

a. Analyze the data to detect a difference in means by using the analysis of variance for a randomized block design. (Use $\alpha = .05$.)
b. What is the relationship between the calculated value of t (Example 9.9) and the calculated value of F?

13.9 Some Comments on Blocking

A randomized block design should not be used when treatments and blocks both correspond to experimental factors. When this is the case, the effect of one factor may depend on the level of the other, and the factors are said to interact. In a randomized block design, it is assumed that the treatment and block factors *do not interact*. When interaction occurs, MSE, the mean squared error, becomes biased upward, resulting in erroneous conclusions concerning the effects of treatments and blocks. Information on designs dealing with several experimental factors can be found in any standard text on the design of experiments.

There are two major steps in designing an experiment that must be kept conceptually separate:
1. The first step encompasses the decision about what treatments to include in the experiment and the number of observations per treatment. Each setting of the independent variable, or each combination of settings of the independent variables if there are more than one, corresponds to a single ─────── .   treatment
2. The second step involves the decision of how to apply the treatments to the experimental _____ . Here is where the choice between a   units completely randomized and a randomized block design should be made.

Blocking produces a gain in information if and only if the between-block variation is (larger, smaller) than the within-block variation. But blocking   larger also costs information because it (increases, reduces) the number of degrees   reduces of freedom associated with the sum of squares for error denoted by SSE. Hence, blocking is beneficial only if the gain in information due to blocking outweighs the loss due to reduced degrees of freedom associated with SSE.

13.10 Selecting the Sample Size

The quantity of information in a sample pertinent to a population parameter can be measured by the width of the confidence interval for the parameter. Since the width of the confidence interval for a single mean or the difference between two means is inversely proportional to the number of observations in a treatment mean, when a bound on the error of estimation is given, the sample sizes can be chosen large enough to achieve the required accuracy.

The selection of the sample size involves the following steps:
1. The experimenter must decide on the parameter or parameters of interest.
2. The experimenter must specify a maximum bound on the error he or she is willing to tolerate.
3. An estimate of σ^2 must be available. This estimate may be obtained from a prior experiment or roughly computed as 1/4 of the expected range of the measurements to be taken.
4. The initial solution, which can be refined if so desired, involves the inequality

$$2\sigma_{\hat{\theta}} \leq B,$$

where 2 is used as an approximate value of $t_{.025}$ or $z_{.025}$ for 95% confidence.

Example 13.15

A completely randomized design is to be used in an experiment to compare the mean response time to a standard dosage of a drug using four different formulations. The experimenter would like to estimate the difference in mean response times correct to within 30 seconds with 95% accuracy. If the range of response times is expected to be about 3 minutes, how many observations should be included in each sample to achieve the required accuracy?

Solution

The 95% confidence interval estimator for $\mu_i - \mu_j$ is given as

$$(\bar{T}_i - \bar{T}_j) \pm t_{.025} \, s \sqrt{\frac{1}{n_i} + \frac{1}{n_j}}.$$

1. Since $t_{.025}$ depends on the degrees of freedom available for estimating σ^2, we will take $t_{.025} \approx$ _____ .
2. When all populations have the same variance, the optimal solution is to take $n_i = n_j =$ _____ .
3. Using the range to approximate σ, we have

$$s \approx R/4 = \underline{\hspace{1cm}}/4 = \underline{\hspace{1cm}}.$$

4. For a maximum bound of 30 seconds = .5 minute, we need to solve

$$t_{.025} \, s \sqrt{\frac{1}{n_i} + \frac{1}{n_j}} \leqslant .5.$$

This is approximately

$$2(.75)\sqrt{2/n} \leqslant .5,$$

$$[2(.75)\sqrt{2}/.5]^2 \leqslant n,$$

or

$$n \geqslant \underline{\hspace{1cm}}.$$

The experimenter should take equal samples of size _____ or more in each of the four groups to achieve the required accuracy.

Example 13.16

An experiment to compare the effect of three stimuli upon reaction times is to be run in a randomized block design using subjects as blocks. Previous experimentation in this area produced a standard deviation of 7.3 seconds. How many subjects should be included in the experiment if the experimenter wishes that the error in estimating the difference in two mean reaction times be less than or equal to 4 seconds with probability .95?

Solution

To estimate the difference in two treatment means in a randomized block design, we use

$$(\bar{T}_i - \bar{T}_j) \pm t_{.025} \, s \sqrt{\frac{1}{b} + \frac{1}{b}}.$$

When $s =$ _____ seconds, $B =$ _____ seconds, and $t_{.025} \approx$ _____, we solve

$$2(7.3)\sqrt{2/b} \leq 4.$$

Solving for b, we find

$$b \geq [2(7.3)\sqrt{2}/4]^2,$$

$b \geq$ _____ . | 26.65

Therefore, the experimenter should include at least _____ subjects in the experiment. | 27

The solutions to these sample-size problems are only approximate because of the approximations used in arriving at the results—for example, $t_{.025} \approx 2$ or $s \approx R/4$. These solutions are very useful nonetheless, since they do give the experimenter the approximate *size* and therefore the approximate _____ of the experiment. For a further review of problems involving the selection of the sample size to achieve specified bounds of error in estimation, refer to Section 7.5 and reread your solutions to the examples given there. | cost

13.11 Assumptions for the Analysis of Variance

To validly apply the testing and estimation procedures in an analysis of variance, the following assumptions concerning the probability distribution of the response x must be satisfied.

Assumptions

1. For any treatment or block combination, the response x is *normally distributed* with a *common variance* σ^2.
2. The observations are selected *randomly* and *independently* so that any two observations are *independent*.

Although it is never known in practice whether these assumptions are satisfied, we should be reasonably sure that violations of these assumptions are moderate. If the random variable under investigation is discrete, it is possible that the distribution for x is mound-shaped and, hence, approximately normal. This would not be the case if the discrete random variable only assumed three or four values.

Although binomial and Poisson random variables have probability distributions that may be approximately normal, these random variables will usually violate the assumption of equal variances. Recall that for a binomial random variable, $\mu =$ _____ and $\sigma^2 =$ _____ $= \mu(1-p)$, while for a Poisson random variable, $\sigma^2 =$ _____ . In either case, if the treatments are effective in changing the means, they will also cause the variances to change. Hence, the homogeneity assumption will be violated. | np; npq
 μ

Even when the response is normally distributed, the variances for the treatment groups may still be unequal. In agricultural variety trials it is not un-

mean

usual for the variability in height of a plant to increase with the average height. Similarly, in economic studies the variability of income within economic groupings is known to increase as the average income increases. Relationships of this sort can be detected by plotting the treatment or group means against the group variances or standard deviations. The coefficient of variation, defined as the ratio of the standard deviation to the _____ expressed as a percentage, can also be used to identify situations in which the standard deviations, and hence the variances, are unequal through their dependence on the mean.

When the data fail to meet the assumptions of normality and equal variances, an appropriate transformation of the data, such as their square roots, logarithms, or some other function of the data values, may be used in order that the transformed values approximately satisfy these assumptions.

independent

When the data consist only of rankings or ordered preferences, appropriate nonparametric testing and estimation procedures can be used. These procedures can also be used when the data fail to satisfy the assumptions of normality and equal variances, since the only requirement needed to use nonparametric techniques is that the observations be _____ within the constraints of the design used.

EXERCISES

1. A large piece of cotton fabric was cut into 12 pieces and randomly partitioned into three groups of four. Three different chemicals designed to produce resistance to stain were applied to the units, one chemical for each group. A stain was applied (as uniformly as possible) over all $n = 12$ units and the intensity of the stain measured in terms of light reflection.
   a. What type of experimental design was employed?
   b. Perform an analysis of variance and construct the ANOVA table for the following data:

	Chemical	
1	2	3
12	14	9
8	9	7
9	11	9
6	10	5

   c. Do the data present sufficient evidence to indicate a difference in mean resistance to stain for the three chemicals?
   d. Give a 95% confidence interval for the difference in means for chemicals 1 and 2.
   e. Give a 90% confidence interval for the mean stain intensity for chemical 2.
   f. Approximately how many observations per treatment would be required to estimate the difference in mean response for two chemicals, correct to within 1.0?

g. Obtain SSE directly for the data of Exercise 1 by calculating the sums of squares of deviations within each of the three treatments and pooling. Compare with the value found by using SSE = Total SS - SST.

2. A substantial amount of variation was expected in the amount of stain applied to the experimental units of Exercise 1. It was decided that greater uniformity could be obtained by applying the stain three units at a time. A repetition of the experiment produced the following results:

| | Chemical | | | |
Application	1	2	3	Total
1	12	15	9	36
2	9	13	9	31
3	7	12	7	26
4	10	15	9	34
Total	38	55	34	127

a. Give the type of design.
b. Conduct an analysis of variance for the data.
c. Do the data provide sufficient evidence to indicate a difference between chemicals?
d. Give the formula for a $(1 - \alpha)$ 100% confidence interval for the difference in a pair of chemical means. Calculate a 95% confidence interval for $(\mu_2 - \mu_3)$.
e. Approximately how many blocks (applications) would be required to estimate $(\mu_1 - \mu_2)$ correct to within .5?
f. We noted that the chemist suspected an uneven distribution of stain when simultaneously distributed over the 12 pieces of cloth. Do the data support this view? (That is, do the data present sufficient evidence to indicate a difference in mean response for applications?)

3. Twenty maladjusted children were randomly separated into four equal groups and subjected to 3 months of psychological treatment. A slightly different technique was employed for each group. At the end of the 3-month period, progress was measured by a psychological test. The scores are shown below (one child in group 3 dropped out of the experiment).

| | Group | | | | |
	1	2	3	4	
	112	111	140	101	
	92	129	121	116	
	124	102	130	105	
	89	136	106	126	
	97	99		119	
Total	514	577	497	567	2155

a. Give the type of design that appears to be appropriate.
b. Conduct an analysis of variance for the data.
c. Do the data present sufficient evidence to indicate a difference in mean response on the test for the four techniques?

d. Find a 95% confidence interval for the difference in mean response on the test for groups 1 and 2.
e. How could one employ blocking to increase the information in this problem? Under what circumstances might a blocking design applied to this problem fail to achieve the objective of the experiment?

4. The Graduate Record Examination scores were recorded for students admitted to three different graduate programs in a university.

	Graduate Programs		
	1	2	3
	532	670	502
	601	590	607
	548	640	549
	619	710	524
	509		542
	627		
	690		

a. Do these data provide sufficient evidence to indicate a difference in mean level of achievement on the GRE for applicants admitted to the three programs?
b. Find a 90% confidence interval for the difference in mean GRE scores for programs 1 and 2.

5. In a study where the objective was to investigate methods of reducing fatigue among employees whose job involved a monotonous assembly procedure, 12 randomly selected employees were asked to perform their usual job under each of three trial conditions. As a measure of fatigue, the experimenter used the total length of time in minutes of assembly line stoppages during a 4-hour period for each trial condition. The data follow.

		Condition	
Employee	1	2	3
1	31	22	26
2	20	15	23
3	26	21	18
4	21	12	22
5	12	16	18
6	13	19	23
7	18	7	16
8	15	9	12
9	21	11	26
10	15	15	19
11	11	14	21
12	18	11	21

a. Perform an analysis of variance for these data, testing for whether there is a significant difference among the mean stoppage times for the three conditions.
b. Is there a significant difference in mean stoppage times for the 12 employees? Was the blocking effective?
c. Estimate the difference in mean stoppage time for conditions 2 and 3 with 95% confidence.

CHAPTER 14
NONPARAMETRIC STATISTICS

14.1 Introduction

In earlier chapters we tested various hypotheses concerning populations in terms of their parameters. These tests represent a group of tests that are called _____ tests, since they specifically involve parameters such as means, variances, or proportions. To apply the techniques of Chapter 7, a large number of observations were required to assure the approximate _____ of the statistics employed in testing. In Chapters 9, 10, 11, and 13, it was assumed that the sampled populations had _____ distributions. Further, if two or more populations were studied in the same experiment, it was necessary to assume that these populations had a common _____ . In this chapter we will be concerned with hypotheses that do not involve population parameters directly but deal rather with the form of the distribution. The hypothesis that two distributions are identical versus the hypothesis that one distribution has typically larger values than the other are nonparametric statements of H_0 and H_a.

parametric

normality
normal

variance

Nonparametric tests are appropriate in many situations where one or more of the following conditions exist:

1. Nonparametric methods can be used when the form of the distribution is unknown, so that descriptive parameters may be of little use.
2. Nonparametric techniques are particularly appropriate if the measurement scale is that of rank ordering.
3. If a response can be measured on a continuous scale, a nonparametric method may nevertheless be desirable because of its relative simplicity when compared to its parametric analogue.
4. Most parametric tests require that the sampled population satisfy certain assumptions. When an experimenter cannot reasonably expect that these assumptions are met, a nonparametric test would be a valid alternative.

The following hypotheses would be appropriate for nonparametric tests:
1. H_0: a given population is normally distributed.
2. H_0: populations I and II have the same distribution.

Since these hypotheses are less specific than those required for parametric tests, we might expect a nonparametric test to be (more, less) efficient than a corresponding parametric test when all the conditions required for the use of the parametric test are met.

less

335

14.2 The Sign Test for Comparing Two Populations

The sign test is based on the _____ of the observed differences. Thus in a paired-difference experiment, we may observe in each pair only whether the first element is larger (or smaller) than the second. If the first element is larger (smaller), we assign a plus (minus) sign to the difference. We will define the test statistic x to be the number of _____ signs observed.

It is worth emphasizing that the sign test *does not* require a numerical measure of a response but merely a statement of which of two responses within a matched pair is larger. Thus, the sign test is a convenient and even necessary tool in many psychological investigations. If within a given pair it is impossible to tell which response is larger (a tie occurs), the pair is omitted. Thus, if 20 differences are analyzed and 2 of them are impossible to classify as plus or minus, we will base our inference on _____ (give number) differences.

Let p denote the probability that a difference selected at random from the population of differences would be given a plus sign. If the two population distributions are identical, the probability of a plus sign for a given pair would equal _____. Then the null hypothesis, "the two populations are identical," could be stated in the form $H_0: p = 1/2$. The test statistic x will have a _____ distribution whether H_0 is true or not. If H_0 is true, then the number of trials n will be the number of pairs in which a difference can be detected and the probability of success (a plus sign) on a given trial will be _____. If the alternative hypothesis is $H_a: p > 1/2$, then (large, small) values of x would be placed in the rejection region. If the alternative hypothesis is $H_a: p < 1/2$, then (large, small) values of x would be used in the rejection region. With $H_a: p \neq 1/2$, the rejection region would include both _____ and _____ values of x.

Example 14.1
Thirty matched pairs of schizophrenic patients were used in an experiment to determine the effect of a certain drug on sociability. In 18 of these pairs, the patient receiving the drug was judged to be more sociable, while in 5 pairs it was not possible to detect a difference in sociability. Test to determine whether or not this drug tends to increase sociability.

Solution
Let p denote the probability that, in a matched pair selected at random, the patient receiving the drug will be more sociable. Further, let x denote the number of pairs in which the drugged patient is more sociable.
1. The null hypothesis that the two populations are identical is stated as $H_0: p =$ _____. The alternative hypothesis that the drugged patients will be more sociable can be written as $H_a:$ _____.
2. The test statistic is x, the number of responses in which the drugged patient was more sociable out of the $n = 25$ pairs in which a difference was detected. For this problem, $x =$ _____.
3. Using $x = 18, 19, \ldots, 25$ as a rejection region yields the value α

Margin notes:

signs

plus

18

1/2

binomial

$p = 1/2$
large
small

large; small

1/2
$p > 1/2$

18

= _____. (Use the binomial tables in the text.) If $x = 17, 18, \ldots,$ | .022
25 is taken as the rejection region, α = _____. Assuming that | .054
$\alpha = .05$ is a satisfactory significance level, we will use the (first, second) | second
rejection region.

4. Since the observed value of x is $x = 18$, we agree to reject H_0 and conclude that the drug tends to increase sociability among schizophrenics.

We observed in Chapter 5 of the text that the normal approximation to binomial probabilities is reasonably accurate when $p = 1/2$ even when n is as small as _____. Thus, the normal distribution can ordinarily be used to approximate α and β for a given rejection region. Furthermore, when n is at least 25, the test can be based on the statistic | 10

$$\frac{x - (n/2)}{.5\sqrt{n}}$$

which will have approximately the _____ _____ | standard normal
distribution when H_0 is true.

Example 14.2

The productivity of 35 students was observed and measured both before and after the installation of new lighting in their classroom. The productivity of 21 of the 35 students was observed to have improved while the productivity of the others appeared to show no perceptible gain as a result of the new lighting. It is necessary to determine whether or not the new lighting was effective in increasing student productivity.

Solution

Let p denote the probability that one of the 35 students selected at random exhibits increased productivity after the installation of new lighting. This constitutes a paired-difference test where the productivity measures are paired on the students. Such pairing tends to block out student variations.

1. The null hypothesis is $H_0: p$ _____. | = 1/2
2. The appropriate one-sided alternative hypothesis is $H_a: p$ _____. | > 1/2
3. If x denotes the number of students who show improved productivity after the installation of the new lighting, then x has a binomial distribution with mean $np = 35(1/2) =$ _____ and variance $npq = 35(1/2)(1/2)$ | 17.5
= _____. Therefore, the test statistic can be taken to be | 8.75

$$z = \frac{x - 17.5}{\sqrt{8.75}}.$$

4. Reject H_0 at the $\alpha = .05$ level of significance if the calculated value of z is greater than $z_{.05}$ = _____. | 1.645
5. Since $x = 21$,

$$z = \frac{21 - 17.5}{\sqrt{8.75}} = \underline{}.$$ | 1.18

338 / Chap. 14: Nonparametric Statistics

would not; has not

Hence, we (would, would not) reject H_0; the new lighting (has, has not) improved student productivity.

14.3 A Comparison of Statistical Tests

The conditions of an experiment are often such that two or more different tests would be valid for testing the hypotheses of interest. How could we compare the efficiencies of two such tests? Statisticians examine the power of a test and use power as a measure of efficiency. The power of a test is defined to be _____ . If β is the probability that H_0 is accepted when H_a is true, then the complement of this event, $1 - \beta$, is the probability that H_0 is _____ when H_a is true. Since the object of a statistical test is to reject H_0 when it is _____ , $1 - \beta$ represents the probability that the test will perform its designated task.

$1 - \beta$

rejected
false

One method of comparing two tests utilizing the same sample size and the same significance level (α) is to compare their powers for alternatives of concern to the experimenter. The most common method of comparing two tests is to find the relative efficiency of one test with respect to the other. Since the sample sizes represent a measure of the costs of the tests in question, we would choose the test requiring (fewer, more) sample observations to achieve the same level of significance (α) and the same power $(1 - \beta)$ as the (more, less) efficient test. If n_A and n_B denote the sample sizes required for tests A and B to achieve the same specified values of α and $1 - \beta$ for a specific alternative hypothesis, then the relative efficiency of test A with respect to test B is _____ . If this ratio is greater than one, test A is said to be (more, less) efficient than B.

fewer

more

n_B/n_A
more

14.4 The Mann-Whitney U Test for Comparing Two Population Distributions

If an experimenter has two independent (not related) random samples in which the measurement scale is at least rank ordering, he can test the hypothesis that the two underlying distributions are identical versus the alternative that they are not identical by using the **Mann-Whitney U test**.

If two independent random samples are drawn from the same population (this is the case if H_0 is true), then we really have one large sample of size $n = n_1 + n_2$. If all measurements were ranked from small (1) to large (n), and each observation from sample 1 replaced with an 1 while each observation from sample 2 was replaced with a 2, we would expect to find the 1's and 2's randomly mixed in the ranking positions. If H_0 is false and the second sample comes from a population whose values tend to be larger than the first population, the 2's would tend to occupy the _____ ranks. However, if the second sample comes from a population whose values tend to be smaller than the first, then the 2's would appear in the _____ rank positions.

higher

lower

A statistic that reflects the positions of the n_1 and n_2 sample values in the total ranking is the sum of the ranks occupied by the first sample or the sum of the ranks occupied by the second sample, denoted by T_1 and T_2, respectively. The stronger the discrepancy between T_1 and T_2, the greater is the evidence to indicate that the samples have been drawn from two _____ populations. The Mann-Whitney U statistic uses this information in testing for a difference in the population frequency distributions giving rise to the sample observations.

different

The Mann-Whitney U statistic is the smaller of U_1 and U_2 where

$$U_1 = n_1 n_2 + \frac{n_1(n_1 + 1)}{2} - T_1$$

$$U_2 = n_1 n_2 + \frac{n_2(n_2 + 1)}{2} - T_2$$

where n_1 and n_2 are the number of observations in samples 1 and 2, respectively and

$$U_1 + U_2 = n_1 n_2.$$

Example 14.3

Five sample observations for each of two samples are given below:

Sample 1: 19 (___) 20 (___) 16 (___) 12 (___) 23 (___)
Sample 2: 17 (___) 21 (___) 22 (___) 25 (___) 18 (___)

5; 6; 2; 1; 9
3; 7; 8; 10; 4

In the space provided, fill in the rank of each of the 10 observations and calculate T_1 and T_2, U_1 and U_2.

$T_1 = 5 + 6 + 2 + 1 + 9 =$ _____

23

$T_2 = 3 + 7 + 8 +$ _____ $+$ _____ $=$ _____

10; 4; 32

Then

$$U_1 = n_1 n_2 + \frac{n_1(n_1 + 1)}{2} - T_1$$

$$= 5(5) + \frac{5(6)}{2} - 23 = \_\_\_\_$$

17

and

$$U_2 = n_1 n_2 + \frac{n_2(n_2 + 1)}{2} - T_2$$

$$= 5(5) + \frac{5(6)}{2} - 32 = \_\_\_\_$$

8

340 / Chap. 14: Nonparametric Statistics

As a check on our calculations, notice that

$$U_1 + U_2 = \underline{\quad n_1 n_2 \quad} = 25$$

In order to use either U_1 or U_2 as the test statistic to test the null hypothesis that the two populations are identical against an alternative that might be one- or two-tailed, it is necessary to determine an appropriate rejection region based on the probability distribution for U_1 or U_2. Table 11 in the text gives the tabled values for $P(U \leq U_0)$ when n_1 and n_2 are ten or less. Since only lower-tailed probabilities are tabulated, it is desirable to design the test statistic to take advantage of this lower-tailed rejection region.

1. When the alternative hypothesis is that the two populations differ in distribution, a two-tailed rejection region is appropriate. In that case we calculate

$$U_1 = n_1 n_2 + (1/2)n_1(n_1 + 1) - T_1$$

and

$$U_2 = n_1 n_2 + (1/2)n_2(n_2 + 1) - T_2.$$

Since U_2 counts the number of times that a 2 precedes a 1, U_1 will be small if U_2 is __large__ and U_1 will be large if U_2 is __small__. Hence, the smaller of the two is used as our sample value of U, and we would reject H_0 if the smaller of U_1 or U_2 is too __small__. In particular, H_0 will be rejected if the smaller of U_1 and U_2 is less than or equal to U_0. Since only the lower portion of the rejection region is used (there is, in fact, another portion of the rejection region in the upper tail of the probability distribution), the tabulated probability, $P(U \leq U_0)$, is doubled and the level of significance is $\alpha = (\underline{\quad 2 \quad}) P(U \leq U_0)$.

2. When the alternative hypothesis is that the distribution of 1's lies to the right of the distribution of 2's, then it will be desirable to choose U_1 as the test statistic. In this alternative situation the 1's will occupy the (lower, __higher__) ranks, T_1 will be (small, __large__), and U_1 will tend to be __small__. Hence, we use U_1 as the test statistic and reject H_0 if $U_1 \leq U_0$ with $\alpha = P(U \leq U_0)$.

3. When the alternative hypothesis is that the distribution of 2's lies to the right of the distribution of 1's, then it will be desirable to choose U_2 as the test statistic. In this situation, if H_a is true, the 2's will occupy the (lower, __higher__) ranks, T_2 will be (small, __large__), and U_2 will tend to be __small__. Hence, we use U_2 as the test statistic and reject H_0 if $U_2 \leq U_0$ with $\alpha = P(U \leq U_0)$.

Example 14.4
Use the data in Example 14.3 to test H_0: the frequency distributions for populations 1 and 2 are identical against H_a: the population frequency distributions are not identical.

Solution

1. From Example 14.3, the Mann-Whitney U statistic is equal to _____, the _____ of U_1 and U_2. | 8
                                                                                                                  | smaller
2. For a two-tailed test the rejection region consists of values of $U \leq U_0$ such that $P(U \leq U_0) \approx \alpha$ with U_0 found in Table 11 of the text when $n_1 = n_2 = $ _____ . An appropriate choice is $U_0 = $ _____ , since that insures that $P(U \leq U_0) = $ _____ , which is approximately equal to $\alpha/2$ for $\alpha = .05$. | 5; 3
   | .0278
3. Since $U = 8$ is greater than $U_0 = 3$, we (reject, do not reject) the null hypothesis of identical population frequency distributions. | do not reject

Example 14.5

Before filling several new teaching positions at the high school level, the principal of the school formed a review board consisting of five teachers who were asked to interview the 12 applicants and rank them in order of merit. Seven of the 12 applicants held college degrees but had limited teaching experience. Of the remaining 5 applicants, all had college degrees and substantial experience. The review board's rankings are given below:

Limited Experience	Substantial Experience
4	1
6	2
7	3
9	5
10	8
11	
12	

Do these rankings indicate that the review board considers experience a prime factor in the selection of the best candidates?

Solution

1. In testing the null hypothesis that the underlying populations are identical versus the alternative hypothesis that the population consisting of applicants having substantial experience is better qualified (will receive low ranks), we require a _____-tailed test. | one
2. In deciding upon the test statistic and the rejection region, we must take care to note that the tables are given with $n_1 \leq n_2$. Hence we take $n_1 = 5$ and $n_2 = 7$ and identify the 5 applicants with substantial experience as 1's and the remaining 7 applicants as 2's. If H_a is true, the 1's will occupy the _____ ranks and U_1, will be (small, large) while U_2, will be (small, large). | lower; large
  | small
3. Using U_2 as the test statistic, with $\alpha \approx .05$, $n_1 = 5$, and $n_2 = 7$, an appropriate rejection region would consist of the values $U \leq $ _____ with $\alpha = $ _____ . | 7
                                                                                                                                                                                                      | .0530

342 / Chap. 14: Nonparametric Statistics

$$U_2 = n_1 n_2 + \frac{n_2(n_2+1)}{2} - T_2$$

59

$$= (5)(7) + \frac{7(8)}{2} - \underline{\hspace{1cm}}$$

4

$$= \underline{\hspace{1cm}}.$$

reject
does

4. Since the observed value of U falls in the rejection region, we (reject, do not reject) H_0 and conclude that the review board (does, does not) consider the applicants with teaching experience to be more highly qualified than those without.

When the sample sizes both exceed ten, Table 11 can no longer be used to locate rejection regions for tests involving the Mann-Whitney U statistic. However, when the sample sizes exceed ten, the distribution of U can be

normal

approximated by a _____ distribution with mean

$$E(U) = n_1 n_2 / 2$$

and variance

$$\sigma_U^2 = n_1 n_2 (n_1 + n_2 + 1)/12.$$

Therefore, as a test statistic we can use

$$z = \frac{U - E(U)}{\sigma_U}$$

with the appropriate one- or two-tailed rejection region expressed in terms of z, the standard normal random variable.

Example 14.6

A manufacturer uses a large amount of a certain chemical. Since there are just two suppliers of this chemical, the manufacturer wishes to test whether the percentage of contaminants is the same for the two sources against the alternative that there is a difference in the percentage of contaminants for the two suppliers. Data from independent random samples are given below:

Supplier	Contaminant Percentages					
1	.86	.69	.72	.65	1.13
	.65	1.18	.45	1.41	.50
	1.04	.41			
2	.55	.40	.22	.58	.16
	.07	.09	.16	.26	.36
	.20	.15			

Solution

1. We combine the obtained contaminant percentages in a single ordered arrangement and identify each percentage by letter.

Percentage	.07	.09	.15	.16	.16	.20	.22	.26
Rank	1	2	3	4.5	4.5	6	7	8
Supplier	2	2	2	2	2	2	2	2

Percentage	.36	.40	.41	.45	.50	.55	.58	.65
Rank	9	10	11	12	13	14	15	16.5
Supplier	2	2	1	1	1	2	2	1

Percentage	.65	.69	.72	.86	1.04	1.13	1.18	1.41
Rank	16.5	18	19	20	21	22	23	24
Supplier	1	1	1	1	1	1	1	1

2. Since the sample sizes of $n_1 = 12$ and $n_2 = 12$ are beyond those given in Table 11, we can use the normal approximation to the distribution of U. The manufacturer, in asking whether there is a difference between the two suppliers, has specified a _____ tailed test. Therefore, we would reject H_0 if U were either too large or too small. (For a two-tailed test using the normal approximation, we are at liberty to use either U_1 or U_2 as the value of U to be tested.)   two-

72; 300

$$U_1 = n_1 n_2 + (1/2)n_1(n_1 + 1) - T_1$$

$$= 144 + 78 - \underline{\qquad}$$ 216

$$= \underline{\qquad},$$ 6

while

$$U_2 = n_1 n_2 + (1/2)n_2(n_2 + 1) - T_2$$

$$= 144 + 78 - \underline{\qquad}$$ 84

$$= \underline{\qquad}.$$ 138

3. The rejection region in terms of $z = (U - E(U))\sigma_U$ would be to reject H_0 if $|z| > \underline{\qquad}$. With $U = U_1 = 6$,   1.96

$$z = (6 - 72)/\sqrt{300} = -66/17.32 = \underline{\qquad}.$$ -3.81

4. Hence we would conclude that there (is, is not) a significant difference in contaminant percentages for the two suppliers.   is

Had we used U_B as the value of U, our result would have been

$$z = (138 - 72)/\sqrt{300} = 66/17.32 = \underline{\qquad}$$ 3.81

and we would have arrived at the same conclusion.

normal; equal

Use of the Mann-Whitney U test eliminates the need for the restrictive assumptions of the Student's t test, which requires that the samples be randomly drawn from _____ populations having _____ variances.

The Mann-Whitney test procedure can be implemented using the MINITAB command MANN-WHITNEY followed by the column numbers identifying the data and the group (sample) subscripts. The value of W, the sum of the ranks for the sample designated as sample one, is used as the test statistic rather than U, the statistic we have been using. The results are equivalent when using either statistic. Notice that the test is one of equality of the population medians, designated on the printout as ETA1 and ETA2. The following MINITAB printout uses the data from Example 14.6.

MTB > MANN-WHITNEY C1 C2

Mann-Whitney Confidence Interval and Test

C1   N =   12   Median =   0.7050
C2   N =   12   Median =   0.2100
Point estimate for ETA1−ETA2 is   0.5000
95.4 pct c.i. for ETA1−ETA2 is (0.2901,0.7800)
W = 216.0
Test of ETA1 = ETA2   vs.   ETA1 n.e. ETA2 is significant at 0.0002
The test is significant at 0.0002 (adjusted for ties)

Self-Correcting Exercises 14A

1. An experiment was designed to compare the durabilities of two highway paints, paint A and paint B, under actual highway conditions. An A strip and a B strip were painted across a highway at each of 30 locations. At the end of the test period, the experimenter observed the following results: At 8 locations paint A showed the least wear; at 17 locations paint B showed the least wear; and at the other 5 locations the paint samples showed the same amount of wear. Can we conclude that paint B is more durable? (Use $\alpha = .05$.)

2. In a deprivation study to test the strength of two physiological drives, ten rats who were fed the same diet according to a feeding schedule were randomly divided into two groups of five rats. Group I was deprived of water for 18 hours and group II was deprived of food for 18 hours. At the end of this time, each rat was put into a maze having the appropriate reward at the exit, and the time required to run the maze was recorded for each rat. The results follow, with time measured in seconds:

Water	Food
16.8	20.8
22.5	24.7
18.2	19.4
13.1	28.9
20.2	25.3

Is there a difference in strength of these two drives as measured by the time required to find the incentive reward? Use the Mann-Whitney U test with $\alpha \approx .05$.

3. Rootstock of varieties A and B was tested for resistance to nematode intrusion. An A and a B were planted side by side in each of ten widely separated locations. At the conclusion of the experiment, all roots were brought into the laboratory for a nematode count. The results are recorded in the following table.

	\multicolumn{10}{c}{Location}
	1	2	3	4	5	6	7	8	9	10
Variety A	463	268	871	730	474	432	538	305	173	592
Variety B	277	130	522	610	482	340	319	266	205	540

Can it be said that varieties A and B differ in their resistance to nematode intrusion? (Use a two-tailed sign test, with $\alpha = .02$.)

4. The score on a certain psychological test, P, is used as an index of status frustration. The scale ranges from $P = 0$ (low frustration) to $P = 10$ (high frustration). This test was administered to independent random samples of seven men and eight women with the following results:

	\multicolumn{8}{c}{Status Frustration Score}
Women	6	10	3	8	8	7	9	
Men	3	5	2	0	3	1	0	4

Use the Mann-Whitney U statistic with α as close to .05 as possible to test whether the distribution of status frustration scores is the same for the two groups against the alternative that the status frustration scores are higher among women.

14.5 The Wilcoxon Signed-Rank Test for a Paired Experiment

One previously discussed nonparametric test that may be used for a paired-difference experiment is the _____ test. While the sign test requires only the direction of the difference within each matched pair, a more efficient test is available if in addition the _____ _____ of the differences can be ranked in order of magnitude. The Wilcoxon signed-rank test employs as a test statistic, T, the (smaller, larger) sum of ranks for differences of the same sign where the differences are ranked in order of their _____ _____. In calculating T, zero differences are _____ and ties in the absolute values of nonzero differences are treated in the same manner as prescribed for the _____ test. Critical values of T are given in Table 12.

sign

absolute values

smaller

absolute values
omitted
Mann-Whitney U

Example 14.7
Twelve matched pairs of brain-damaged children were formed for an experiment to determine which of two forms of physical therapy is the more effective. One child was chosen at random from each pair and treated over a

period of several months using therapy A, while the other child was treated during this period using therapy B. There was judged to be no difference within two of the matched pairs at the end of the treatment period. The results are summarized in the following table:

Pair	Difference Favorable to Treatment	Rank for the Absolute Value of the Difference
1	A	9
2	A	5
3	B	1.5
4	A	4
5	A	1.5
6	*	*
7	A	7
8	A	8
9	*	*
10	A	10
11	A	6
12	A	3

*zero difference

8; 1.5

reject

For a two-sided test with $\alpha = .05$, we should reject H_0, "treatments equally effective," when $T \leq$ _____. The sample value of T is _____. Hence we _____ H_0.

It can be shown that the expected value and variance of T are

$$E(T) = n(n+1)/4$$

$$\sigma_T^2 = n(n+1)(2n+1)/24.$$

pairs

where n is the number of _____ in the experiment. When n is at least 25, we may employ the test statistic

$$z = \frac{T - E(T)}{\sigma_T}$$

standard normal

which will have approximately the _____ _____ distribution when H_0 is true.

Example 14.8
A drug was developed for reducing the cholesterol level in heart patients. The cholesterol levels before and after drug treatment were obtained for a random sample of 25 heart patients, with the following results:

Sect. 14.5: The Wilcoxon Signed-Rank Test for a Paired Experiment / 347

Patient	Cholesterol Level Before	Cholesterol Level After	Patient	Cholesterol Level Before	Cholesterol Level After
1	257	243	13	364	343
2	222	217	14	210	217
3	177	174	15	263	243
4	258	260	16	214	198
5	294	295	17	392	388
6	244	236	18	370	357
7	390	383	19	310	299
8	247	233	20	255	258
9	409	410	21	281	276
10	214	216	22	294	295
11	217	210	23	257	227
12	340	335	24	227	231
			25	385	374

It is necessary to determine whether or not this drug has an effect on the cholesterol level of heart patients.

Solution
Differences, Before − After, arranged in order of their absolute values are shown below together with the corresponding ranks. Fill in the missing ranks.

Difference	Rank	Difference	Rank	
−1	2	7	14	
−1	2	−7	14	
−1	2	7	14	
−2	4.5	8	16	
−2	4.5	11	_____	17.5
3	6.5	11	_____	17.5
−3	6.5	13	19	
−4	8.5	14	_____	20.5
4	8.5	14	_____	20.5
5	11	16	_____	22
5	11	20	23	
5	11	21	24	
		30	25	

Suppose that the alternative hypothesis of interest to the experimenter is the statement "the drug has the effect of reducing cholesterol levels in heart patients." Thus, the appropriate rejection region for $\alpha = .05$ is $z < -1.645$, where, in calculating z, we take T to be the smaller sum of ranks (the sum of ranks of the _____ differences).   negative

When H_0 is true,

$$E(T) = n(n+1)/4 = _____$$ 325

and

$$\sigma_T^2 = n(n+1)(2n+1)/24 = _____ .$$ 1,381.25

348 / Chap. 14: Nonparametric Statistics

Thus, we will reject H_0 at the $\alpha = .05$ significance level if

$$z = \frac{T - 325}{\sqrt{1{,}381.25}} < -1.645.$$

Summing the ranks of the negative differences, we obtain $T =$ _____ [44]
and hence, $z =$ _____ [−7.56]. Comparing z with its critical value, we
_____ [reject] H_0 in favor of the alternative hypothesis that the drug has the effect of reducing cholesterol levels in heart patients.

It is interesting to see what conclusion is obtained by using the sign test. Recall that x is equal to the number of positive differences and that the test statistic

$$z = \frac{x - n/2}{\sqrt{n/2}}$$

has approximately the _____ _____ [standard normal] distribution when n is greater than ten and $H_0: p = 1/2$ is true. With $\alpha = .05$, the rejection region for z is z _____ [> 1.645]. But $x = 17$, so that $z =$ _____ [1.8]. Thus, we obtain the same conclusion as before, though the sample value of the test statistic does not penetrate as deeply into the rejection region as when the Wilcoxon signed rank test was used. Since the Wilcoxon signed rank test makes fuller use of the information available in the experiment, we say that the Wilcoxon signed rank test is more _____ [efficient] than the sign test.

The MINITAB command WTEST, followed by the column containing the differences to be analyzed, will implement Wilcoxon's signed-rank procedure. The Wilcoxon statistic reported is the rank sum for the positive differences. One-tailed tests are indicated by using the subcommand ALTERNATIVE, followed by a −1 for a left-tailed test, or a 1 for a right-tailed test. The following printout used the data from Example 14.8. Since the sum of the ranks of the positive differences and the sum of the ranks of the negative differences equal $n(n + 1)/2$, the rank sum of the positive differences is

$$\begin{aligned} T^+ &= 25(26)/2 - 44 \\ &= 325 - 44 \\ &= 281 \end{aligned}$$

as reported on the printout.

```
MTB > WTEST C1;
SUBC> ALTERNATIVE 1.

TEST OF MEDIAN = 0.000000 VERSUS MEDIAN G.T. 0.000000

              N FOR  WILCOXON            ESTIMATED
         N    TEST   STATISTIC  P-VALUE   MEDIAN
   C1    25    25      281.0     0.001    6.000
```

Self-Correcting Exercises 14B

1. The sign test is not as efficient as the Wilcoxon signed rank test for data of the type presented in Exercise 3, Self-Correcting Exercises 14A. Analyze the data of Exercise 3, Self-Correcting Exercises 14A, by using the two-tailed Wilcoxon test with $\alpha = .02$. Can it be said that varieties A and B differ in their resistance to nematode intrusion?
2. The sign test is sometimes used as a "quick and easy" substitute for more powerful tests that require lengthy computations. The following differences were obtained in a paired-difference experiment: -.93, .95, .52, -.26, -.75, .25, 1.08, 1.47, .60, 1.20, -.65, -.15, 2.50, 1.22, .80, 1.27, 1.46, 3.05, -.43, 1.82, -.56, 1.08, -.16, 2.64. Use the sign test with $\alpha = .05$ to test $H_0: \mu_d = 0$ against the one-sided alternative $H_a: \mu_d > 0$.
3. Refer to Exercise 2. Use the large-sample Wilcoxon signed rank test with $\alpha = .05$ to test $H_0: \mu_d = 0$ against the alternative hypothesis $H_a: \mu_d > 0$. Compare (in efficiency and in computational requirements) the sign test and the Wilcoxon signed rank test as substitute tests in a paired-difference experiment.

14.6 The Kruskal-Wallis H Test for Completely Randomized Designs

In comparing several populations based upon independent samples from these populations, the Kruskal-Wallis H test, which uses the rank sums for each sample, is an extension of the Mann-Whitney U test. The test statistic for comparing k populations is

$$H = \frac{12}{n(n-1)} \sum_{i=1}^{k} \frac{T_i^2}{n_i} - 3(n+1)$$

for T_i, the rank sum of the n_i observations in the ith sample, based upon the total ranking of all $n = n_1 + n_2 + \ldots + n_k$ observations.

The hypothesis to be tested using the Kruskal-Wallis H test is

H_0: all k population distributions are identical

H_a: at least one of the k population distributions is different

If H_0 is true, and all the samples are being drawn from the same population, there should be (little, large) variation in the rank sums, $T_1, T_2, \ldots, T_k$. However, if one or more samples are from different populations, the rank sums will exhibit (little, large) variation and the value of H will increase. This statistic then always uses an upper-tailed rejection region. Furthermore, when the null hypothesis is true, the test statistic H has an approximate Chi-square distribution with $(k - 1)$ degrees of freedom. Hence, Table 5 can be used to determine the appropriate rejection region for the test.

little

large

Example 14.9
Three job-training programs were tested on 15 new employees by randomly assigning 5 employees to participate in each program. After completing the

programs and having performed on the job for one week, the fifteen were ranked according to their ability to perform the task for which they had been trained, with a high rank indicating a low ability.

	Program		
	A	B	C
	2	6	10
	13	7	15
	1	9	8
	5	3	12
	4	11	14
Sums	25; 36; 59	___	___

Do these rankings indicate that one program is better than another at the $\alpha = .05$ level of significance?

Solution:
1. We are interested in testing whether these three samples of 5 measurements come from the same population, against the alternative that at least one sample comes from a population different from the others.
2. Since the data are given directly as ranks, we need but calculate the statistic H using $T_1 = 25$, $T_2 = 36$, and $T_3 = 59$ with $n_1 = n_2 = n_3 = 5$ and $n =$ _____ . [15]

$$H = \frac{12}{n(n+1)} \sum_{i=1}^{3} \frac{T_i^2}{n_i} - 3(n+1)$$

$$= \frac{12}{15(16)} \left(\frac{25^2}{5} + \frac{36^2}{5} + \frac{59^2}{5} \right) - 3(16)$$

$$= \frac{12}{15(16)} \left(\frac{5402}{5} \right) - \underline{48}$$

$$= \underline{54.02} - 48$$

$$= \underline{6.02}$$

3. Using Table 5 with $(k-1) = 2$ degrees of freedom and $\alpha = .05$, the rejection region consists of values of $H \geq \chi^2_{.05} = 5.99$. Since the calculated value of H exceeds 5.99, we (<u>reject</u>, do not reject) H_0 and conclude that a difference (<u>exists</u>, does not exist) among the three job-training programs. Although this is all that can be said statistically, by looking at the rank sums for the three programs, program A seems to produce employees with higher job abilities.

Example 14.10
In Example 14.4, we considered a comparison of the length of time required for kindergarten-age children to assemble a device when the children had been

instructed for four different lengths of time. Four children were randomly assigned to each instructional group, but two were eliminated during the experiment due to sickness. The length of time to assemble the device was recorded for each child in the experiment.

Training Periods (in hours)			
0.5	1.0	1.5	2.0
8 (9.5)	9 (11.5)	4 (1.5)	4 (1.5)
14 (14)	7 (7)	6 (5)	7 (___)
9 (___)	5 (___)	7 (7)	5 (3.5)
12 (13)		8 (___)	

7
11.5; 3.5
9.5

Use the Kruskal-Wallis H test to determine whether the data present sufficient evidence to indicate a difference in the distribution of times for the four different lengths of instructional time. Use $\alpha = .01$.

Solution

1. The data are first ranked according to their magnitude. The ranks of the combined sample of $n = 20$ are shown in parentheses above. Fill in the missing entries. The rank sums are

$$T_1 = 48 \qquad T_3 = \underline{}$$

23

$$T_2 = 22 \qquad T_4 = 12$$

with $n_1 = n_3 = 4$ and $n_2 = n_4 = 3$.

2. The test statistic is calculated as

$$H = \frac{12}{n(n+1)} \sum_{i=1}^{4} \frac{T_i^2}{n_i} - 3(n+1)$$

$$= \frac{12}{14(15)} \left(\frac{48^2}{4} + \frac{22^2}{3} + \frac{23^2}{4} + \frac{12^2}{3} \right) - 3(15)$$

$$= \underline{} - 45 = \underline{}$$

52.4333; 7.4333

3. Using Table 5 with $(k - 1) = 3$ degrees of freedom and $\alpha = .01$, the rejection region is $H \geq \chi^2_{.01} = 11.3449$. Hence, we (reject, do not reject) H_0. There (is, is not) sufficient evidence to indicate that there is a difference in the distributions of times for the four groups.

do not reject
is not

The Kruskal-Wallis test can also be implemented using the MINITAB command KRUSKAL-WALLIS, followed by the column designations for the data and the group subscripts. The following printout used the data from Example 14.10. Notice that the Kruskal-Wallis statistic can be adjusted for any ties that may occur in the rankings. Although we have not discussed this problem, the adjustment has the effect of increasing the value of the statistic, and hence, decreasing its *p*-value. Differences in the adjusted and unadjusted statistics are small if the number of ties is not large.

```
MTB > PRINT C1 C2       MTB > KRUSKAL-WALLIS C1 C2

ROW   C1   C2      LEVEL     NOBS     MEDIAN    AVE. RANK    Z VALUE

 1     8    1        1         4      10.500      12.0         2.55
 2    14    1        2         3       7.000       7.3        -0.08
 3     9    1        3         4       6.500       5.8        -0.99
 4    12    1        4         3       5.000       4.0        -1.63
 5     9    2     OVERALL     14                   7.5
 6     7    2
 7     5    2     H = 7.43    d.f. = 3    p = 0.060
 8     4    3     H = 7.57    d.f. = 3    p = 0.056 (adj. for ties)
 9     6    3
10     7    3     * NOTE    * One or more small samples
11     8    3
12     4    4
13     7    4
14     5    4
```

Self-Correcting Exercises 14C

1. In Exercise 1, Self-Correcting Exercises 13A, the dressed weights of five chickens fed from birth on one of three rations were recorded. The data are reproduced below.

	Rations	
1	*2*	*3*
7.1	4.9	6.7
6.2	6.6	6.0
7.0	6.8	7.3
5.6	4.6	6.2
6.4	5.3	7.1

Use the Kruskal-Wallis H test to determine whether the data present sufficient evidence to indicate a difference in the distribution of weights for the three rations. Use $\alpha = .05$.

2. In Exercise 2, Self-Correcting Exercises 13A, we considered an investigation of a citizens committee's complaint about the availability of fire protection within the county. The distance in miles to the nearest fire station was measured for each of 5 randomly selected residences in each of four areas. The data are reproduced below.

		Areas		
	1	*2*	*3*	*4*
	7	1	7	4
	5	4	9	6
	5	3	8	3
	6	4	7	7
	8	5	8	5

Suppose that the experimenter was not willing to assume that the distribution of distances for each of the four areas was normal. Use the Kruskal-Wallis H test to determine if there is sufficient evidence to indicate a difference in the distributions of distances for the four areas. Use $\alpha = .01$.

14.7 The Friedman F_r Test for Randomized Block Designs

The randomized block design is used in situations where the individual experimental units may be quite different, but where blocks of relatively homogeneous experimental units may be formed. In this situation, the Friedman test provides an extension of the Wilcoxon Signed Rank test and makes use of the ranks of the individual measurements. Instead of determining the ranks of the observations with respect to the entire data set as was done in the case of the Kruskal-Wallis test, the ranks of the observations are determined within each block. Tied measurements in the same block are each given the average of the ranks the measurements would have received if there had been slight differences in the observed values.

The hypothesis to be tested is

H_0: the distributions for the k treatment populations are identical

H_a: at least one of the distributions differs from the other $k - 1$.

The Friedman F_r statistic is given by

$$F_r = \frac{12}{bk(k+1)} \left(\sum_{i=1}^{k} T_i^2 \right) - 3b(k+1)$$

where k is the number of _____, b is the number of _____ and $T_1, T_2, \ldots, T_k$ are the _____ _____ for the k treatments. Wide discrepancies in the values of $T_1, T_2, \ldots, T_k$ occur if high or low ranks tend to be concentrated in certain treatments, which indicates that the population distributions are not identical. Discrepancies in the values of the rank sums cause F_r to be large. For this reason, the rejection region consists of _____ values of F_r. As with the Kruskal-Wallis test, the sampling distribution of the statistic F_r is approximately _____ with _____ degrees of freedom. Thus, for significance level α, the null hypothesis that the k treatments have the same population frequency distributions should be rejected if $F_r > $ _____ with $k - 1$ degrees of freedom.

treatments; blocks
rank sums

large
χ^2
$k - 1$

χ_α^2

Example 14.11
In an applicant-screening interview, four applicants were ranked by six panel members with the following results:

354 / Chap. 14: Nonparametric Statistics

	Applicant			
Panel Member	A_1	A_2	A_3	A_4
1	3	4	1	2
2	4	2	3	1
3	4	3	2	1
4	3	4	1	2
5	4	3	1	2
6	4	2	1	3

Do these rankings indicate agreement among the panel members with respect to their ranking of the four applicants at the $\alpha = .05$ level of significance?

Solution:

1. Since the observations are in fact rank orderings, to test the hypothesis that the panel members' rankings represent a random rank assignment versus the alternative that there is "agreement," we need to find the rank sum for each applicant.

22; 18; 9

$$T_1 = \underline{\hspace{1cm}} ; T_2 = \underline{\hspace{1cm}} ; T_3 = \underline{\hspace{1cm}} ;$$

11

$$T_4 = \underline{\hspace{1cm}}$$

2. The value of F_r is found as

$$F_r = \frac{12}{bk(k+1)} \sum_{i=1}^{4} T_i^2 - 3b(k+1)$$

$$= \frac{12}{6(4)(5)} (22^2 + 18^2 + 9^2 + 11^2) - 3(6)(5)$$

90

$$= \frac{12}{120}(1010) - \underline{\hspace{1cm}}$$

11

$$= \underline{\hspace{1cm}}$$

3. Using Table 5 with $(k - 1) = 3$ degrees of freedom and $\alpha = .05$, the rejection region consists of values of $F_r \geq \chi^2_{.05} = 7.81$. Since the calculated

reject
does

value of F_r exceeds 7.81, we (reject, do not reject) H_0 and conclude that a difference (does, does not) exist among the four applicants.

Example 14.12

In a study where the objective was to investigate methods of reducing fatigue among employees whose jobs involved a monotonous assembly procedure, twelve randomly selected employees were asked to perform their usual job under each of three trial conditions. As a measure of fatigue, the experimenter used the number of assembly-line stoppages during a four-hour period for each trial condition. Do the following data indicate that employee fatigue as measured by stoppages differ for these three conditions? Use $\alpha = .05$.

Sect. 14.7: The Friedman F_r Test for Randomized Block Designs

		Conditions		
Employees	1	2	3	
1	31 (_____)	22 (_____)	26 (_____)	3; 1; 2
2	20 (2)	15 (1)	23 (3)	
3	26 (3)	21 (2)	18 (1)	
4	31 (2)	22 (1)	32 (3)	
5	12 (_____)	16 (_____)	18 (_____)	1; 2; 3
6	22 (1)	29 (2)	34 (3)	
7	28 (_____)	17 (_____)	26 (_____)	3; 1; 2
8	15 (3)	9 (1)	12 (2)	
9	41 (2)	31 (1)	46 (3)	
10	19 (_____)	19 (_____)	25 (_____)	1.5; 1.5; 3
11	31 (1)	34 (2)	41 (3)	
12	18 (2)	11 (1)	21 (3)	
Rank Sums	$T_1 =$ _____	$T_2 =$ _____	$T_3 =$ _____	24.5; 16.5; 31

Solution:
1. In order to use the Friedman test as an alternative to the parametric analysis of variance, we need to rank the responses within employees and find the rank totals for each of the three conditions. Complete any missing rank entries and find the rank sums in the table above.
2. In testing for significant differences among the three conditions, we shall use Friedman's statistic, given as

$$F_r = \frac{12}{bk(k+1)} \left(\sum_{i=1}^{k} T_i^2 \right) - 3b(k+1)$$

For $b =$ _____, $k =$ _____, $T_1 = 24.5$, $T_2 = 16.5$, $T_3 = 31$, | 12; 3

$$F_r = \frac{12}{(12)(3)(4)} [24.5^2 + 16.5^2 + 31^2] - 3(12)(4)$$

$= 152.79 -$ _____ | 144

$=$ _____ | 8.79

3. We shall use the Chi-square approximation in finding a rejection region. With $k - 1 =$ _____ degrees of freedom, $\chi^2_{.05} =$ _____. Since the observed value of $F_r = 8.79$ is _____ than 5.99, we (reject, do not reject) H_0 and conclude that there (are, are not) significant differences among the three conditions at the $\alpha = .05$ level of significance. | 2; 5.99 greater; reject are

In using the MINITAB package to implement the Friedman test procedure, the entry of the information *must* be in the following order: FRIEDMAN DATA, TREATMENT SUBSCRIPTS, BLOCK SUBSCRIPTS. Of course, this information will be stored in three columns, and the number of the columns given in the order prescribed. Analyzing the data of Example 14.12 using the MINITAB command FRIEDMAN produced the following printout.

```
MTB > FRIEDMAN C1 C2 C3

Friedman test of C1 by C2 blocked by C3

S = 8.79    d.f. = 2    p = 0.013
S = 8.98    d.f. = 2    p = 0.011 (adjusted for ties)

                     Est.      Sum of
    C2      N      Median     RANKS
    1       12     24.750      24.5
    2       12     19.917      16.5
    3       12     27.083      31.0

Grand median    =    23.917
```

Self-Correcting Exercises 14D

1. Refer to Exercise 1, Self-Correcting Exercises 13B, in which the growth in pounds for pigs fed on four feed additives were compared, using pig litters as blocks. The data are reproduced below.

| | \multicolumn{4}{c}{Additive} |
Litter	1	2	3	4
1	78	69	78	85
2	66	64	70	70
3	81	78	72	83
4	76	66	77	74
5	61	66	69	70

Use Friedman's test to determine whether there is a significant difference among the distributions of growth for pigs fed on the four additives. Use $\alpha = .01$.

2. In a brand identification experiment involving four brands, 10 subjects in each of 5 geographic areas were asked to listen to an advertising jingle associated with each of the four brands and identify the brand through its jingle. The length of time in seconds until correct identification was averaged for the 10 people with the following results.

| | \multicolumn{4}{c}{Brands} |
Areas	1	2	3	4
1	3.7	3.9	4.2	4.0
2	4.2	4.8	4.6	4.7
3	2.9	3.5	3.0	3.4
4	5.0	5.4	5.0	5.5
5	3.3	4.3	4.1	3.9

Use Friedman's test to determine whether there is a significant difference in the distribution of times among the four brands. Use $\alpha = .05$.

14.8 Rank Correlation Coefficient, r_s

The Spearman rank correlation coefficient, r_s, is a numerical measure of the association between two variables y and x. As implied in the name of the test statistic, r_s makes use of _____ and hence the exact value of numerical measurements on y and x need not be known. Conveniently, r_s is computed in exactly the same manner as _____, Chapter 10.

To determine whether two variables y and x are related, we select a _____ sample of n experimental units (or items) from the population of interest. Each of the n items is ranked first according to the variable x and then according to the variable _____. Thus for each item in the experiment, we obtain two _____. (Tied ranks are treated as in other parts of this chapter.) Let x_i and y_i denote the respective ranks assigned to item i. Then, as in Chapter 10,

$$r_s = S_{xy}/\sqrt{S_{xx} S_{yy}}.$$

ranks

r

random

y
ranks

Example 14.13

An investigator wished to determine whether "leadership ability" is related to the amount of a certain hormone present in the blood. Six individuals were selected at random from the membership of the Junior Chamber of Commerce in a large city and ranked on the characteristic leadership ability. A determination of hormone content for each individual was made from blood samples. The leadership ranks and hormone measurements are recorded in the following table. Fill in the missing hormone ranks. Note that no difference in leadership ability could be detected for individuals 2 and 5.

Individual	Leadership Ability Rank (y_i)	Hormone Content	Hormone Rank (x_i)
1	6	131	1
2	3.5	174	____
3	1	189	____
4	2	200	6
5	3.5	186	____
6	5	156	____

3
5

4
2

To calculate r_s we form an auxiliary table. Fill in the missing quantities.

Individual	y_i	y_i^2	x_i	x_i^2	$x_i y_i$
1	6	____	i	1	6
2	3.5	12.25	3	9	10.5
3	1	1	5	___	5
4	2	4	36	12	6
5	___	12.25	4	16	14
6	5	25	2	4	___
Total	___	90.5	21	91	57.5

36
10.5
25
6
3.5
10

21

Thus,

$$S_{xy} = 57.5 - (21)(21)/6$$

-16

$$= \underline{\hspace{1cm}},$$

$$S_{xx} = 91 - (21)^2/6$$

17.5

$$= \underline{\hspace{1cm}},$$

$$S_{yy} = 90.5 - (21)^2/6$$

17

$$= \underline{\hspace{1cm}}.$$

Finally,

17

$$r_s = \frac{-16}{\sqrt{(17.5)(\underline{\hspace{1cm}})}} = -.93.$$

Thus, high leadership ability (reflected in low rank) seems to be associated with higher amounts of hormone.

The Spearman rank correlation coefficient may be employed as a test statistic to test an hypothesis of _____ _____ between two characteristics. Critical values of r_s are given in Table 14. The tabulated quantities are values of r_0 such that $P(r_s > r_0) = .05, .025, .01,$ or $.005$, as indicated. For a lower-tail test, reject H_0; "no association between the two characteristics," when $r_s < \underline{\hspace{1cm}}$.

no association

$-r_0$

Two-tailed tests require doubling the stated values of α, and hence, critical values for two-tailed tests may be read from Table 14 if $\alpha = \underline{\hspace{1cm}}$, _____ , _____ , or _____ .

.10
.05; .02; .01

Example 14.9

Continuing Example 14.8, we may wish to test whether leadership ability is associated with hormone level. If the experimenter had designed the experiment with the objective of demonstrating that low leadership ranks (high leadership abilities) are associated with high hormone levels, the appropriate test would be (a lower, an upper) tail test.

a lower
.829
-.829
does; do

For $\alpha = .05$ the critical value of r_s is $r_0 = \underline{\hspace{1cm}}$. Hence, we reject H_0 if $r_s < \underline{\hspace{1cm}}$. Since the sample value of r_s found in Example 14.8 (does, does not) fall in the rejection region, we (do, do not) reject H_0.

Self-Correcting Exercises 14E

1. An interviewer was asked to rank seven applicants as to their suitability for a given position. The same seven applicants took a written examination that was designed to rate an applicant's ability to function in the given position. The interviewer's ranking and the examination score for each applicant are given below.

Applicant	Interview Rank	Examination Score
1	4	49
2	7	42
3	5	58
4	3	50
5	6	33
6	2	65
7	1	67

Calculate the value of Spearman's rank correlation coefficient for these data. Test for a significant negative rank correlation at the $\alpha = .05$ level of significance.

2. A manufacturing plant is considering building a subsidiary plant in another location. Nine plant sites are currently under consideration. After considering land and building costs, zoning regulations, available local work force, and transportation facilities associated with each possible plant site, two corporate executives have independently ranked the nine possible plant sites as follows:

	Site								
	1	2	3	4	5	6	7	8	9
Executive 1	2	7	1	5	3	9	4	8	6
Executive 2	1	4	3	6	2	9	7	8	5

a. Calculate Spearman's rank correlation coefficient between the two sets of rankings.
b. Is there reason to believe the two executives are in basic agreement regarding their evaluation of the nine plant sites? (Use $\alpha = .05$.)

EXERCISES

1. For each of the following tests, state whether the test would be used for related samples or for independent samples: sign test, Mann-Whitney U test, Wilcoxon signed rank test.
2. About 1.2% of our combat forces in a certain area develop combat fatigue. To find identifying characteristics of men who are predisposed to this breakdown, the level of a certain adrenal chemical was measured in samples from two groups: men who had developed battle fatigue and men who had adjusted readily to combat conditions. The following determinations were recorded:

Battle fatigue group	23.35	21.08	22.36	20.24
	21.69	21.54	21.26	20.71
	20.00	23.40	21.43	21.54
	22.21			
Well-adjusted group	21.66	21.85	21.01	20.54
	20.19	19.26	21.16	19.97
	20.40	19.92	20.52	19.78
	21.15			

Use a large-sample, one-tailed, Mann-Whitney test with α approximately equal to .05 to test whether the distributions of levels of this chemical are the same in the two groups against the alternative that the mean level is higher in the combat fatigue group.

3. An experiment was designed to determine whether exposure to cigarette smoke has an effect on the length of life of beagle dogs. Twenty beagles of the same age were used in the experiment. The animals were assigned at random to one of two groups. Ten of the dogs were subjected to conditions equivalent to smoking up to 12 cigarettes each day. The other 10 acted as a control group. Recorded in the table is the number of days until death for the dogs in both groups. Since the experiment was concluded when the last of the experimental dogs died, an L is recorded for each of the dogs still living.

Experimental Group	Control Group
45	315
112	474
251	727
340	894
412	L
533	L
712	L
790	L
845	L
974	L

Use the Mann-Whitney U and $\alpha = .0526$ to obtain a one-tailed test of whether the experimental group has a shorter mean life than the control group.

4. The value of r (defined in Chapter 10) for the following data is .636:

x	y
.05	1.08
.14	1.15
.24	1.27
.30	1.33
.47	1.41
.52	1.46
.57	1.54
.61	2.72
.67	4.01
.72	9.63

Calculate r_s for these data. What advantage of r_s was brought out in this example?

5. A ranking of the quarterbacks in the top eight teams of the National Football League was made by polling a number of professional football coaches and sportswriters. This "true ranking" is shown below together with my ranking.

   a. Calculate r_s.

b. Do the data provide evidence at the $\alpha = .05$ level of significance to indicate a positive correlation between my ranking and that of the experts?

	Quarterback							
	A	B	C	D	E	F	G	H
True ranking	1	2	3	4	5	6	7	8
My ranking	3	1	4	5	2	8	6	7

6. Eight recent college graduates have interviewed for positions within the marketing department of a large industrial organization. The organization's vice-president for marketing and the personnel director rated each candidate independently on a 0–10 assumed interval scale. Their ratings are shown below. Use a two-tailed sign test to determine if the vice-president and the personnel manager differ in their evaluations of the eight candidates. (Use $\alpha \leq .10$.)

	Candidate							
	1	2	3	4	5	6	7	8
Vice-president	3	7	6	9	7	4	3	8
Personnel manager	2	4	5	5	9	1	6	6

7. Refer to Exercise 6 and analyze the data using the two-tailed Wilcoxon test with α as close as possible to the significance level used in Exercise 6. Explain any differences in conclusions arrived at using the Wilcoxon versus the sign test. Which conclusion should we believe?

8. It is of interest to determine whether the efficiency of a certain machine operator is superior to the efficiency of another. To examine this question, the percentages of defective items produced daily by machine operators A and B are recorded over a period of time. More recorded data are available from operator A due to a recent illness experienced by operator B. The data are shown below.

	Percentage of Output Defective								
Operator A	3	2	7	6	5	5	3	7	4	6
Operator B	6	5	8	4	8	7	9	10		

The plant analyst realizes that distributions of percentages do not follow a normal distribution and is hesitant to use a t test in the analysis. Use the most powerful nonparametric test at your disposal to determine whether operator A produces a lower percentage defective than operator B. (Use $\alpha = .05$.)

9. Daily lost production from three production lines in a manufacturing operation were recorded for a ten-day period.

	Line		
Day	1	2	3
1	15	11	8
2	9	9	6
3	6	8	4
4	7	6	5
5	16	13	9
6	23	25	14
7	12	9	7
8	10	12	9
9	12	10	11
10	16	10	9

a. What type of experimental design has been used?

b. If the assumptions required for the parametric analysis of variance have been violated, use an alternative nonparametric test to determine whether there is a difference in the distributions of lost daily production for the three production lines. Use $\alpha = .01$.

10. Refer to Exercise 3, Chapter 13. Four different treatments were used for maladjusted children, and psychological scores were recorded at the end of a three-month period. The data are reproduced below.

Treatments			
1	2	3	4
112	111	140	101
92	129	121	116
124	102	130	105
89	136	106	126
97	99		119

a. Give the type of design which has been used in this experiment.

b. Use an appropriate nonparametric test to determine whether there is a significant difference in the distribution of psychological scores for the four treatments. Use $\alpha = .05$.

APPENDIX
USEFUL MATHEMATICAL RESULTS

A.1 Introduction

To fulfill the objective of statistics, information must be extracted from measurements contained in a sample. Formulas for extracting this information are typically expressed in summation notation. A second useful notation, functional notation, is important because it is used in the development of summation notation and in presenting formulas that we will encounter later in the text. Consequently, this Appendix is devoted to familiarizing the student with functional notation and summation notation.

A.2 Functions and Functional Notation

A function consists of two sets of elements and a rule which associates one and only one element of the second set with each of the elements in the first set. It can be displayed graphically, showing the elements of the first and second sets as points inside respective enclosures and the rule of association indicated by joining associated points (elements) with lines.

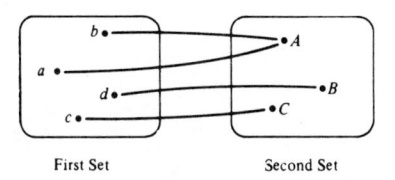

First Set   Second Set

The function visualized above might also be exhibited as a collection of ordered pairs. In each pair, the right member is an element of the second set, and *no two pairs* have the same (first, second) element. Thus our function can also be displayed as the collection $\{(a,A),(b,A),(d,B),(\underline{})\}$ of ordered pairs. Though two of these pairs have the same second element, we note that no two of these pairs have the same \underline{} element.
   A third manner of representing a function is to employ functional notation.

first

c,C

first

363

364 / Appendix Useful Mathematical Results

The element of the second set corresponding to the element x of the first set is denoted by an expression of the type $f(x)$.

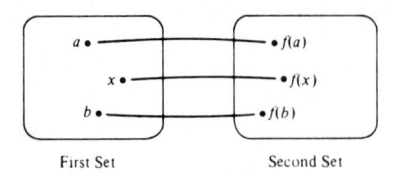

[Diagram: First Set with elements a, x, b mapped to Second Set elements $f(a)$, $f(x)$, $f(b)$]

The functional value $f(x)$ is most commonly expressed by a formula when the first and second sets are sets of numbers.

Example A.1
Write down the collection of ordered pairs represented in functional notation by:

$$f(x) = x^2, \quad x = 1, 2, 3.$$

Solution

$\{(1,1), (2,4), (\underline{})\}$. Here, the first set is the set of numbers $\{1, 2, 3\}$, while the second set is the set of numbers $\{\underline{}\}$.

3,9
1, 4, 9

Example A.2

$$f(x) = 2x^2 - 3x + 1, \quad x = 0, 1, 2, 3.$$

Supply each of the following functional values:

$f(0) = \underline{}$. [1]

$f(1) = \underline{}$. [0]

$f(2) = \underline{}$. [3]

$f(3) = \underline{}$. [10]

This function is the collection of ordered pairs:

$$\{(0,1), (1,0), (2,3), (\underline{})\}.$$

Some Useful Functions
Functions of the type $f(x) = (1/2)^x$, $x = 0, 1, 2, \ldots, n$, are used in this text. We shall be consistent with mathematical convention and define $a^0 = 1$ if a is not zero. Thus if $f(x) = (1/2)^x$, then $f(0) = 1$. Also $f(2) = 1/4$ and $f(4) = \underline{}$. [1/16] The letter used to represent a typical value from the first set (here we have used the letter x) is called the independent variable.

$f(x)$ is then said to be the _____ variable since the value taken on by $f(x)$ depends on the value used for x.   dependent

The factorial function, $f(n) = n!, n = 0, 1, 2, \ldots$, is used in the expression of certain important probability distributions. Again we follow mathematical convention and define $f(0) = 0!$ to be 1 ($0! = 1$). To complete the definition of $n!$ we require that $n! = n(n - 1)!$ for $n = 1, 2, 3, \ldots$. Hence,

$1! = (1)(0)! = 1$,

$2! = (2)(1)! = 2$,

$3! = (3)(2)! = 3 \cdot 2 \cdot 1 =$ _____,   6

$4! = (4)(3)! = 4 \cdot 3 \cdot 2 \cdot 1 =$ _____,   24

and so on.

Example A.3

Do not be disturbed if letters other than x are used to represent independent variables or if symbols other than $f(x)$ are used to represent dependent variables. Supply each missing entry in the following table:

Value of Independent Variable	Formula for Dependent Variable	Value of Dependent Variable	
a. $x = 2$	$f(x) = x^2 - 1$	$f(2) =$ _____	3
b. $y = -1$	$g(y) = 2y + 1$	$g(-1) =$ _____	-1
c. $x = 2$	$f(x) = 3^x - 1$	$f(2) =$ _____	8
d. $t = 0$	$f(t) = 3^t - 1$	$f(0) =$ _____	0
e. $u = 3$	$h(u) = u!$	$h(3) =$ _____	6
f. $u = 0$	$h(u) = u! + 1$	$h(0) =$ _____	2
g. $i = 5$	$g(i) = x_i$	$g(5) =$ _____	x_5

Self-Correcting Exercises A

1. $f(x) = 2x + 3$. Find $f(4)$.
2. $g(t) = t^2 - 2$. Find $g(0)$ and $g(4)$.
3. $h(x) = x!$ Find $h(0)$ and $h(4)$.
4. $g(u) = u!$ Find $g(0) + g(4)$.

A.3 Numerical Sequences

If the functional values corresponding to the values $y = 1, y = 2, \ldots$ for the independent variable are arranged so that $f(1)$ is in the first position, $f(2)$ is in the second position, $f(3)$ is in the third position, and so on, we say that the ordered array

$\{f(1), f(2), f(3), \ldots\}$

is a sequence. To select or identify a specific term in a sequence, we need only specify its position in the sequence. For example, if the first four terms of a sequence are 3, 6, 9, 12, the fifth and sixth terms are 15 and _____ , respectively. When the elements of a sequence correspond to the functional values of $f(y)$ for $y = 1, 2, 3, \ldots$, then y is often called a position variable, since the value given to y determines the position of the element $f(y)$ in the sequence.

18

Example A.4
Find a formula expressing the typical element of the sequence

$$3, 6, 9, 12, \ldots$$

as a function of its position in the sequence.

Solution
We note that the element in the first position is 3(1), the element in the second position is 3(2), the third is _____ , and so on. Hence, a formula expressing the typical element as a function of its position is

3(3)

$$f(y) = \underline{\qquad}, \quad y = 1, 2, 3, \ldots.$$

3y

Example A.5
Find a formula expressing the typical element of the sequence

$$5, 8, 11, 14, \ldots$$

as a function of its position in the sequence.

Solution
1. The terms of this sequence appear to be related to the sequence in Example A.4, since $5 = 3(1) + 2$, $8 = 3(2) + 2$, $11 = 3(\underline{\qquad}) + 2$, and so on. Hence, we could use the function

3

$$f(x) = \underline{\qquad}, \quad x = 1, 2, 3, \ldots$$

3x + 2

2. But the sequence could also be described by considering $5 = 3(2) - 1$, $8 = 3(3) - 1$, $11 = 3(4) - 1$, and so on, so that we might try

$$f(x) = 3(x + 1) - 1, \quad x = 1, 2, 3, \ldots$$

However, upon algebraic simplification, we find that

$$3(x + 1) - 1 = 3x + 3 - 1 = \underline{\qquad}$$

3x + 2

as in step 1.

3. We see, then, that if two students find two seemingly different functions using position as the independent variable and these two functions generate

the same sequence, algebraic manipulation will show that the two functions are identical.

Statisticians generally work with sequences of measurements and therefore need a convenient way of referring to the first, second, ..., and last measurement observed. Suppose that a statistician were working with measurements that were the construction costs per mile for ten different sections of an interstate highway. Wishing to use the variable c to designate cost, he would probably write c_1 (c-sub-one) to designate the cost for the first section, c_2 to designate the cost for the second section, and so on. Hence, he would refer to his measurements as

$$c_1, c_2, c_3, \ldots, c_9, c_{10}.$$

In this case, we would write a typical element of this sequence as

$$f(i) = \underline{\qquad}, \quad i = 1, 2, \ldots, 10, \qquad\qquad c_i$$

using i as the position variable. Hence, $f(3) = \underline{\qquad}$, the cost of the third section. $\qquad c_3$

A.4 Summation Notation

Consider the expression

$$\sum_{x=2}^{4} f(x).$$

1. The Greek letter Σ (capital sigma) is an instruction to perform the operation of _____. $\qquad\qquad$ addition
2. $f(x)$ is the xth term of the sequence $\{f(1), f(2), f(3), \ldots\}$ and is called the typical _____ of summation. $\qquad\qquad$ element
3. The notation "$x = 2$" found below the symbol Σ indicates two things:
   a. The letter x is to be used as the position variable or the *variable of summation*.
   b. The first term in the sum is to be $f(2)$, the _____ term in the sequence. $\qquad\qquad$ second

   The number of the first term in the sum is usually referred to as the *lower limit of summation*. In this problem, the lower limit of summation is _____. $\qquad\qquad$ 2
4. The number "4" above the symbol Σ indicates that the last term in the sum is to be $f(4)$. In general this number is called the *upper limit of summation*.
5. The total expression $\sum_{x=2}^{4} f(x)$ is the instruction to add the second, third, and fourth terms of the sequence $\{f(1), f(2), \ldots\}$.

Example A.6

Evaluate $\sum_{y=1}^{4} y^2$.

Solution

$$\sum_{y=1}^{4} y^2 = 1^2 + 2^2 + 3^2 + 4^2$$

$$= 1 + 4 + 9 + 16$$

$$= \underline{}.$$

30

Example A.7

Evaluate $\sum_{y=2}^{4} (y + 2)$.

Solution

$$\sum_{y=2}^{4} (y + 2) = 4 + 5 + \underline{} = \underline{}.$$

6; 15

You must pay close attention to parentheses in problems involving summations. Notice the difference between the next two problems:

$$\sum_{x=1}^{3} (x^2 - 1) = (1^2 - 1) + (2^2 - 1) + (3^2 - 1) = \underline{}.$$

11

$$\sum_{x=1}^{3} x^2 - 1 = (1^2 + 2^2 + 3^2) - 1 = \underline{}.$$

13

In the first problem, the typical element of summation is $(x^2 - 1)$, which means that within each term one is subtracted from x^2. In the second problem, the typical element of summation is \underline{}, so that after summing, one is subtracted from $\sum_{x=1}^{3} x^2$. The placement or absence of a set of parentheses is crucial in defining the typical element of summation.

x^2

Example A.8
Evaluate the following summations:

a. $\sum_{x=3}^{5} (x^2 + 2x) = 15 + 24 + 35 = \underline{}.$

74

b. $\sum_{z=1}^{3} z/3 + 10 = (1/3 + 2/3 + 3/3) + 10 =$ _____ . | 12

c. $\sum_{i=1}^{4} (i^2 - i + 1) = 1 + 3 + 7 + 13 =$ _____ . | 24

d. $\sum_{j=1}^{4} (j^2 - j) + 1 = (0 + 2 + 6 + 12) + 1 =$ _____ . | 21

Notice that any symbol can be used as a valid variable of summation. In fact, the letters i and j are very commonly used as position variables in statistical problems.

Example A.9

The following measurements represent the cost in cents of five different bunches of broccoli in the produce department of a particular grocery store:

$$x_1 = 119,\ x_2 = 98,\ x_3 = 79,\ x_4 = 89,\ x_5 = 95.$$

Using these values, evaluate the following summations:

a. $\sum_{i=1}^{5} x_i.$

b. $\sum_{i=1}^{5} x_i/5.$

c. $\sum_{i=1}^{5} (x_i - 96).$

d. $\sum_{i=1}^{5} (x_i - 96)^2.$

e. $\sum_{i=1}^{5} x_i^2 - \frac{\left(\sum_{i=1}^{5} x_i\right)^2}{5}.$

Solution

In each of these five problems, the letter i is used as a position variable to designate the ith measurement (cost) in the group of five measurements.

a.

$$\sum_{i=1}^{5} x_i = x_1 + x_2 + x_3 + x_4 + x_5$$

$$= 119 + 98 + 79 + 89 + 95$$

$$= \underline{\qquad}.$$

| 480

370 / Appendix Useful Mathematical Results

b. From part a, $\sum_{i=1}^{5} x_i = 480$, so

96

$$\sum_{i=1}^{5} x_i/5 = 480/5 = \underline{}.$$

c. This problem can be solved directly as

$$\sum_{i=1}^{5} (x_i - 96) = (119 - 96) + (98 - 96) + (79 - 96) + (89 - 96)$$
$$+ (95 - 96)$$

7; 1

$$= 23 + 2 - 17 - \underline{} - \underline{}$$

0

$$= \underline{}.$$

d. Using the results of part c,

$$\sum_{i=1}^{5} (x_i - 96)^2 = (23)^2 + (2)^2 + (-17)^2 + (-7)^2 + (-1)^2$$

872

$$= \underline{}.$$

e. To compute this summation, we first need to find

$$\sum_{i=1}^{5} x_i^2 = 119^2 + 98^2 + 79^2 + 89^2 + 95^2$$

$$= 46{,}952.$$

480

$$\sum_{i=1}^{5} x_i^2 - \frac{\left(\sum_{i=1}^{5} x_i\right)^2}{5} = 46{,}952 - \frac{(\underline{})^2}{5}$$

46,080

$$= 46{,}952 - \underline{}$$

872

$$= \underline{}.$$

(It is no accident that the answers to parts d and e are identical.)

Self-Correcting Exercises B

1. $h(y) = 2(1/3)^y$. Find $\sum_{y=1}^{3} h(y)$.

2. $g(x) = 2x^2 - 5$. Find $\sum_{x=1}^{4} g(x)$.

Use the following set of measurements to answer Exercises 3 through 5.

i	1	2	3	4	5	6	7	8	9	10
x_i	-1	2	1	0	4	-3	1	6	-5	-2

3. $\sum_{i=1}^{10} x_i$.

4. $\sum_{i=1}^{10} x_i^2$.

5. $\sum_{i=1}^{10} x_i/10$.

A.5 Summation Theorems

We will now review three theorems involving summations. These theorems will prove useful when evaluating statistical descriptive measures that characterize a distribution of measurements.

Theorem A.1 $\quad \sum_{x=1}^{n} c = nc.$

We say that c is a constant because c does not depend upon the variable of summation. You may encounter sums with a lower limit different from one. A useful device for arriving at the correct answer in this case is illustrated by the following example.

Example A.10

Evaluate $\sum_{x=4}^{10} 5$.

Solution

10; 3

1. $\sum_{x=1}^{10} 5$ has _____ terms while $\sum_{x=1}^{3} 5$ has _____ terms.

2. But $\sum_{x=4}^{10} 5 = \sum_{x=1}^{10} 5 - \sum_{x=1}^{3} 5.$

7; 35

3. $\sum_{x=4}^{10} 5$ has _____ terms and hence, $\sum_{x=4}^{10} 5 = (10 - 3)5 =$ _____.

Theorem A.2 $\quad \sum_{x=1}^{n} cf(x) = c \sum_{x=1}^{n} f(x).$

An equivalent form of this theorem is $\sum_{i=1}^{n} cx_i = c \sum_{i=1}^{n} x_i.$

6; 150

Thus, $\sum_{x=1}^{3} 25x = 25 \sum_{x=1}^{3} x = 25(\underline{\quad}) = \underline{\quad}.$

Theorem A.3 $\quad \sum_{x=1}^{n} [f(x) \pm g(x)] = \sum_{x=1}^{n} f(x) \pm \sum_{x=1}^{n} g(x).$

An equivalent form of this theorem is

$$\sum_{i=1}^{n} (x_i \pm y_i) = \sum_{i=1}^{n} x_i \pm \sum_{i=1}^{n} y_i$$

Thus,

$$\sum_{x=1}^{6} (x - 5) = \sum_{x=1}^{6} x - \sum_{x=1}^{6} 5$$

6(5)

$= 21 - \underline{\quad}$

-9

$= \underline{\quad}.$

Theorems A.1, A.2, and A.3 may be used in combination, as in the following example.

Example A.11

Evaluate $\sum_{x=1}^{5} (3x^2 + 5x - 2)$.

Solution
1. Using Theorem A.3,

$$\sum_{x=1}^{5} (3x^2 + 5x - 2) = \sum_{x=1}^{5} 3x^2 + \sum_{x=1}^{5} 5x - \sum_{x=1}^{5} 2.$$

2. The first two terms can be simplified using Theorem A.2 to give

$$\sum_{x=1}^{5} (3x^2 + 5x - 2) = 3\sum_{x=1}^{5} x^2 + 5\sum_{x=1}^{5} x - \sum_{x=1}^{5} 2.$$

3. Finally, using Theorem A.1 to simplify $\sum_{x=1}^{5} 2 = 5(2)$, we have

$$\sum_{x=1}^{5} (3x^2 + 5x - 2) = 3(55) + 5(15) - 10$$

230

Example A.12

Evaluate $\sum_{x=1}^{5} (x - 4)^2$.

Solution
1. If the element of summation is not written as a simple sum or difference of terms, it may be possible to convert it to a sum or difference by algebraic manipulation.
2. The typical element of summation is $(x - 4)^2$. But $(x - 4)^2 = x^2 - 8x +$ _____, so that

16

$$\sum_{x=1}^{5}(x-4)^2 = \sum_{x=1}^{5}(x^2 - 8x + 16)$$

$$= \sum_{x=1}^{5} x^2 - 8\sum_{x=1}^{5} x + 5(16)$$

15; 80

$$= 55 - 8(\underline{}) + \underline{}$$

15

$$= \underline{}.$$

Example A.13
Using summation theorems, rewrite the following summation as sums or differences of terms.

$$\sum_{i=1}^{n}(x_i - c)^2$$

Solution
1. Following the technique given in Example A.12, the typical element of summation is $(x_i - c)^2$, which when algebraically squared equals $(x_i^2 - 2cx_i + c^2)$.
2. Using the summation theorems, write

nc^2

$$\sum_{i=1}^{n}(x_i - c)^2 = \sum_{i=1}^{n} x_i^2 - 2c\sum_{i=1}^{n} x_i + \underline{}.$$

3. Note that until the values of $x_1, x_2, \ldots, x_n$ and c are given, we cannot proceed any further with this problem.

Example A.14
Use the results of Example A.13 to find $\sum_{i=1}^{5}(x_i - 3)^2$ if

$$x_1 = 2, \ x_2 = 4, \ x_3 = 1, \ x_4 = 3, \ x_5 = 4.$$

Solution
1. From Example A.13,

$$\sum_{i=1}^{5}(x_i - 3)^2 = \sum_{i=1}^{5} x_i^2 - 6\sum_{i=1}^{5} x_i + 5(9).$$

Therefore, we need to find $\sum_{i=1}^{5} x_i^2$ and $\sum_{i=1}^{5} x_i$.

2. $\sum_{i=1}^{5} x_i^2 = 2^2 + 4^2 + 1^2 + 3^2 + 4^2 =$ _____ . 46

$\sum_{i=1}^{5} x_i = 2 + 4 + 1 + 3 + 4 =$ _____ . 14

3. Collecting results we have

$\sum_{i=1}^{5} (x_i - 3)^2 =$ _____ $- 6($_____$) + 45$ 46; 14

$=$ _____ .   7

Although this method may seem slow and sluggish and you feel you would rather use $\sum_{i=1}^{5} (x_i - 3)^2$ directly for this problem, when the number of observations, n, is larger than 5 and the measurements are not small whole numbers, the expanded form of the summation will prove to be the simpler of the two to use in practice.

Self-Correcting Exercises C

1. Evaluate $\sum_{x=1}^{10} 3$.

2. Evaluate $\sum_{i=7}^{14} 4$.

3. Refer to Self-Correcting Exercises B, Exercises 3, 4, and 5. Using the set of measurements given there, evaluate the following summations:

   a. $\sum_{i=3}^{8} (x_i - 4)$.

b. $\sum_{i=1}^{10} (x_i - 5)^2.$

c. $\sum_{i=1}^{5} (x_i^2 - x_i).$

d. $\sum_{i=1}^{10} x_i^2 - \dfrac{\left(\sum_{i=1}^{10} x_i\right)^2}{10}.$

A.6 The Binomial Theorem

The following identity is proved in most high school algebra books:

$$(a + b)^n = \sum_{x=0}^{n} \frac{n!}{x!(n-x)!} a^x b^{n-x}$$

where a and b are any real numbers and n is a positive integer. This identity is known as the binomial theorem. The coefficients in the sum are known as binomial coefficients and we can write

$$\frac{n!}{x!(n-x)!} = C_x^n$$

to represent the coefficient of $a^x b^{n-x}$ in the expansion of $(a + b)^n$.

For the special case when $n = 2$ we find

$$(a + b)^2 = \sum_{x=0}^{2} C_x^2 a^x b^{2-x}$$

$$= a^2 + 2ab + b^2.$$

When $n = 3$, we have

$$(a + b)^3 = \sum_{x=0}^{3} C_x^3 a^x b^{3-x}$$

$$= a^3 + 3a^2 b + 3ab^2 + b^3.$$

If a is replaced by a probability p and b is replaced by the probability $q = 1 - p$, then the binomial theorem implies that

$$\sum_{x=0}^{n} C_x^n p^x q^{n-x} = (p+q)^n.$$

Note the following points about this binomial identity involving p and q when $0 < p < 1$:
1. The terms $C_x^n p^x q^{n-x}$ are each positive.
2. The sum of the terms is $(p+q)^n = 1$.
3. Since the total can be no greater than any of its parts, then

$$0 \leqslant C_x^n p^x q^{n-x} \leqslant 1.$$

Thus the function

$$p(x) = C_x^n p^x q^{n-x}, \quad x = 0, 1, 2, \ldots, n,$$

satisfies the two requirements of a probability function for a discrete random variable, namely,

$$0 \leqslant p(x) \leqslant 1 \quad \text{and} \quad \sum_x p(x) = 1.$$

EXERCISES

1. $f(y) = 3y - 2u$. Find $f(u)$.
2. $G(u) = 3u^3 - 2u$. Find $G(a)$.
3. $H(y) = 3(1/2)^y$. Find $H(0)$ and $H(2)$.
4. $f(u) = 2u^2 - 3u + 5$. Find $f(1/x)$.
5. $F(u) = 2u^2 - 3/u$. Find $F(1/v)$.
6. Expand $\sum_{i=3}^{6} (x_i - a)$.
7. Express the following in a more compact form by using summation notation:

$$(x_3 - m)^2 + (x_4 - m)^2 + (x_5 - m)^2 + (x_6 - m)^2.$$

8. Evaluate $\sum_{u=1}^{3} (3x + u!)$.

9. Evaluate $\sum_{x=0}^{3} \frac{(.75)}{x! \, (3-x)!}$.

10. Evaluate $\sum_{x=7}^{8} 4x$.

11. If $\sum_{i=1}^{25} y_i = 50$, find $\sum_{i=1}^{25} (2y_i - 3)$.

12. If $\sum_{i=1}^{25} y_i = 50$ and $\sum_{i=1}^{25} y_i^2 = 250$, find $\sum_{i=1}^{25} (y_i - 10)^2$.

SOLUTIONS TO SELF-CORRECTING EXERCISES

Set 2A

1. a. range = 55 − 30 = 25.
   b.–c. Each student will obtain slightly different results. Dividing the range by 5 produces intervals of length 5. A convenient choice is to use intervals of length 5, beginning at 29.5. Note that this forces the use of 6 intervals.

Class	Class Boundaries	Tally	f_i
1	29.5–34.5	⊮⊮ ⊮⊮ II	12
2	34.5–39.5	⊮⊮ ⊮⊮ ⊮⊮	15
3	39.5–44.5	⊮⊮ ⊮⊮ III	13
4	44.5–50.5	⊮⊮ IIII	9
5	50.5–54.5		0
6	54.5–59.5	I	1

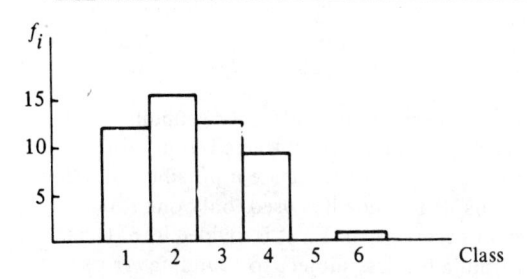

   d. The distribution is skewed, with a "piling up" of the data in the earlier (younger) age classes.

2. a. An extra column in the tabulation is used to calculate relative frequency.

Class	Class Boundaries	Tally	f_i	f_i/n
1	5.55–7.55	⊮⊮	5	5/32
2	7.55–9.55	⊮⊮	5	5/32
3	9.55–11.55	⊮⊮ ⊮⊮ II	12	12/32
4	11.55–13.55	⊮⊮	5	5/32
5	13.55–15.55	III	3	3/32
6	15.55–17.55	I	1	1/32
7	17.55–19.55	I	1	1/32

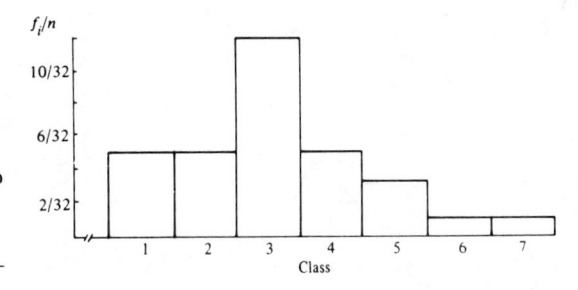

   b. $(1/32) + (1/32) = 2/32$.
   c. $(5/32) + (5/32) = 10/32$.
   d. $(12/32) + (5/32) + (3/32) = 20/32$.

Set 2B

1. The stems range from 1 to 5, and the leaves have been reordered in order of ascending magnitude.

1	8 9
2	1 1 3 3 3 5 6 7 7 8 8 9
3	0 0 1 2 2 2 3 4 4 5 5 6 7 8 9
4	1 1 2 3 4 4 6 9
5	0 1 2 5 9

2. a. The stems range from 0 to 10. One large measurement was placed in the HI category.

0	0 0 2 3 3 4 4 6 7 7 8 8 8
1	3 4 5
2	1 2 4 5 7
3	3 5 5 7 8 9
4	1 1 7 7 8 9
5	4 5 7
6	5 9 9
7	1
8	1 4 6 7
9	0 1 6
10	5 9
HI	186.0

380 / Solutions to Self-Correcting Exercises

b. The measurement 186.0 is a possible outlier. Most of the number of farms per state lie within 1,000 and 11,000.
c. The stem-and-leaf display might be improved by combining the stems 0 and 1, 2 and 3, 4 and 5, 6 and 7, 8 and 9, and 10 and 11.

Set 2C

1. Arrange the set of data in order of ascending magnitude:

6	9	11	13	16
8	10	12	13	17
9	10	12	15	19

median = 12; $\bar{x} = \sum_{i=1}^{n} x_i/n = 180/15 = 12$.

2. Arrange the data in order of ascending magnitude.

x_i	$(x_i - \bar{x})$	$(x_i - \bar{x})^2$
0	-3	9
1	-2	4
2	-1	1
2	-1	1
2	-1	1
3	0	0
3	0	0
4	1	1
4	1	1
5	2	4
7	4	16
33	0	38

a. median = 3, $\bar{x} = 33/11 = 3$.
b. range = 7 − 0 = 7.
c. $s^2 = \Sigma(x_i - \bar{x})^2/(n-1) = 38/10 = 3.8$; $s = \sqrt{3.8} = 1.95$.

Set 2D

1. Display the data in a table as follows:

x_i	x_i^2
36	1296
48	2304
45	2025
29	841
49	2401
35	1225
65	4225
307	14317

$$s^2 = \frac{\Sigma x_i^2 - \frac{(\Sigma x_i)^2}{n}}{n-1} = \frac{14{,}317 - (307)^2/7}{6} = 142.14286$$

Then the standard deviation is $s = \sqrt{s^2} = 11.92$.

2. Display the data in a table as follows:

x_i	x_i^2
0	0
1	1
2	4
2	4
2	4
3	9
3	9
4	16
4	16
5	25
7	49
33	137

$$s^2 = \frac{\Sigma x_i^2 - (\Sigma x_i)^2/n}{n-1} = \frac{137 - (33)^2/11}{10} = \frac{137 - 99}{10} = \frac{38}{10} = 3.8.$$

3. If $\bar{x}$ has been rounded off, then rounding error occurs each time $\bar{x}$ is subtracted from x_i in formula a. Hence there are n possible rounding errors. If formula b is used, only one rounding error occurs when $(\Sigma x_i)^2$ is divided by n. Hence formula b is less subject to rounding errors and results in a more accurate computation.

4. $\Sigma x_i = 29.7$, $\Sigma x_i^2 = 129.19$,

$$s^2 = \frac{\Sigma x_i^2 - (\Sigma x_i)^2/n}{n-1} = \frac{129.19 - (29.7)^2/7}{6}$$

$$= \frac{129.19 - 126.0129}{6} = \frac{3.1771}{6} = .5295,$$

$s = \sqrt{.5295} = .73$.

5. $\Sigma x_i = 356$, $\Sigma x_i^2 = 25{,}362$,

$$s^2 = \frac{25{,}362 - (356)^2/5}{4} = \frac{25{,}362 - 25{,}347.2}{4}$$

$$= \frac{14.8}{4} = 3.7,$$

$$\bar{x} = \frac{356}{5} = 71.2.$$

Set 2E

1. a. Since $n = 8$, the median is average of the fourth and fifth ordered measurements, which are given below.

   3, 4, 4.5, 5, 7.5, 8.5, 12, 30

   Then $m = \frac{5 + 7.5}{2} = 6.25$. The lower quartile has position $.25(n + 1) = 2.25$ while the upper quartile has position $.75(n + 1) = 6.75$.

   Then $Q_L = 4 + .25(4.5 - 4) = 4.125$,

   while $Q_U = 8.5 + .75(12 - 8.5) = 11.125$.

   b. Calculate $\Sigma x_i = 74.5$, $\Sigma x_i^2 = 1242.75$,

   $\bar{x} = \frac{74.5}{8} = 9.3125$, $s^2 = 78.4241$. The z-scores are calculated as $z = \frac{x - \bar{x}}{s}$ and are shown below for the ordered data.

   $z_1 = -.71$ $z_5 = -.20$
   $z_2 = -.60$ $z_6 = -.09$
   $z_3 = -.54$ $z_7 = .30$
   $z_4 = -.49$ $z_8 = 2.24$

   c. The largest observation, $x = 30$, is a suspected outlier.

2. Since $n = 42$, the median is the average of the 21st and 22nd ordered observations. Then $m = 33.5$. Calculate $(n + 1)/2 = 21.5$, so that $d(M) = 21$ and the hinge position is

   $$\frac{d(M) + 1}{2} = 11.$$

   lower hinge $= 27$

   upper hinge $= 42$

   H-spread $= 15$.

   The inner fences are

   $27 - 1.5(15) = 4.5$

   $42 + 1.5(15) = 64.5$

   The outer fences are

   $27 - 3(15) = -18$

   $42 + 3(15) = 87$

   The box plot is shown below. There are no outliers.

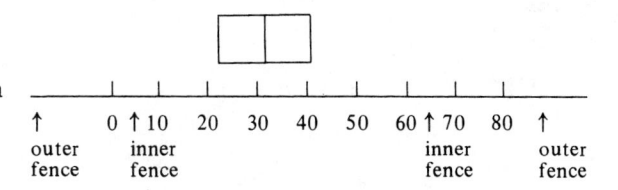

   ↑   0 ↑10 20 30 40 50 60 ↑70 80 ↑
   outer inner   inner   outer
   fence fence   fence   fence

3. Since $n = 50$, the median is the average of the 25th and 26th ordered measurements. Calculate $(n + 1)/2 = 25.5$, so that $d(M) = 25$ and the hinge position is

   $$\frac{d(M) + 1}{2} = 13.$$

   Then,

   lower hinge $= 8.9$

   upper hinge $= 69.0$

   H-spread $= 69.0 - 8.9 = 60.1$.

   The inner fences are

   $8.9 - 1.5(60.1) = -81.25$

   $69.0 + 1.5(60.1) = 159.15$.

   The outer fences are

   $8.9 - 3(60.1) = -171.4$

   $69.0 + 3(60.1) = 249.3$.

   The value $x = 186$ is a suspect outlier.

Set 3A

1. a. One camera is drawn at random from five, after which a second camera is chosen. Each is tested and found to be defective or nondefective.

   b. Since it is important whether the first or second camera is defective (see event D), there are 20 simple events to be listed. The first element in the pair denotes the first camera chosen.

   $E_1: G_1G_2$ $E_8: G_3D_1$ $E_{15}: G_3G_2$
   $E_2: G_1G_3$ $E_9: G_3D_2$ $E_{16}: D_1G_2$

382 / Solutions to Self-Correcting Exercises

$E_3: G_1 D_1$ $E_{10}: D_1 D_2$ $E_{17}: D_2 G_2$
$E_4: G_1 D_2$ $E_{11}: G_2 G_1$ $E_{18}: D_1 G_3$
$E_5: G_2 G_3$ $E_{12}: G_3 G_1$ $E_{19}: D_2 G_3$
$E_6: G_2 D_1$ $E_{13}: D_1 G_1$ $E_{20}: D_2 D_1$
$E_7: G_2 D_2$ $E_{14}: D_2 G_1$

c. $A: \{E_{10}, E_{20}\}$
$B: \{E_1, E_2, E_5, E_{11}, E_{12}, E_{15}\}$
$C: \{E_3, E_4, E_6, E_7, E_8, E_9, E_{10}, E_{13}, E_{14}, E_{16}, E_{17}, E_{18}, E_{19}, E_{20}\}$
$D: \{E_{10}, E_{13}, E_{14}, E_{16}, E_{17}, E_{18}, E_{19}, E_{20}\}$.

d. $P(A) = 2/20 = 1/10$ $P(B) = 6/20 = 3/10$
$P(C) = 14/20 = 7/10$ $P(D) = 8/20 = 2/5$.

2. Denote the four good items as G_1, G_2, G_3, G_4 and the two defectives as D_1 and D_2.
a. $E_1: G_1 G_2$ $E_6: G_2 G_3$ $E_{11}: G_3 D_1$
$E_2: G_1 G_3$ $E_7: G_2 G_4$ $E_{12}: G_3 D_2$
$E_3: G_1 G_4$ $E_8: G_2 D_1$ $E_{13}: G_4 D_1$
$E_4: G_1 D_1$ $E_9: G_2 D_2$ $E_{14}: G_4 D_2$
$E_5: G_1 D_2$ $E_{10}: G_3 G_4$ $E_{15}: D_1 D_2$

b. "At least one defective" implies one or two defectives, while "no more than one defective" implies zero or one defective.
$A: \{E_4, E_5, E_8, E_9, E_{11}, E_{12}, E_{13}, E_{14}, E_{15}\}$
$B: \{E_4, E_5, E_8, E_9, E_{11}, E_{12}, E_{13}, E_{14}\}$
$C: \{E_1, E_2, E_3, E_4, E_5, E_6, E_7, E_8, E_9, E_{10}, E_{11}, E_{12}, E_{13}, E_{14}\}$

c. Each simple event is assigned equal probability; that is, $P(E_i) = 1/15$.
$P(A) = 9/15 = 3/5$, $P(B) = 8/15$, $P(C) = 14/15$.

3. The experiment consists of choosing two photographs from a total of four, and the six simple events are
E_1: 1,2 E_2: 1,3 E_3: 1,4 E_4: 2,3 E_5: 2,4 E_6: 3,4
Then: $A: \{E_1\}$; $B: \{E_2, E_4, E_6\}$; C: no simple events; $D: \{E_1, E_2, E_4, E_6\}$;
$P(A) = 1/6$; $P(B) = 3/6 = 1/2$; $P(C) = 0$;
$P(D) = 4/6 = 2/3$.

4. a. E_1: HHH E_3: NHN E_5: HHN E_7: NHH
E_2: HNN E_4: NNH E_6: HNH E_8: NNN

b. $P(A) = 3/8$; $P(B) = 4/8 = 1/2$; $P(C) = 1/8$;
$P(D) = 4/8 = 1/2$.

Set 3B

1. a. $A \cup C$: same as C; AD: same as A; CD: same as D; $A \cup D$: same as D.
b. $P(A \cup C) = 7/10$, $P(AD) = 1/10$, $P(CD) = 2/5$, $P(A \cup D) = 2/5$.

2. a. E_1: FFFF E_9: MFFM
E_2: MFFF E_{10}: MFMF
E_3: FMFF E_{11}: MMFF
E_4: FFMF E_{12}: FMMM
E_5: FFFM E_{13}: MFMM
E_6: FFMM E_{14}: MMFM
E_7: FMFM E_{15}: MMMF
E_8: FMMF E_{16}: MMMM

b. $A: \{E_6, E_7, E_8, E_9, E_{10}, E_{11}\}$
$B: \{E_1\}$
$C: \{E_2, E_3, E_4, E_5, E_6, E_7, E_8, E_9, E_{10}, E_{11}, E_{12}, E_{13}, E_{14}, E_{15}, E_{16}\}$

c. AB: { no simple events }; $B \cup C$:
$\{E_1, E_2, \ldots, E_{16}\} = S$;
$AC \cup BC: \{E_6, E_7, E_8, E_9, E_{10}, E_{11}\}$;
$\overline{C}: \{E_1\}$;
$\overline{AC}: \{E_1, E_2, \ldots, E_5, E_{12}, E_{13}, \ldots, E_{16}\}$.

d. Since each simple event is equally likely,

$P(A) = 6/16 = 3/8$ $P(B \cup C) = 1$
$P(B) = 1/16$ $P(AC \cup BC) = 3/8$
$P(C) = 15/16$ $P(\overline{C}) = 1/16$
$P(AB) = 0$ $P(\overline{AC}) = 10/16 = 5/8$

e. $P(A|C) = \dfrac{P(AC)}{P(C)} = \dfrac{6/16}{15/16} = \dfrac{6}{15}$ while $P(A) = \dfrac{3}{8}$.

A and C are dependent but are not mutually exclusive.

f. $P(B|C) = P(BC)/P(C) = 0 \Big/ \dfrac{15}{16} = 0$ while $P(B) = \dfrac{1}{16}$ and $P(BC) = 0$. B and C are dependent and mutually exclusive.

3. a. $P(A) = P$ (the executive represents a small corporation) $= 75/200 = 3/8$;
$P(F) = P$ (the executive favors gas rationing) $= 15/200 = 3/40$; $P(AF) = 3/200$;
$P(A \cup G) = P$ (executive favors conversion or represents a small corporation or both)
$= P(A) + P(G) - P(AG) = (75 + 22 - 10)/200$
$= 87/200$;
$P(AD) = P$ (executive represents a small corporation and favors car pooling) $= 20/200$;
$P(\bar{F}) = 1 - P(F) = 1 - (3/40) = 37/40$.

b. $P(A|F) = \dfrac{P(AF)}{P(F)} = \dfrac{3/200}{15/200} = \dfrac{3}{15}$,

$P(A|D) = \dfrac{20/200}{55/200} = \dfrac{20}{55}$. Neither A and F nor A and D are mutually exclusive. A and F and A and D are both dependent.

4. Define A: company A shows an increase; B: company B shows an increase; C: company C shows an increase. It is given that $P(A) = .4$, $P(B) = .6$, $P(C) = .7$, and A, B, and C are independent events.

a. $P(ABC) = P(A) P(B) P(C) = (.4)(.6)(.7) = .168$.

b. $P(\bar{A}\,\bar{B}\,\bar{C}) = P(\bar{A}) P(\bar{B}) P(\bar{C})$
$= [1 - P(A)] [1 - P(B)] [1 - P(C)]$
$= (.6)(.4)(.3) = .072$

c. P (at least one shows profit)
$= 1 - P$ (none show profit)
$= 1 - P(\bar{A}\,\bar{B}\,\bar{C}) = 1 - .072 = .928$.

5. Define H: marksman hits the target; $\bar{H}$: marksman misses the target. It is given that $P(H) = .9$. Therefore, $P(\bar{H}) = 1 - .9 = .1$.

a. Since his shots are independent, $P(HHH)$
$= P(H) P(H) P(H) = (.9)^3 = .729$.

b. The event of interest is A: marksman hits with at least one of the next three shots. Then $\bar{A}$: marksman misses all three shots. $P(A)$
$= 1 - P(\bar{A}) = 1 - P(\bar{H}\bar{H}\bar{H}) = 1 - (.1)^3 = .999$.

6. Define A_1: person passes on first try; A_2: person passes on second try; A_3: person passes on third try. It is given that $P(A_1) = .7$, $P(A_2|\bar{A}_1) = .8$, $P(A_3|\bar{A}_1 \bar{A}_2) = .9$.

a. P(person passes on second try) $= P(\bar{A}_1 A_2)$
$= P(\bar{A}_1) P(A_2|\bar{A}_1) = .3(.8) = .24$.

b. P(person passes on third try) $= P(\bar{A}_1 \bar{A}_2 A_3)$
$= P(\bar{A}_1) P(\bar{A}_2|\bar{A}_1) P(A_3|\bar{A}_1 \bar{A}_2) = .3(.2)(.9)$
$= .054$.

c. P(person passes) $= P$(passes on first, second, or third tries) $= P(A_1) + .24 + .054$
$= .7 + .24 + .054 = .994$.

Set 3C

1. Define the events F: a favorable seismic outcome results; O: oil is actually present; $\bar{O}$: oil is not present. $P(F|O) = .8$; $P(F|\bar{O}) = .3$; $P(O) = .5$; $P(\bar{O}) = 1 - P(O) = .5$;

$P(O|F) = \dfrac{P(F|O) P(O)}{P(F|O) P(O) + P(F|\bar{O}) P(\bar{O})}$

$= \dfrac{(.8)(.5)}{(.8)(.5) + (.3)(.5)} = \dfrac{8}{11}$.

2. Define the events D: a defective item is passed by an inspector; A_1: inspector 1 inspected the item; A_2: inspector 2 inspected the item. $P(A_1) = .6$; $P(A_2) = .4$; $P(D|A_1) = .01$; $P(D|A_2) = .05$;

$P(A_1|D) = \dfrac{P(D|A_1) P(A_1)}{P(D|A_1) P(A_1) + P(D|A_2) P(A_2)}$

$= \dfrac{.01(.6)}{.01(.6) + .05(.4)} = \dfrac{.006}{.026} = .23$.

Set 3D

1. a. Since the pairs are ordered (see part b, Exercise 1, Self-Correcting Exercises 3A), the number of simple events is $P_2^5 = (5)(4) = 20$.

b. Event A: $P_2^2 = (2)(1) = 2$; event B: $P_2^3 = (3)(2) = 6$; event C: $2 P_1^2 P_1^3 + P_2^2 = 2(2)(3) + 2 = 14$ since the defective item may be chosen either first or second; event D: $P_1^2 P_1^3 + P_2^2$
$= 6 + 2 = 8$ since once the first camera is found to be defective, the second can be either defective or good.

c. See Exercise 1d, Self-Correcting Exercises 3A.

2. Since the awards are different, order is important.
a. $P_3^5 = 5!/2! = 60$.
b. $P_3^3 = 3! = 3(2) = 6$.
c. P(ctiy officials receive three awards)
$= 6/60 = 1/10$

3. Since a random sample of size 6 does not involve order, we have the following:
a. $C_6^{20} = 38{,}760$.
b. Drawing 6 males from a total of 10 males can be done in $C_6^{10} = 210$ ways. Similarly, 6 females can be drawn in $C_6^{10} = 210$ ways.
c. Three females or 3 males can be drawn in $C_3^{10} = 120$ ways. Using the mn rule, 3 men *and* 3 women can be drawn in $C_3^{10} C_3^{10} = 120^2$
$= 14{,}400$ ways.

d. P(all persons of same sex) = P(all men) + P(all women) = (210 + 210)/38,760 = .0108.
e. P(3 males and 3 females) = 14,400/38,760 = .3715.
f. Yes, 37% is a fairly high probability.

Set 3E

1. The simple events are $(A, B), (A, C), (A, D), (B, C), (B, D), (C, D)$. If there is no discrimination in the selection, each simple event has probability 1/6. Collecting information,

Simple Events	x	$p(x)$
(C, D)	0	1/6
$(A, C), (A, D), (B, C), (B, D)$	1	4/6
(A, B)	2	1/6

2. a. $p(0) = -2/10$ is not between 0 and 1
   b. $\sum_x p(x) = 7/10 \ne 1$.

3. Let U be the event that the primary wage earner is unemployed. There are 16 simple events with unequal probabilities.

   $p(0) = P[\bar{U}\bar{U}\bar{U}\bar{U}] = (.93)^4 = .7481$
   $p(1) = 4P(\bar{U})^3 P(U) = 4(.93)^3(.07) = .2252$
   $p(2) = 6P(\bar{U})^2 P(U)^2 = 6(.93)^2(.07)^2 = .0254$
   $p(3) = 4P(\bar{U})P(U)^3 = 4(.93)(.07)^3 = .0013$
   $p(4) = P(UUUU) = (.07)^4 = .000024$.

Set 3F

1. a. Since $\sum_x p(x) = 1, p(0) = 1 - (6/9) = 3/9$.
   b. $\mu = \sum_x x p(x) = -2(1/9) + (-1)(1/9) + 0(4/9) + 1(3/9)$
   $= (-2/9) - (1/9) + (3/9) = 0$.
   c. $\sigma^2 = E(x^2) - \mu^2 = (-2)^2(1/9) + (-1)^2(1/9) + 0^2(4/9) + 1^2(3/9) - 0^2$
   $= (4/9) + (1/9) + (3/9) = 8/9 = .8889$
   $\sigma = \sqrt{.8889} = .94$.

2. a. $\mu = E(x) = \sum_x x p(x) = 0(.1) + 1(.6) + 2(.2) + 3(.1) = .6 + .4 + .3 = 1.3$.

b. $\sigma^2 = E((x - \mu)^2) = (0 - 1.3)^2(.1) + (1 - 1.3)^2(.6) + (2 - 1.3)^2(.2) + (3 - 1.3)^2(.1) = .61$, or
$\sigma^2 = E(x^2) - \mu^2 = 0^2(.1) + 1^2(.6) + 2^2(.2) + 3^2(.1) - (1.3)^2$
$= .6 + .8 + .9 - 1.69 = .61$.

c. $\mu \pm 2\sigma = 1.3 \pm 2(.61) = 1.3 \pm 1.22$, or .08 to 2.52. Referring to the initial probability distribution, $P(.08 < x < 2.52) = .6 + .2 = .8$, which agrees with Tchebysheff's Theorem.

3. $P(x \geq 2) = p(2) + p(3) = .2 + .1 = .3$.

4. Let x be the gain to the insurance company and let r be the premium charged by the company.

x	$p(x)$
r	.9900
$-40,000 + r$	.0075
$-80,000 + r$	.0025

In order to break even, $E(x) = 0$, or
$E(x) = \sum_x x p(x) = .99r + .0075(-40,000 + r) + (-80,000 + r)(.0025) = 0$
$r - 300 - 200 = 0$
$r = \$500$.

Set 4A

1. Let x be the number of apartment dwellers who move within a year. Then $p = P$(move within a year) = .2 and $n = 7$.
   a. $P(x = 2) = C_2^7 (.2)^2 (.8)^5 = .27525$.
   b. $P(x \leq 1) = C_0^7 (.2)^0 (.8)^7 + C_1^7 (.2)^1 (.8)^6$
   $= .209715 + .367002 = .57617$.

2. Let x be the number of letters delivered within 4 days. Then $p = P$(letter delivered within 4 days) = .7 and $n = 20$.
   a. $P(x \geq 15) = 1 - P(x \leq 14) = 1 - .584 = .416$.
   b. $P(x \geq 10) = 1 - P(x \leq 9) = 1 - .017 = .983$. Notice that if 10 or fewer letters arrive later than 4 days, then 20 - 10 = 10 or more will arrive within 4 days.

3. Use Table 1 in the text, indexing $n = 15$, $p = 1/2$.

a. $P(x = 4) = P(x \leq 4) - P(x \leq 3) = .059 - .018 = .041$.

b. $P(x \geq 4) = 1 - P(x \leq 3) = 1 - .018 = .982$.

4. Let x be the number of minority members on a list of 80, so that $p = .20$ and $n = 80$. Then $\mu = 80(.2) = 16$ and $\sigma = \sqrt{npq} = \sqrt{12.8} = 3.58$. The limits are $\mu \pm 2\sigma = 16 \pm 7.16$, or 8.84 to 23.16. We expect to see between 9 and 23 minority group members on the jury lists.

5. x = number watching the TV program; $p = .4$; $n = 400$; $\mu = 400(.4) = 160$; $\sigma^2 = 400(.4)(.6) = 96$; $\sigma = \sqrt{96} = 9.80$. Calculate $\mu \pm 2\sigma = 160 \pm 2(9.8) = 160 \pm 19.6$, or 140.4 to 179.6. Since we would expect the number watching the show to be between 141 and 179 with probability .95, it is highly unlikely that only 96 people would have watched the show *if* the 40% claim is correct. It is more likely that the percentage of viewers for this particular show is less than 40%.

6. a. $P(\text{acceptance}) = C_0^4 p^0 q^4 + C_1^4 p^1 q^3$ for various values of p. When $p = 0$, $P(\text{acceptance}) = 1$; $p = .3$, $P(\text{acceptance}) = (.7)^4 + 4(.3)(.7)^3 = .2401 + .4116 = .6517$. When $p = 1$, $P(\text{acceptance}) = 0$. The graph follows the procedures used in Example 4.13 and is omitted here.

   b. Keep $n = 4$, take $a > 1$; keep $a = 1$, take $n < 4$.

7. a.

p	0	.1	.3	.5	1.0
$n = 10, a = 1$	1	.736	.149	.011	0
$n = 25, a = 1$	1	.271	.002	.000	0

   b.-c. If the student will graph the 3 O.C. curves as given in part a and Exercise 6, he will see that increasing n has the effect of decreasing the probability of acceptance.

8. a.

p	0	.1	.3	.5	1.0
$n = 25, a = 3$	1	.764	.033	.000	0
$n = 25, a = 5$	1	.967	.193	.002	0

   b.-c. Increasing a has the effect of increasing the probability of acceptance.

Set 4B

1. Let x be the number of fires observed so that $p = P[\text{fire}] = .005$ and $n = 1000$. The random variable is binomial; however, since n is large and p is small with $\mu = np = 5$, the Poisson approximation is appropriate.

   a. $P[x = 0] = \dfrac{\mu^0 e^{-\mu}}{0!} = e^{-5} = .006738$

   b. $P[x \leq 3] = \dfrac{5^0 e^{-5}}{0!} + \dfrac{5^1 e^{-5}}{1!} + \dfrac{5^2 e^{-5}}{2!} + \dfrac{5^3 e^{-5}}{3!}$

   $= .006738(1 + 5 + 12.5 + 20.833) = .2650$

2. Let x be the number of defective panels with $\mu = 2$.

   a. $P[x = 3] = \dfrac{2^3 e^{-2}}{3!} = 1.33(.135335) = .180$

   b. $P[x \geq 2] = 1 - P[x \leq 1] = 1 - \dfrac{2^0 e^{-2}}{0!} - \dfrac{2^1 e^{-2}}{1!}$

   $= 1 - e^{-2}(1 + 2) = 1$
   $= 1 - 3(.135335)$
   $= .594$

3. We are now concerned with the random variable x, the number of defective panels in a bundle of 200, with $\mu = 2(2) = 4$. Then

   $P[x \leq 4] = \dfrac{4^0 e^{-4}}{0!} + \dfrac{4^1 e^{-4}}{1!} + \dfrac{4^2 e^{-4}}{2!} + \dfrac{4^3 e^{-4}}{3!} + \dfrac{4^4 e^{-4}}{4!}$

   $= .018316(1 + 4 + 8 + 10.67 + 10.67)$
   $= .629$

4. Four employees will be chosen from 15, nine of whom are men and six of whom are women. Hence, $N = 15$, $k = 9$, $n = 4$, $N - k = 6$.

   a. $P[\text{two or more men}]$

   $= P[x \geq 2] = \dfrac{C_2^9 C_2^6}{C_4^{15}} + \dfrac{C_3^9 C_1^6}{C_4^{15}} + \dfrac{C_4^9 C_0^6}{C_4^{15}}$

   $= \dfrac{36}{1365} + \dfrac{84(6)}{1365} + \dfrac{126}{1365} = \dfrac{666}{1365} = .488$

   b. $P[\text{exactly three women}] = P[\text{exactly one man}]$

   $= P[x = 1] = \dfrac{C_1^9 C_3^6}{C_4^{15}}$

   $= \dfrac{9(20)}{1365} = .132$

5. Define x to be the number of defective units chosen. Then $N = 50$, $k = 3$, $N - k = 47$, $n = 5$.

$$P[x=0] = \frac{C_0^3 C_5^{47}}{C_5^{50}} = \frac{47!5!45!}{5!42!50!}$$

$$= \frac{47(46)(45)(44)(43)}{50(49)(48)(47)(46)} = .724$$

Set 4C

1. a. Since all ranks are equally likely, r has a uniform distribution on the integers 1 to n.
   b. The mean of a uniform random variable is
   $$\mu = \frac{n+1}{2}$$
   while the standard deviation is
   $$\sigma = \sqrt{\frac{n^2-1}{12}}$$
   c. For $n = 10$,
   $$\mu = \frac{10+1}{2} = 5.5$$
   and
   $$\sigma = \sqrt{\frac{10^2-1}{12}} = 2.87.$$
   Using Tchebysheff's Theorem with $k = 2$, the interval
   $$\mu \pm 2\sigma = 5.5 \pm 2(2.87) = 5.5 \pm 5.74$$
   or from $-.24$ to 11.24 will contain *at least* $1 - (\frac{1}{2})^2 = .75$, or 75% of the observations.

2. The random variable x has a uniform distribution on the integers 1 to 6. Therefore, the mean is
   $$\mu = \frac{6+1}{2} = 3.5$$
   while
   $$\sigma = \sqrt{\frac{6^2-1}{12}} = 1.71.$$

3. a. The number of potential buyers until the first sale has a geometric distribution with $p = .2$ and $q = .8$, so that
   $$p(x) = (.2)(.8)^{x-1} \quad x = 1, 2, \ldots$$

   b. The average number of potential buyers until the first sale is
   $$\mu = \frac{1}{p} = \frac{1}{.2} = 5.$$

   c. To use Tchebysheff's Theorem, we need to calculate
   $$\sigma = \sqrt{\frac{q}{p^2}} = \sqrt{\frac{.8}{(.2)^2}} = 4.47.$$
   Therefore, with $k = 2$, at least 75% of her sales would occur between
   $$5 \pm 2(4.47) = 5 \pm 8.94$$
   or from -3.94 to 13.94. Hence, at least 75% of the time, her first sale would occur before or at her 14th potential buyer.

Set 5A

1. Since $b - a = .05 - (-.05) = .10$, $f(x) = 10$ for $-.05 \le x \le .05$.

   a. $P(x \ge .025) = (.05 - .025)/.10 = .25$
   b. $P(|x| \ge .025) = P(x \ge .025) + P(x \le -.025)$
   $$= .25 + .25 = .50$$
   c. For the uniform distribution $\mu = \frac{-.05 + .05}{2} = 0$
   and $\sigma = .10/\sqrt{12} = .02887$
   so that $\mu \pm \sigma = 0 \pm .02887$ or $-.02887$ to $.02887$
   Then,
   $$P(\mu - \sigma \le x \le \mu + \sigma) = \frac{2(.02887)}{.10} = .577$$

2. Since $\mu = 8$, $\lambda = 1/8$ and $f(x) = \frac{1}{8}e^{-x/8}$.

   a. $P(x < 8) = 1 - P(x \ge 8) = 1 - e^{-8/8}$
   $$= 1 - e^{-1} = .63212$$

   b. It is necessary to find x_0 such that
   $$P(x > x_0) = .05$$
   $$e^{-x_0/8} = .05$$

Taking the natural logarithm of both sides,

$-x_0/8 = \ln .05$

$x_0 = 8(2.9957) = 23.97$

c. It is necessary to find the median, m, such that

$P(x > m) = .5$

$e^{-m/8} = .5$

$\dfrac{m}{8} = -\ln .5$

$m = 5.545$

3. Since $\mu = 10$, $\lambda = 0.1$ and $f(x) = 0.1e^{-x/10}$

   a. $P(x > 15) = e^{-15/10} = e^{-1.5} = .223$

   b. $P(x < 5) + P(x > 15) = 1 - e^{-5/10} + e^{-1.5}$
      $= .616$

   c. It is necessary to find x_0 such that

   $P(x < x_0) = .90$ or $P(x \geq x_0) = .10$

   Then, $P(x \geq x_0) = e^{-x_0/10} = .10$

   $\dfrac{x_0}{10} = -\ln .10$

   $x_0 = 23.03$

4. a. $\mu = \dfrac{0+2}{2} = 1$; $\sigma = \dfrac{2-0}{\sqrt{12}} = .577$

   b. $P(x \leq 1.5) = \dfrac{1.5}{2} = .75$

   c. $P(.25 < x \leq 2) = (2 - .25)/2 = .875$

   d. It is necessary to find x_0 such that

   $P(x \geq x_0) = .95$

   $\dfrac{2 - x_0}{2} = .95$

   $x_0 = 2 - 2(.95) = .10$ hour or 6 minutes

Set 5B

Note: The student should illustrate each problem with a diagram and list all pertinent information before attempting the solution. Diagrams are omitted here in order to conserve space.

1. a. $P(z > 2.1) = .5000 - A(2.1) = .5000 - .4821$
      $= .0179$.

   b. $P(z < -1.2) = .5000 - A(1.2) = .5000 - .3849$
      $= .1151$.

   c. $P(.5 < z < 1.5) = P(0 < z < 1.5)$
      $\qquad\qquad\qquad\quad - P(0 < z < .5)$
      $= A(1.5) - A(.5) = .4332$
      $- .1915 = .2417$.

   d. $P(-2.75 < z < -1.70) = A(2.75) - A(1.7)$
      $= .4970 - .4554$
      $= .0416$.

   e. $P(-1.96 < z < 1.96) = A(1.96) + A(1.96)$
      $= 2(.4750) = .95$.

   f. $P(z > 1.645) = .5000 - A(1.645)$
      $= .5000 - .4500 = .05$.

   Linear interpolation was used in part f. That is, since the value $z = 1.645$ is halfway between two table values, $z = 1.64$ and $z = 1.65$, the appropriate area is taken to be halfway between the two table areas, $A(1.64) = .4495$ and $A(1.65) = .4505$. As a general rule, values of z will be rounded to two decimal places, except for this particular example, which will occur frequently in our calculations.

2. a. We know that $P(z > z_0) = .10$, or $.5000 - A(z_0) = .10$, which implies that $A(z_0) = .4000$. The value of z_0 that satisfies this equation is $z_0 = 1.28$, so that $P(z > 1.28) = .10$.

   b. $P(z < z_0) = .01$ so that $.5000 - A(z_0) = .01$ and $A(z_0) = .4900$. The value of z_0 that satisfies this equation is $z_0 = -2.33$, so that $P(z < -2.33) = .01$. The student who draws a diagram will see that z_0 must be negative, since it must be in the left-hand portion of the curve.

   c. $P(-z_0 < z < z_0) = A(z_0) + A(z_0) = .95$ so that $A(z_0) = .4750$. The necessary value of z_0 is $z_0 = 1.96$ and $P(-1.96 < z < 1.96) = .95$.

   d. $P(-z_0 < z < z_0) = 2A(z_0) = .99$ so that $A(z_0) = .4950$. The necessary value of z_0 is $z_0 = 2.58$ and $P(-2.58 < z < 2.58) = .99$.

3. The random variable of interest has a standard normal distribution and hence may be denoted as z.

   a. $P(z > 1) = .5000 - A(1) = .5000 - .3413$
      $= .1587$.

   b. $P(z > 1.5) = .5000 - A(1.5) = .5000 - .4332$
      $= .0668$.

   c. $P(-1 < z < -.5) = A(1) - A(.5) = .3413$
      $- .1915 = .1498$.

d. The problem is to find a value of z, say z_0, such that $P(-z_0 < z < z_0) = .95$. This was done in 2c and $z_0 = 1.96$. Hence 95% of the billing errors will be between $-1.96 and $1.96.

e. Undercharges imply negative errors. Hence the problem is to find z_0 such that $P(z < z_0) = .05$. That is, $.5000 - A(z_0) = .05$, or $A(z_0) = .4500$. The value of z_0 is $z_0 = -1.645$ (see 1f) and hence 5% of the undercharges will be at least $1.65.

Set 5C

1. We have $\mu = 10$, $\sigma = \sqrt{2.25} = 1.5$.

a. $P(x > 8.5) = P\left(\dfrac{x - \mu}{\sigma} > \dfrac{8.5 - 10}{1.5}\right)$

$= P(z > -1)$

$= .5000 + A(1) = .5000 + .3413$

$= .8413$.

b. $P(x < 12) = P\left(z < \dfrac{12 - 10}{1.5}\right)$

$= P(z < 1.33) = .5000 + A(1.33)$

$= .5000 + .4082 = .9082$.

c. $P(9.25 < x < 11.25)$

$= P\left(\dfrac{9.25 - 10}{1.5} < z < \dfrac{11.25 - 10}{1.5}\right)$

$= P(-.5 < z < .83) = .1915 + .2967 = .4882$.

d. $P(7.5 < x < 9.2) = P(-1.67 < z < -.53)$

$= .4525 - .2019 = .2506$.

e. $P(12.25 < x < 13.25)$

$= P(1.5 < z < 2.17) = .4850 - .4332 = .0518$.

2. The random variable of interest is x, the length of life for a standard household light bulb. It is normally distributed with $\mu = 250$ and $\sigma = \sqrt{2,500} = 50$.

a. $P(x > 300) = P\left(z > \dfrac{300 - 250}{50}\right) = P(z > 1)$

$= .5000 - .3413 = .1587$.

b. $P(190 < x < 270) = P(-1.2 < z < .4)$

$= .3849 + .1554 = .5403$.

c. $P(x < 260) = P(z < .2) = .5000 + .0793$

$= .5793$

d. It is necessary to find a value of x, say x_0, such that $P(x > x_0) = .90$. Now

$P(x > x_0) = P\left(\dfrac{x - \mu}{\sigma} > \dfrac{x_0 - 250}{50}\right) = .90$,

so that

$P\left(z > \dfrac{x_0 - 250}{50}\right) = .5 + A\left(\dfrac{x_0 - 250}{50}\right)$

$= .90$, or

$A\left(\dfrac{x_0 - 250}{50}\right) = .40$.

By looking at a diagram, the student will notice that the value satisfying this equation must be negative. From Table 3, this value, $(x_0 - 250)/50$, is $(x_0 - 250)/50 = -1.28$, or $x_0 = -1.28(50) + 250 = 186$. Ninety percent of the bulbs have a useful life in excess of 186 hours.

e. Similar to d. It is necessary to find x_0 such that $P(x < x_0) = .95$. Now

$P(x < x_0) = P\left(z < \dfrac{x_0 - 250}{50}\right) = .95$,

$.5000 + A\left(\dfrac{x_0 - 250}{50}\right) = .95$, and

$A\left(\dfrac{x_0 - 250}{50}\right) = .45$.

Hence, $(x_0 - 250)/50 = 1.645$, or $x_0 = 332.25$. That is, 95% of all bulbs will burn out before 332.25 hours.

3. The random variable is x, scores on a trade school entrance examination, and has a normal distribution with $\mu = 50$ and $\sigma = 5$.

a. $P(x > 60) = P\left(z > \dfrac{60 - 50}{5}\right)$

$= P(z > 2) = .5000 - .4772$

$= .0228$.

b. $P(x < 45) = P(z < -1) = .5000 - .3413$

$= .1587$.

c. $P(35 < x < 65) = P(-3 < z < 3)$

$= .4987 + .4987 = .9974$.

d. It is necessary to find x_0 such that $P(x < x_0)$ = .95. As in 2e, $A(x_0 - 50/5) = .45$. $(x_0 - 50)/5 = 1.645$, or $x_0 = 58.225$.

Set 5D

1. a. Using Table 1 with $n = 20$, $p = .7$,
   $P(10 \leq x \leq 16) = P(x \leq 16) - P(x \leq 9)$
   $= .893 - .017 = .876$.
   b. The probabilities associated with the values $x = 10, 11, \ldots, 16$ are needed; hence the area of interest is the area to the right of 9.5 and to the left of 16.5. Further, $\mu = np = 14$, $\sigma^2 = npq = 4.2$.

   $P(10 \leq x \leq 16) \approx P(9.5 < x < 16.5)$
   $= P\left(\dfrac{9.5 - 14}{\sqrt{4.2}} < z < \dfrac{16.5 - 14}{\sqrt{4.2}}\right)$
   $= P(-2.20 < z < 1.22)$
   $= .4861 + .3888 = .8749$.

2. Let $p = P$ (income is less than 22,000) = 1/2, since 22,000 is the median income. Also, $n = 100$, $\mu = np = 50$, $\sigma^2 = npq = 25$.

   $P(x \leq 37) \approx P(x < 37.5) = P\left(z < \dfrac{37.5 - 50}{5}\right)$
   $= P(z < -2.5) = .5000 - .4938 = .0062$.

   The observed event is highly unlikely under the assumption that $22,000 is the median income. The $22,000 figure does not seem reasonable.

3. x = number of white seeds, $p = P$ (white seed) = .4, $n = 100$; $\mu = np = 40$, $\sigma^2 = npq = 24$, $\sigma = 4.90$.

   a. $P(x \leq 50) \approx P(x < 50.5)$
   $= P\left(z < \dfrac{50.5 - 40}{4.9}\right)$
   $= P(z < 2.14)$
   $= .5 + .4848 = .9838$.

   b. $P(x \leq 35) \approx P(x < 35.5) = P(z < -.92)$
   $= .5 - .3212 = .1788$.

   c. $P(25 \leq x \leq 45) \approx P(24.5 < x < 45.5)$
   $= P(-3.16 < z < 1.12)$
   $= .5 + .3686 = .8686$.

4. $\mu \pm 2\sigma = 40 \pm 2(4.9) = 40 \pm 9.8$, or 31 to 49.

Set 6A

1. The weights of the rats will result from a combination of random factors, such as initial weight of the rat, genetic make-up of the rat, amount of food consumed, and so on. Hence, the weights behave as a sum of independent random variables, and as such would be normally distributed according to the Central Limit Theorem.

2. Since $\bar{x}$ is normally distributed with mean $\mu = 60$ and with standard deviation $\sigma/\sqrt{n} = 10/\sqrt{30} = 1.8257$, the probability of interest is

   $P[\bar{x} > 65] = P\left[z > \dfrac{65 - 60}{1.8257}\right] = P[z > 2.74]$
   $= .5 - .4969 = .0031$

Set 6B

1. a. $P(.8 \leq \hat{p} \leq .9) = P(16 \leq x \leq 18) = 1 - .994 = .006$ from Table 1.
   b. Using the fact that $\mu_{\hat{p}} = p = .5$ and $\sigma_{\hat{p}} = \sqrt{\dfrac{pq}{n}} = \sqrt{\dfrac{.5(.5)}{20}} = .1118$,

   $P(.8 \leq \hat{p} \leq .9) = P\left(\dfrac{.8 - .5}{.1118} < z < \dfrac{.9 - .5}{.1118}\right) =$
   $P(2.68 < z < 3.58) = .5 - .4963 = .0037$

2. Let $p = P$[favor the canal]. Then $p = .3$ and $n = 50$, while $\mu_{\hat{p}} = p = .3$ and $\sigma_{\hat{p}} = \sqrt{\dfrac{pq}{n}} = \sqrt{\dfrac{.3(.7)}{50}} = .0648$. The probability of interest is

   $P(\hat{p} > .50) = P\left(z > \dfrac{.5 - .3}{.0648}\right) = P(z > 3.09) =$
   $.5 - .4990 = .001$

Set 6C

1. The random variable $\bar{x}_1 - \bar{x}_2$ has an approximately normal distribution with mean $\mu_1 - \mu_2 =$ -5 and variance $\dfrac{\sigma_1^2}{n_1} + \dfrac{\sigma_2^2}{n_2} = \dfrac{150}{35} + \dfrac{100}{35} = 7.14286$.

a. $P[\bar{x}_1 - \bar{x}_2 > 1] = P\left[z > \dfrac{1-(-5)}{\sqrt{7.14286}}\right] =$

$P[z > 2.24] = .5 - .4875 = .0125$

b. $P[0 \leq (\bar{x}_1 - \bar{x}_2) \leq 6] = P\left[\dfrac{0-(-5)}{\sqrt{7.14286}} < z < \dfrac{6-(-5)}{\sqrt{7.14286}}\right] = P[1.87 < z < 4.12] =$

$.5 - .4693 = .0307$

c. $P[(\bar{x}_1 - \bar{x}_2) > 2] + P[\bar{x}_1 - \bar{x}_2) < -2] =$

$P[z > 2.62] + P[z < 1.12] = .5 - .4956 +$

$.5 + .3686 = .8730$

2. The sampling distribution of $\hat{p}_1 - \hat{p}_2$ is approximately normal with mean $p_1 - p_2 = -.1$ and variance $\dfrac{p_1 q_1}{n_1} + \dfrac{p_2 q_2}{n_2} = \dfrac{.3(.7)}{100} + \dfrac{.4(.6)}{100} = .0045.$

a. $P[(\hat{p}_1 - \hat{p}_2) > .25] + P[(\hat{p}_1 - \hat{p}_2) < -.25] =$

$P\left[z > \dfrac{.25 - (-.1)}{\sqrt{.0045}}\right] + P\left[z > \dfrac{-.25 - (-.1)}{\sqrt{.0045}}\right] =$

$P[z > 5.22] + P[z < -2.24] = .5 - .4875 = .0125$

b. $P[\hat{p}_1 > \hat{p}_2] = P[\hat{p}_1 - \hat{p}_2 > 0] =$

$P\left[z > \dfrac{0 - (-.1)}{\sqrt{.0045}}\right] = P[z > 1.49] =$

$.5 - .4319 = .0681$

Set 7A

1. $\bar{x} = 67.5$ with approximate bound on error $1.96s/\sqrt{n} = 1.96(8.2)/\sqrt{93} = 1.67.$

2. $\hat{p}_1 - \hat{p}_2 = \dfrac{x_1}{n_1} - \dfrac{x_2}{n_2} = \dfrac{31}{204} - \dfrac{41}{191} = .15 - .21 = -.06.$

Approximate bound on error:

$1.96\sqrt{\dfrac{\hat{p}_1 \hat{q}_1}{n_1} + \dfrac{\hat{p}_2 \hat{q}_2}{n_2}}$

$= 1.96\sqrt{.000625 + .000869}$

$= 1.96(.039) = .076$

3. $\hat{p} = x/n = 8/50 = .16$ with approximate bound on error $1.96\sqrt{\hat{p}\hat{q}/n} = 1.96\sqrt{.16(.84)/50} = 1.96(.0518) = .102$

4. $\bar{x}_1 - \bar{x}_2 = 150.5 - 160.2 = -9.7$ with approximate bound on error

$1.96\sqrt{\dfrac{s_1^2}{n_1} + \dfrac{s_2^2}{n_2}} = 1.96\sqrt{\dfrac{23.72}{35} + \dfrac{36.37}{35}}$

$= 1.96\sqrt{1.7169} = 2.57$

Set 7B

1. $\hat{p} \pm z_{\alpha/2}\sqrt{\hat{p}\hat{q}/n}$, where $\hat{p} = 30/65 = .46$;

$.46 \pm 2.33\sqrt{.46(.54)/65}$; $.46 \pm 2.33(.0618)$;

$.46 \pm .14$, or $.32$ to $.60$.

2. $\hat{p}_1 = 50/100 = .50$; $\hat{p}_2 = 60/200 = .30$.

$(\hat{p}_1 - \hat{p}_2) \pm 1.96\sqrt{\dfrac{\hat{p}_1 \hat{q}_1}{n_1} + \dfrac{\hat{p}_2 \hat{q}_2}{n_2}}$

$(.50 - .30) \pm 1.96\sqrt{\dfrac{.5(.5)}{100} + \dfrac{.3(.7)}{200}}$

$.2 \pm 1.96(.0596)$ $.2 \pm .12$, or $.08$ to $.32$.

3. a. $(\bar{x}_1 - \bar{x}_2) \pm z_{\alpha/2}\sqrt{\dfrac{s_1^2}{n_2} + \dfrac{s_2^2}{n_2}}$

$(28{,}520 - 27{,}210) \pm 2.58\sqrt{\dfrac{(1{,}510)^2}{90} + \dfrac{(950)^2}{60}}$

$1{,}310 \pm 2.58(200.938)$

$1{,}310 \pm 518.42$ or 791.58 to $1{,}828.42$.

b. If the two schools belonged to populations having the same mean annual income, then $\mu_1 = \mu_2$, or $\mu_1 - \mu_2 = 0$. This value of $\mu_1 - \mu_2$ does not fall in the confidence interval obtained above. Hence it is unlikely that the two schools belong to populations having the same mean annual income.

4. $\bar{x} \pm z_{\alpha/2}(s)/\sqrt{n}$; $22 \pm 1.645(4)/\sqrt{39}$;

$22 \pm (6.58/6.245)$; 22 ± 1.05, or 20.95 to 23.05.

Set 7C

1. The estimator of μ is $\bar{x}$, with standard deviation $\sigma/\sqrt{n}$. Hence, solve

   $1.96\sigma/\sqrt{n} = B$, $1.96(8)/\sqrt{n} = 3$, $\sqrt{n} = 5.227$,

   $n = 27.32$.

   The experimenter should obtain $n = 28$ measurements.

2. For each additive, the range of the measurements is 80, so that $\sigma_1 = \sigma_2 \approx \text{range}/4 = 20$. Assuming equal sample sizes are acceptable, solve

   $1.96\sqrt{\dfrac{\sigma_1^2}{n} + \dfrac{\sigma_2^2}{n}} = 10$, $1.96\sqrt{\dfrac{2(20)^2}{n}} = 10$,

   $\sqrt{n} = 5.543$, $n = 30.73$.

   Hence, 31 subjects should be included in each group.

3. Maximum variation occurs when $p_1 = p_2 = .5$. Again assuming equal sample sizes, solve

   $1.96\sqrt{\dfrac{p_1 q_1}{n} + \dfrac{p_2 q_2}{n}} = .01$,

   $1.96\sqrt{\dfrac{2(.5)(.5)}{n}} = .01$, $\sqrt{n} = \dfrac{1.96\sqrt{.5}}{.01}$,

   $n = 19{,}208$.

4. Maximum variation occurs when $p = .5$. Since 90% accuracy is involved, solve

   $1.645\sqrt{pq/n} = .05$, $1.645\sqrt{(.5)(.5)/n} = .05$,

   $\sqrt{n} = 1.645(.5)/.05 = 16.45$,

   $n = 270.6$; 271 patient records should be sampled.

Set 8A

1. $H_0: \mu_1 - \mu_2 = 0$; $H_a: \mu_1 - \mu_2 > 0$; test statistic:

   $z = \dfrac{(\bar{x}_1 - \bar{x}_2) - 0}{\sqrt{\dfrac{s_1^2}{n_1} + \dfrac{s_2^2}{n_2}}}$.

   With $\alpha = .05$, reject H_0 if $z > 1.645$. Calculate

   $z = \dfrac{720 - 693}{\sqrt{\dfrac{104}{50} + \dfrac{85}{50}}} = \dfrac{27}{1.94} = 13.92$.

   Reject H_0. There is a significant difference in the mean scores. Men score higher on the average than women.

2. $H_0: \mu = 6.35$; $H_a: \mu \neq 6.35$; test statistic:

   $z = \dfrac{\bar{x} - \mu_0}{s/\sqrt{n}}$. With $\alpha = .05$,

   reject H_0 if $|z| > 1.96$. Calculate

   $\bar{x} = 2{,}464.40/400 = 6.161$;

   $s^2 = (16156.728 - 15183.168)/399 = 2.4400$;

   $s = \sqrt{2.44} = 1.56$;

   $z = \dfrac{\bar{x} - \mu_0}{s/\sqrt{n}} = \dfrac{6.161 - 6.35}{1.56/20} = -2.42$.

   Reject H_0. Mean revenue is different from $6.35.

3. $H_0: p = 1/2$; $H_a: p \neq 1/2$; test statistic:

   $z = \dfrac{\hat{p} - p_0}{\sqrt{p_0 q_0/n}}$.

   With $\alpha = .10$, reject H_0 if $|z| > 1.645$. Since $\hat{p} = 480/900 = .533$,

   $z = \dfrac{.533 - .5}{\sqrt{.5(.5)/900}} = \dfrac{.033}{.5/30} = 1.98$.

   Reject H_0. Half the cases are not male.

4. $H_0: p_1 - p_2 = 0$; $H_a: p_1 - p_2 \neq 0$; test statistic:

   $z = \dfrac{(\hat{p}_1 - \hat{p}_2) - 0}{\sqrt{\hat{p}\hat{q}\left(\dfrac{1}{n_1} + \dfrac{1}{n_2}\right)}}$.

   Note that if $p_1 = p_2$ as proposed under H_0, the best estimate of this common value of p is

   $\hat{p} = (x_1 + x_2)/(n_1 + n_2) = (16 + 6)/(100 + 50)$

   $= .15$.

   With $\alpha = .05$, reject H_0 if $|z| > 1.96$. Calculate

   $z = \dfrac{(16/100) - (6/50)}{\sqrt{.15(.85)\left(\dfrac{1}{100} + \dfrac{1}{50}\right)}} = \dfrac{.04}{.0618} = .65$.

   Do not reject H_0. There is insufficient evidence to detect a difference in the performances of the machines.

Set 8B

1. The level of significance for a two-tailed test is
   p-value $= P[|z| > 2.42] = 2P[z > 2.42] =$
   $2(.5 - .4922)$
   $= .0156$. Since this is a relatively small α value, the researcher would probably reject H_0.

2. The level of significance is p-value $= P[|z| > .65]$
   $= 2P[z > .65] = 2(.5 - .2422) = .5156$. This is a very large value, and the researcher would not reject H_0 unless he could tolerate an α as large as .5156.

3. $H_0: p = .1$; $H_a: p < .1$;

   Test statistic: $z = \dfrac{\hat{p} - .1}{\sqrt{\dfrac{.1(.9)}{n}}} = \dfrac{.08 - .1}{\sqrt{\dfrac{.1(.9)}{100}}} = -.67$.

   Then p-value $= P[z < -.67] = .5 - .2486 = .2514$. This is a large level of significance. The researcher would not reject H_0 unless he could tolerate an α as large as .2514.

Set 9A

1. $H_0: \mu = 35$; $H_a: \mu > 35$; test statistic: $t = \dfrac{\bar{x} - \mu}{s/\sqrt{n}}$.

   With $\alpha = .05$, and $n - 1 = 19$ degrees of freedom, reject H_0 if $t > t_{.05} = 1.729$. Calculate

   $t = \dfrac{42 - 35}{6.2/\sqrt{20}} = \dfrac{7}{1.386} = 5.05$.

   Reject H_0. The mean riding time is greater than 35 minutes. From Table 4 with 19 degrees of freedom, the observed $t = 5.05$ exceeds $t_{.005} = 2.861$. Hence, p-value $< .005$.

2. $\bar{x} \pm t_{.025} \dfrac{s}{\sqrt{n}}$, $42 \pm 2.093 (1.386)$, 42 ± 2.90, or 39.1 to 44.9.

3. a. $\Sigma x_i = 387$; $\Sigma x_i^2 = 17695$; $\bar{x} = 43$;

   $s^2 = \dfrac{17695 - 16641}{8} = 131.75$.

   $\bar{x} \pm t_{.05} \dfrac{s}{\sqrt{n}}$, $43 \pm 1.86 \sqrt{\dfrac{131.75}{9}}$, 43 ± 7.12 or 35.88 to 50.12.

   b. $H_0: \mu = 45$; $H_a: \mu \neq 45$; test statistic: $t = \dfrac{\bar{x} - \mu}{s/\sqrt{n}}$

   $= \dfrac{43 - 45}{\sqrt{\dfrac{131.75}{9}}} = -.52$. With $\alpha = .05$ and 8

   degrees of freedom, reject H_0 if $|t| > 2.306$. Do not reject H_0.

Set 9B

1. See Section 9.4, paragraph 1.

2. $H_0: \mu_1 - \mu_2 = 0$; $H_a: \mu_1 - \mu_2 \neq 0$; test statistic:

   $t = \dfrac{(\bar{x}_1 - \bar{x}_2) - D_0}{s\sqrt{\dfrac{1}{n_1} + \dfrac{1}{n_2}}}$.

   With $\alpha = .05$ and $n_1 + n_2 - 2 = 16 + 9 - 2 = 23$ degrees of freedom, reject H_0 if $|t| > t_{.025} = 2.069$. Calculate

   $s^2 = \dfrac{(n_1 - 1)s_1^2 + (n_2 - 1)s_2^2}{n_1 + n_2 - 2}$

   $= \dfrac{15(2.8)^2 + 8(5.3)^2}{23} = \dfrac{117.6 + 224.72}{23}$

   $= 14.8835$ and $s = 3.86$.

   $t = \dfrac{5.2 - 8.7}{3.86\sqrt{\dfrac{1}{16} + \dfrac{1}{9}}} = \dfrac{-3.5}{3.86(.42)} = -2.16$.

   Reject H_0. The mean distance to the health center differs for the two groups.

3. $(\bar{x}_1 - \bar{x}_2) \pm t_{.025} s \sqrt{\dfrac{1}{n_1} + \dfrac{1}{n_2}}$,

   $-3.5 \pm 2.069(1.62)$, -3.5 ± 3.35, or

   -6.85 to -0.15 miles.

4. $H_0: \mu_1 - \mu_2 = 0$; $H_a: \mu_1 - \mu_2 \neq 0$; test statistic:

   $t = \dfrac{(\bar{x}_1 - \bar{x}_2) - D_0}{s\sqrt{\dfrac{1}{n_1} + \dfrac{1}{n_2}}}$.

   With $\alpha = .05$ and $n_1 + n_2 - 2 = 34$ degrees of freedom, reject H_0 if $|t| > t_{.025} = 1.96$. Calculate

   $s^2 = \dfrac{(n_1 - 1)s_1^2 + (n_2 - 1)s_2^2}{n_1 + n_2 - 2}$

$$= \frac{17(23.2) + 17(19.8)}{34} = 21.5.$$

$$t = \frac{81.7 - 77.2}{\sqrt{21.5 \left(\frac{2}{18}\right)}} = \frac{4.5}{1.5456} = 2.91.$$

Reject H_0. There is a difference in mean scores for the two methods. From Table 4 with 34 degrees of freedom, $t = 2.91$ exceeds $t_{.005} = 2.576$. Hence, p-value $< 2(.005) = .01$.

Set 9C

1. $H_0: \mu_d = \mu_P - \mu_H = 0$; $H_a: \mu_d = \mu_P - \mu_H < 0$;

    test statistic: $t = \dfrac{\bar{d} - \mu_d}{s_d/\sqrt{n}}$.

With $\alpha = .05$ and $n - 1 = 10 - 1 = 9$ degrees of freedom, reject H_0 if $t < -t_{.05} = -1.833$. The 10 differences and the calculation of the test statistic follow.

d_i	d_i^2
-5	25
-2	4
4	16
-3	9
-3	9
1	1
-1	1
0	0
-5	25
1	1
-13	91

$$\bar{d} = \frac{-13}{10} = -1.3,$$

$$s_d^2 = \frac{91 - (-13)^2/10}{9} = \frac{74.1}{9} = 8.2333,$$

$$s_d = \sqrt{8.2333} = 2.869,$$

$$t = \frac{-1.3 - 0}{2.869/\sqrt{10}} = -1.43.$$

Do not reject H_0. With 9 degrees of freedom, $t = -1.43$ falls between $-t_{.10}$ and $-t_{.05}$. Hence, $.05 < p$-value $< .10$.

2.

d_i	d_i^2
4.6	21.16
1.8	3.24
-1.0	1.00
2.2	4.84
1.0	1.00
1.2	1.44
2.6	6.76
2.8	7.84
2.0	4.00
3.0	9.00
20.2	60.28

$$\bar{d} = \frac{20.2}{10} = 2.02,$$

$$s_d^2 = \frac{60.28 - 40.804}{9} = \frac{19.476}{9} = 2.164,$$

$$s_d = 1.47.$$

a. $\bar{d} \pm t_{.025} \dfrac{s_d}{\sqrt{n}}$, $2.02 \pm 2.262 \dfrac{1.47}{\sqrt{10}}$,

2.02 ± 1.05, or $.97 < \mu_d < 3.07$.

b. $H_0: \mu_P - \mu_H = 0$; $H_a: \mu_P - \mu_H > 0$;

    test statistic: $t = \dfrac{\bar{d} - \mu_d}{s_d/\sqrt{n}}$.

With $\alpha = .05$ and $n - 1 = 9$ degrees of freedom, reject H_0 if $t > t_{.05} = 1.833$. Calculate

$t = 2.02/.465 = 4.34$.

Reject H_0. Per-unit scale increases production.

Set 9D

1. $H_0: \sigma = 3$ $(\sigma^2 = 9)$; $H_a: \sigma < 3$ $(\sigma^2 < 9)$; test statistic:

$$\chi^2 = (n-1)s^2/\sigma_0^2.$$

With $\alpha = .05$ and $n - 1 = 4$ degrees of freedom, reject H_0 if $\chi^2 < \chi^2_{.95} = .710721$. Calculate

$$s^2 = \frac{\Sigma x_i^2 - (\Sigma x_i)^2/n}{n-1} = \frac{1,930.5 - 1,905.152}{4}$$

$$= 6.337.$$

$\chi^2 = 25.348/9 = 2.816$. Do not reject H_0.

2. $H_0: \sigma^2 = 100$; $H_a: \sigma^2 < 100$; test statistic:
$\chi^2 = (n-1)s^2/\sigma_0^2$.

With $\alpha = .05$ and $n - 1 = 29$ degrees of freedom, reject H_0 if $\chi^2 < \chi^2_{.95} = 17.7083$. Calculate

$\chi^2 = 29(8.9)^2/100 = 22.97$. Do not reject H_0. From Table 5 with 29 degrees of freedom, $\chi^2 = 22.97$ exceeds $\chi^2_{.90}$. Hence, $p > .10$.

3. a. $H_0: \sigma = 5$; $H_a: \sigma > 5$; test statistic:
$\chi^2 = (n-1)s^2/\sigma_0^2$.

With $\alpha = .05$, reject H_0 if $\chi^2 > 42.5569$. Since $\chi^2 = 29(7.3)^2/25 = 61.81$, reject H_0. The new technique is less sensitive.

b. $\dfrac{(n-1)s^2}{\chi^2_U} < \sigma^2 < \dfrac{(n-1)s^2}{\chi^2_L}$,

$\dfrac{29(7.3)^2}{45.7222} < \sigma^2 < \dfrac{29(7.3)^2}{16.0471}$,

$33.80 < \sigma^2 < 96.30$, $5.81 < \sigma < 9.81$.

Set 9E

1. $H_0: \sigma_1^2 = \sigma_2^2$; $H_a: \sigma_1^2 \neq \sigma_2^2$; test statistic:

$F = s_1^2/s_2^2$, where population 1 is the population of distances for people wanting the center closed. With $\alpha = .02$ and $\nu_1 = 8$, $\nu_2 = 15$, reject H_0 if $F > 4.00$ from Table 9.

$F = (5.3)^2/(2.8)^2 = 28.09/7.84 = 3.58$.

Do not reject H_0. Assumption has been met.

2. $H_0: \sigma_1^2 = \sigma_2^2$; $H_a: \sigma_1^2 \neq \sigma_2^2$; test statistic:

$F = s_1^2/s_2^2$, where population 1 is the population of thresholds for females. With $\alpha = .10$, $\nu_1 = 12$, $\nu_2 = 9$, reject H_0 if $F > 3.07$. Calculate

$F = 26.9/11.3 = 2.38$. Do not reject H_0. The two groups exhibit the same basic variation. From Tables 6-10, the following values for F_α are found:

α	F_α
.10	2.12
.05	2.64
.025	3.20
.01	4.00
.005	4.67

Since the observed value of $F = 2.38$ falls between $F_{.10}$ and $F_{.05}$, $2(.05) < p$-value $< 2(.10)$ or $.10 < p$-value $< .20$.

Set 10A

1. a.

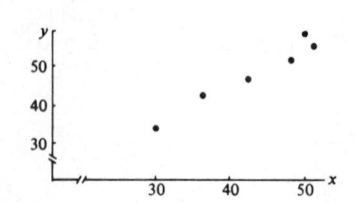

$\Sigma x_i = 256$ $\Sigma x_i y_i = 12{,}608$
$\Sigma y_i = 286$ $n = 6$
$\Sigma x_i^2 = 11{,}294$ $\Sigma y_i^2 = 14{,}096$

The trend appears to be linear.

b. $S_{xy} = 12{,}608 - (256)(286)/6 = 405.3333$
$S_{xx} = 11{,}294 - (256)^2/6 = 371.3333$
$\hat{\beta}_1 = S_{xy}/S_{xx} = 405.3333/371.3333 = 1.09$
$\hat{\beta}_0 = (286/6) - 1.09(256/6)$
$= 47.6667 - 46.5067 = 1.16$

c. $\hat{y} = 1.16 + 1.09(50) = 55.66$, or 5,566 students.

2. a. $\Sigma x_i = 31.6$ $\Sigma x_i y_i = 624.6$
$\Sigma y_i = 135$ $n = 7$
$\Sigma x_i^2 = 149.82$ $\Sigma y_i^2 = 2{,}645$

$S_{xy} = 624.6 - (31.6)(135)/7 = 15.1714$
$S_{xx} = 149.82 - (316)^2/7 = 7.1686$
$\hat{\beta}_1 = S_{xy}/S_{xx} = 15.1714/7.1686 = 2.12$
$\hat{\beta}_0 = 19.29 - 2.12(4.51) = 9.73$.

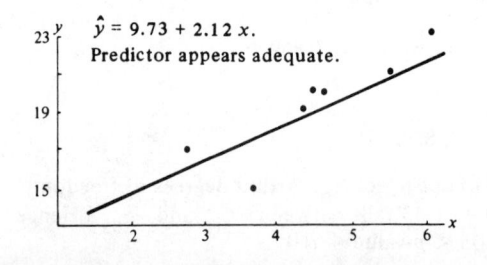

$\hat{y} = 9.73 + 2.12 x$.
Predictor appears adequate.

3. a. $\Sigma x_i = 96$ $\quad$ $\Sigma x_i y_i = 1{,}799$

$\Sigma y_i = 135$ $\quad$ $n = 7$

$\Sigma x_i^2 = 1{,}402$ $\quad$ $\Sigma y_i^2 = 2{,}645$

$S_{xy} = 1{,}799 - (96)(135)/7 = -52.4286$

$S_{xx} = 1{,}402 - (96)^2/7 = 85.4286$

$\hat{\beta}_1 = S_{xy}/S_{xx} = -52.4286/85.4286 = -0.61$

$\hat{\beta}_0 = 19.29 - (-.61)(13.71) = 27.65$

$\hat{y} = 27.65 - .61x$

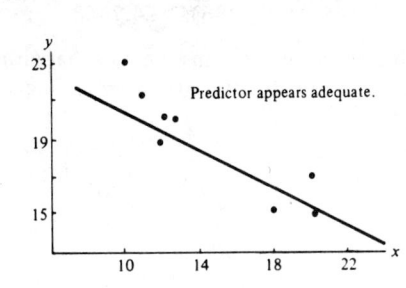

b.

Predictor appears adequate.

Set 10B

1. $SSE = S_{yy} - \hat{\beta}_1 S_{xy} = S_{yy} - (S_{xy})^2/S_{xx}$

$= 14{,}096 - [(286)^2/6]$

$\quad - [(405.3333)^2/371.3333]$

$= 463.33 - 442.45 = 20.88.$

Note that $\hat{\beta}_1 S_{xy} = (S_{xy})^2/S_{xx}$ has been used for the sake of accuracy.

$s^2 = 20.88/4 = 5.22;\ s = \sqrt{5.22} = 2.28.$

a. $H_0: \beta_1 = 0;\ H_a: \beta_1 \neq 0.$

Reject H_0 if $|t| > t_{.025,4} = 2.776.$

Test statistic: $t = \dfrac{\hat{\beta}_1 - \beta_1}{s/\sqrt{S_{xx}}} = \dfrac{1.09}{\sqrt{5.22/371.33}}$

$= \dfrac{1.09}{.12} = 9.08.$ Reject H_0.

b. $\hat{\beta}_1 \pm t_{.025} s/\sqrt{S_{xx}} = 1.09 \pm 2.776(.12)$

$= 1.09 \pm .33,$ or

$.76 < \beta_1 < 1.42.$

2. $SSE = 2645 - (1/7)(135)^2 - (15.1714)^2/7.1686$

$= 41.43 - 32.11 = 9.32.$

$s^2 = 9.32/5 = 1.86;\ s = \sqrt{1.86} = 1.36.$

$H_0: \beta_1 = 0;\ H_a: \beta_1 \neq 0.$

Rejection region: With 5 degrees of freedom and $\alpha = .05$, reject H_0 if $|t| > t_{.025} = 2.571.$
Test statistic:

$t = \dfrac{\hat{\beta}_1 - 0}{s/\sqrt{S_{xx}}} = \dfrac{2.12}{\sqrt{1.86/7.17}} = \dfrac{2.12}{.51} = 4.16.$ Reject H_0.

3. $SSE = 2645 - (135)^2/7 - (-52.4286)^2/85.4286$

$= 41.43 - 32.18 = 9.25$

$s^2 = 9.25/5 = 1.85;\ s = 1.36.$

$H_0: \beta_1 = 0;\ H_a: \beta_1 \neq 0.$ Reject H_0 if

$|t| > t_{.025,5} = 2.571.$

$t = \dfrac{-.61 - 0}{\sqrt{1.85/85.43}} = \dfrac{-.61}{.15} = -4.067.$ Reject H_0.

Set 10C

1. $H_0: E(y|x=0) = 0;\ H_a: E(y|x=0) \neq 0.$
Test statistic:

$t = \dfrac{\hat{y} - E_0}{s\sqrt{\dfrac{1}{n} + \dfrac{(x_0 - \bar{x})^2}{S_{xx}}}}$

Rejection region: With $\alpha = .05$, reject H_0 if $|t| > t_{.025} = 2.776.$ Calculate

$\hat{y} = 1.16 + 1.09(0) = 1.16$

$t = \dfrac{1.16 - 0}{\sqrt{5.22\left[\dfrac{1}{6} + \dfrac{(42.67)^2}{371.33}\right]}} = \dfrac{1.16}{\sqrt{26.47}} = \dfrac{1.16}{5.14} = .226.$

Do not reject H_0. Note that $\hat{y} = 1.16 + 1.09x$ does not pass through the origin, even though we could not reject the hypothesis, $H_0: E(y|x=0) = \beta_0 = 0$. We could not reject H_0 because of the small number of observations and the large variation.

2. $(\hat{y}|x = 4.5) = 9.73 + 2.12(4.5) = 9.73 + 9.54$

$= 19.27,$

$V(\hat{y}|x) = 1.87\left[\dfrac{1}{7} + \dfrac{(4.50 - 4.51)^2}{7.17}\right]$

$= 1.87(.14) = .2618.$

The 90% confidence interval is $\hat{y} \pm t_{.05}\sqrt{V(\hat{y}|x)}$
$= 19.27 \pm 2.015\sqrt{.2618} = 19.27 \pm 1.03,$ or

$18.24 < E(y|x = 4.5) < 20.30$. Since $x = 250$ is outside the limits for the observed x, one should not predict for that value.

3. $(\hat{y}|x = 12) = 27.65 - .61(12) = 20.33$,

$$V(\hat{y}|x = 12) = 1.85\left[\frac{1}{7} + \frac{(12 - 13.71)^2}{85.43}\right]$$

$= 1.85(.18) = .33$.

The 90% confidence interval is $20.33 \pm 2.015\sqrt{.33}$ $= 20.33 \pm 2.015(.57) = 20.33 \pm 1.15$, or $19.18 < E(y|x = 12) < 21.48$. Note that this interval predicts a slightly higher expected yield.

4. $(\hat{y}|x = 40) = 1.16 + 1.09(40) = 44.76$;

$$V(\hat{y}|x) = 5.22\left[\frac{1}{6} + \frac{(40 - 42.67)^2}{371.33}\right]$$

$= 5.22(.19) = .99$. The 95% confidence interval is $44.76 \pm 2.776\sqrt{.99} = 44.76 \pm 2.776$, or $41.98 < E(y|x = 40) < 47.54$. Enrollment will be between 4,198 and 4,754 with 95% confidence.

Set 10D

1. a. $r = S_{xy}/\sqrt{S_{xx} S_{yy}}$

   $= 15.1714/\sqrt{(7.1686)(41.43)}$

   $= 15.1714/17.2335 = .88$.

   b. $r^2 = (.88)^2 = .7744$. Total variation is reduced by 77.44% by using number of cotton bolls to aid in prediction.

   c. $H_0: \rho = 0; H_a: \rho > 0$
      Test statistic:

   $$t = \frac{r\sqrt{n-2}}{\sqrt{1-r^2}} = \frac{.88\sqrt{5}}{\sqrt{1-.7744}} \approx 4.14$$

   Rejection region: With 5 degrees of freedom and $\alpha = .05$, reject H_0 if $t > t_{.05} = 2.015$. Reject H_0. Note discrepancy in t value in SCE 10B, #2 due to rounding error.

2. a. $r = -52.4286/\sqrt{(85.4286)(41.43)}$

   $= -52.4286/59.4921 = -.88$.

   Total variation is reduced by 77.44% by using number of damaging insects to aid in prediction.

   b. The predictors are equally effective.

3.

x_1 (Bolls)	x_2 (Insects)
5.5	11
2.8	20
4.7	13
4.3	12
3.7	18
6.1	10
4.5	12

$\Sigma x_1 = 31.6$

$\Sigma x_1^2 = 149.82$

$n = 7$

$\Sigma x_2 = 96$

$\Sigma x_2^2 = 1,402$

$\Sigma x_1 x_2 = 410.80$

$$r = \frac{410.8 - (31.6)(96)/7}{\sqrt{(7.1686)(85.4286)}} = \frac{-22.5714}{24.7468} = -.91.$$

High correlation explains the fact that either variable is equally effective in predicting cotton yield.

Set 11A

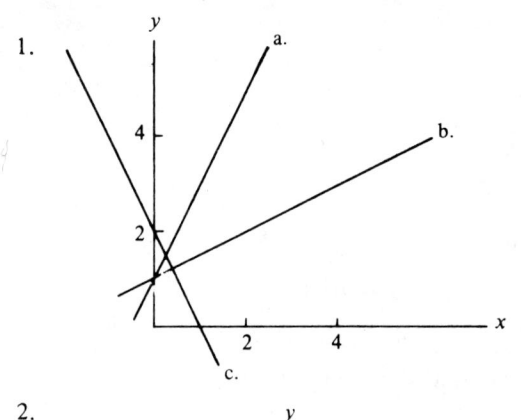

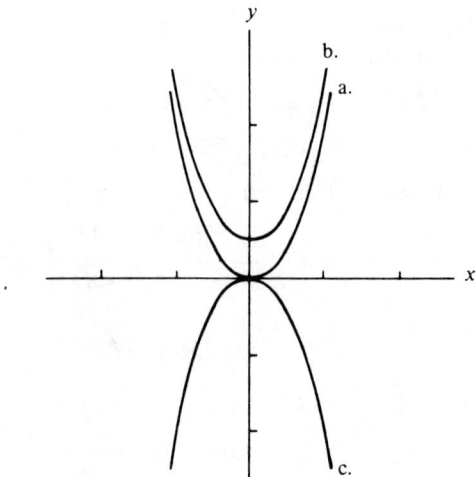

1. [graph showing lines a, b, c on xy-axes]

2. [graph showing curves a, b, c on xy-axes]

3.

a.

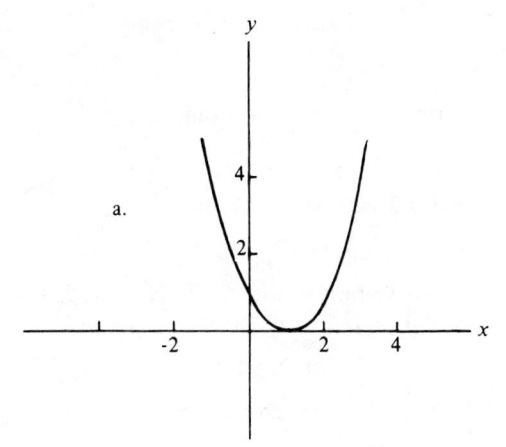

b. The addition of $-2x$ to the equation has the effect of moving the parabola one unit to the right along the x-axis.

c. If the term $2x$ were added to the equation, the parabola would be moved one unit to the left along the x-axis.

4. a. When $x_2 = 0$, $E(y) = 2 + 3x_1$. When $x_2 = 1$, $E(y) = 1 + 3x_1$ and when $x_2 = 2$, $E(y) = 3x_1$.
b. The three lines are parallel.

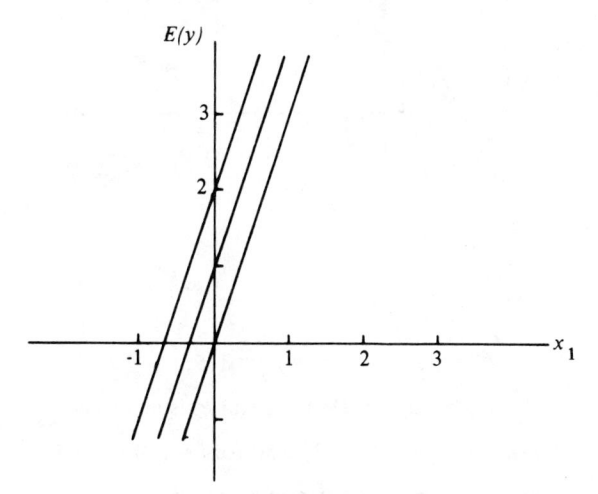

c. When $x_1 = 0$, $E(y) = 2 - x_2$. When $x_1 = 1$, $E(y) = 5 - x_2$ and when $x_1 = 2$, $E(y) = 8 - x_2$.
d. The lines are again parallel.

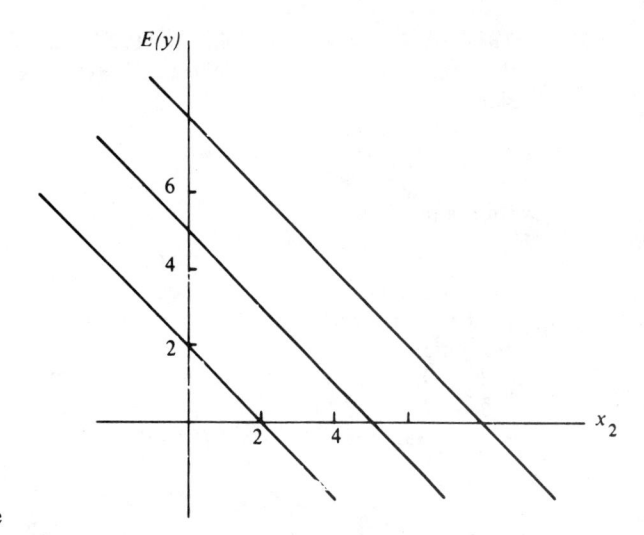

Set 11B

1. a. Refer to the *Analysis of Variance* section of the printout. In the column labeled SS, find $S_{yy} = 431.60$, SSE = 76.38 and SSR = 355.22. Then s^2 = MSE = 10.91 in the MS column. $s = \sqrt{s^2}$ is found on the printout as 3.303.

b. The entry R-sq appears under the individual variables section and is given as 82.3%, meaning that 82.3% of the variation in the response y can be explained by regression.

c. $R^2 = \left(\dfrac{\text{SSR}}{S_{yy}}\right) 100\% = \left(\dfrac{355.22}{431.60}\right) 100\%$

   $= 82.3\%$,

   which agrees with the value given on the printout.

d. $H_0: \beta_1 = \beta_2 = 0$; H_a: at least one nonzero β_i;

   test statistic: $F = \dfrac{\text{MSR}}{\text{MSE}} = 16.28$ with level of significance equal to 0.002. H_0 is rejected. The model contributes significant information for predicting y.

e. $H_0: \beta_2 = 0$; $H_a: \beta_2 \neq 0$; test statistic:

   $t = \dfrac{\hat{\beta}_2}{s_{\hat{\beta}_2}} = \dfrac{4.4343}{.8002} = 5.54$

   with level of significance 0.000. Reject H_0 and conclude that $\beta_2 \neq 0$.

f. Correct to three digit accuracy, the *regression equation* is

   $\hat{y} = -8.18 + 0.292x_1 + 4.43x_2$.

g. When $x_1 = 22$ and $x_2 = 4$, $\hat{y} = -8.177 + 6.427 + 17.737 = 15.987$.

2. a. The entry R-sq = 85.9% implies that 85.9% of the total variation can be explained by regression.
   b. $R^2 = \dfrac{SSR}{S_{yy}} = \dfrac{19.90431}{23.1800} = .8587$ which agrees with the printout.
   c. $H_0: \beta_1 = \beta_2 = 0$; test statistic: $F = \dfrac{MSR}{MSE}$
   = 18.23 with level of significance .0028. Reject H_0. The model contributes significant information.
   d. $H_0: \beta_2 = 0; H_a: \beta_2 \neq 0$; test statistic:
   $t = \dfrac{\hat{\beta}_2}{s_{\hat{\beta}_2}} = -4.87$ with level of significance .0028. Reject H_0. There is significant curvature.
   e. When $x = 1$, $\hat{y} = 11.43463 + .34 - .20519 = 11.56944$.

Set 11C

1. a. $E(y) = \beta_0 + \beta_1 x_1 + \beta_2 x_2$.
   b. When x_2 is constant, the equation is that of a straight line with slope β_1 and intercept $\beta_0 + \beta_2 x_2$.
   c. When x_1 is constant, the equation is that of a straight line with slope β_2 and intercept $\beta_0 + \beta_1 x_1$.
   d. If the term $\beta_3 x_1 x_2$ is added to the model, the slope and intercept will vary as x_2, since the model will be $E(y) = (\beta_0 + \beta_2 x_2) + (\beta_1 + \beta_3 x_2) x_1$. The line has slope $\beta_1 + \beta_3 x_2$ and intercept $\beta_0 + \beta_2 x_2$.
2. a. $E(y) = \beta_0 + \beta_1 x_1 + \beta_2 x_2 + \beta_3 x_3$.
   b. $E(y) = \beta_0 + \beta_1 x_1 + \beta_2 x_2 + \beta_3 x_3 + \beta_4 x_1 x_2 + \beta_5 x_1 x_3 + \beta_6 x_2 x_3 + \beta_7 x_1^2 + \beta_8 x_2^2 + \beta_9 x_3^2$.
   c. The quadratic terms allow for curvature.
   d. The interaction terms are $\beta_4 x_1 x_2, \beta_5 x_1 x_3, \beta_6 x_2 x_3$. They allow for warping or twisting of the response surface.
3. $E(y) = \beta_0 + \beta_1 x_1 + \beta_2 x_2 + \beta_3 x_1 x_2$. Note that if $x_2 = 0$, $E(y) = \beta_0 + \beta_1 x_1$, while if $x_2 = 1$, $E(y) = (\beta_0 + \beta_2) + (\beta_1 + \beta_3) x_1$. The relationship between x_1 and $E(y)$ is linear for both areas, though not the same in slope or intercept.
4. Refer to exercise 3. The third area involves the addition of only one interaction term. That is, $E(y) = \beta_0 + \beta_1 x_1 + \beta_2 x_2 + \beta_3 x_1 x_2 + \beta_4 x_3 + \beta_5 x_1 x_3$.

Then, for area 1, $x_2 = 0, x_3 = 0$ and
$E(y) = \beta_0 + \beta_1 x_1$
For area 2, $x_2 = 1, x_3 = 0$ and
$E(y) = (\beta_0 + \beta_2) + (\beta_1 + \beta_3) x_1$
For area 3, $x_2 = 0, x_3 = 1$ and
$E(y) = (\beta_0 + \beta_4) + (\beta_1 + \beta_5) x_1$.

5. Refer to exercise 4, with $\beta_0 = 2, \beta_1 = 1, \beta_2 = 1, \beta_3 = 3, \beta_4 = 2, \beta_5 = 1$. The three lines are
$E(y) = 2 + x_1$ for area 1,
$E(y) = 3 + 4x_1$ for area 2,
$E(y) = 4 + 2x_1$ for area 3.

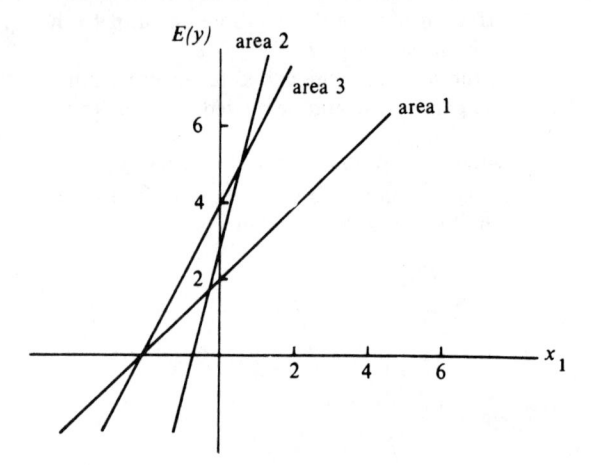

Set 12A

1. $H_0: p_1 = p_2 = p_3 = p_4 = p_5 = 1/5$;
   H_a: at least one of these equalities is incorrect.
   $E(n_i) = np_i = 250(1/5) = 50$ for $i = 1, 2, \ldots, 5$.
   With $k - 1 = 5 - 1 = 4$ degrees of freedom, reject H_0 if $X^2 > \chi^2_{.05} = 9.49$. Test statistic:
   $X^2 = \sum \dfrac{[n_i - E(n_i)]^2}{E(n_i)}$
   $= [(62-50)^2 + (48-50)^2 + (56-50)^2 + (39-50)^2 + (45-50)^2]/50$
   $= \dfrac{144 + 4 + 36 + 121 + 25}{50} = 6.6$.
   Do not reject H_0. We cannot say there is a preference for color.

2. $H_0: p_1 = .41, p_2 = .12, p_3 = .03, p_4 = .44$;
H_a: at least one equality is incorrect.

	Blood Type			
	A	B	AB	O
Observed n_i	90	16	10	84
Expected $E(n_i)$	82	24	6	88

With $k - 1 = 3$ degrees of freedom, reject H_0 if $X^2 > \chi^2_{.05} = 7.81$. Test statistic:

$$X^2 = \frac{(90-82)^2}{82} + \frac{(16-24)^2}{24} + \frac{(10-6)^2}{6} + \frac{(84-88)^2}{88}$$

$= .7805 + 2.6667 + 2.6667 + .1818 = 6.30$.
Do not reject H_0. There is insufficient evidence to refute the given proportions.

Set 12B

1. H_0: independence of classifications; H_a: classifications are not independent. Expected and observed cell counts are given in the table.

Party Affiliation	Income			Total
	Low	Average	High	
Republican	33 (30.57)	85 (74.77)	27 (39.66)	145
Democrat	19 (30.78)	71 (75.29)	56 (39.93)	146
Other	22 (12.65)	25 (30.94)	13 (16.41)	60
Total	74	181	96	351

With $(r-1)(c-1) = 2(2) = 4$ degrees of freedom, reject H_0 if $X^2 > \chi^2_{.05} = 9.49$. Test statistic:

$$X^2 = \frac{(2.43)^2}{30.57} + \frac{(10.23)^2}{74.77} + \ldots + \frac{(-3.41)^2}{16.41} = 25.61.$$

Reject H_0. There is a significant relationship between income levels and political party affiliation.

2. H_0: opinion independent of sex; H_a: opinion dependent on sex. Expected and observed cell counts are given in the table.

	Opinion		Total
Sex	For	Against	
Male	114 (116.58)	60 (57.42)	174
Female	87 (84.42)	39 (41.58)	126
Total	201	99	300

With $(r-1)(c-1) = 1$ degree of freedom and $\alpha = .05$, reject H_0 if $X^2 > \chi^2_{.05} = 3.84$. The test statistic is

$$X^2 = \frac{(-2.58)^2}{116.58} + \frac{(2.58)^2}{57.42} + \frac{(2.58)^2}{84.42} + \frac{(-2.58)^2}{41.58}$$

$= .4119$.

Do not reject H_0. There is insufficient evidence to show that opinion is dependent on sex.

Set 12C

1. $H_0: p_1 = p_2 = p_3 = p_4 = p$; $H_a: p_i \neq p$ for at least one $i = 1, 2, 3, 4$.

	Ward				Total
	1	2	3	4	
Favor	75 (66.25)	63 (66.25)	69 (66.25)	58 (66.25)	265
Against	125 (133.75)	137 (133.75)	131 (133.75)	142 (133.75)	535
Total	200	200	200	200	800

With $(r-1)(c-1) = 3$ degrees of freedom and $\alpha = .05$, reject H_0 if $X^2 > \chi^2_{.05} = 7.81$. Test statistic:

$$X^2 = \frac{8.75^2}{66.25} + \frac{(-3.25)^2}{66.25} + \ldots + \frac{(8.25)^2}{133.75}$$

$$= \frac{162.75}{66.25} + \frac{162.75}{133.75} = 3.673.$$

Do not reject H_0. There is insufficient evidence to suggest a difference from ward to ward.

2. H_0: independence of classifications; H_a: dependence of classifications.

With $(r-1)(c-1) = 6$ degrees of freedom, reject H_0 if $X^2 > \chi^2_{.05} = 12.59$. Test statistic:

$$X^2 = \frac{(-10)^2}{50} + \frac{10^2}{50} + \ldots + \frac{4^2}{60} = 8.73.$$

Do not reject H_0.

Set 12D

1. With $e^{-2} = .135335$, $p(0) = .135335$, $p(1) = 2e^{-2} = .270670$, $p(2) = 2e^{-2} = .27067$, and $P(x \geq 3) = 1 - p(0) - p(1) - p(2) = .323325$. The observed and expected cell counts are shown below.

n_i	4	15	16	15
$E(n_i)$	6.77	13.53	13.53	16.17

With $k - 1 = 3$ degrees of freedom, reject H_0 if $X^2 > \chi^2_{.05} = 7.81$. Test statistic:

$$X^2 = \frac{(-2.77)^2}{6.77} + \frac{(1.47)^2 + (2.47)^2}{13.53} + \frac{(-1.17)^2}{16.17}$$

$= 1.133 + .611 + .085 = 1.829$.

Do not reject the model.

2. Expected numbers are $E(n_i) = np_i = 100 p_i$, where $p_i = P$(observation falls in cell i | score drawn from the standard normal distribution). Hence, using Table 3,

$p_1 = P(z < -1.5) = .0668$

$p_2 = P(-1.5 < z < -.5) = .2417$

$p_3 = P(-.5 < z < .5) = 2(.1915) = .3830$

$p_4 = .2417$

$p_5 = .0668$

Observed and expected cell counts are shown below.

n_i	8	20	40	29	3
$E(n_i)$	6.68	24.17	38.30	24.17	6.68

With $k - 1 = 4$ degrees of freedom, reject H_0 if $X^2 > \chi^2_{.05} = 9.49$. Test statistic:

$$X^2 = \frac{1.32^2}{6.68} + \frac{(-4.17)^2}{24.17} + \ldots + \frac{(-3.68)^2}{6.68}$$

$= .2608 + .7194 + .0755 + .9635 + 2.0273$

$= 4.047$.

Do not reject the model.

Set 13A

1. $\sigma_1 = 6$ and $\sigma_2 = 4$. Hence, 3/5 of the units should be assigned to population 1 and 2/5 of the units to population 2. Then

$n_1 = (3/5)(50) = 30$ and $n_2 = 20$.

2. a. The ten values of x are 120, 140, 160, 180, 200, 220, 240, 260, 280, 300, with $\Sigma x_i = 2{,}100$ and $\Sigma x_i^2 = 474{,}000$. Then $\Sigma (x_i - \bar{x})^2 = 474{,}000 - 441{,}000 = 33{,}000$ and

$\sigma_{\hat{\beta}_1} = 10/\sqrt{33{,}000} = 1/\sqrt{330}$.

b. $\Sigma x_i = 2{,}100$, $\Sigma x_i^2 = 522{,}000$,

$\Sigma (x_i - \bar{x})^2 = 522{,}000 - 441{,}000 = 81{,}000$, and

$\sigma_{\hat{\beta}_1} = 10/\sqrt{81{,}000} = 1/\sqrt{810}$.

c. $\sqrt{810}/\sqrt{330} = 1.57$ times as large.

d. Allocation b is optimal.

Set 13B

1. CM $= (93.8)^2/15 = 586.5627$;

Total SS $= 596.26 - $ CM $= 9.6973$;

$$\text{SST} = \frac{32.3^2 + 28.2^2 + 33.3^2}{5} - \text{CM}$$

$= 589.484 - $ CM $= 2.9213$;

SSE $= 9.6973 - 2.9213 = 6.7760$.

ANOVA

Source	d.f.	SS	MS	F
Treatments	2	2.9213	1.4607	2.5867
Error	12	6.7760	.5647	
Total	14	9.6973		

a. $H_0: \mu_1 = \mu_2 = \mu_3$; H_a: at least one of the equalities is incorrect. Test statistic: $F = $ MST/MSE. With $\nu_1 = 2$ and $\nu_2 = 12$ degrees of freedom, reject H_0 if $F > F_{.05} = 3.89$. Since $F = 2.5867$, do not reject H_0. We cannot find a significant difference.

b. $(\bar{T}_2 - \bar{T}_3) \pm t_{.025} \sqrt{\text{MSE}\left(\frac{1}{n_2} + \frac{1}{n_3}\right)}$

$$= (5.64 - 6.66) \pm 2.179 \sqrt{\frac{2(.5647)}{5}}$$

$$= -1.02 \pm 2.179 \, (.4753)$$

$$= -1.02 \pm 1.04.$$

$-2.06 < \mu_2 - \mu_3 < .02$ with 95% confidence.

2. $CM = (112)^2/20 = 627.2$; Total SS = 708 − CM = 80.80;

$$SST = \frac{31^2 + 17^2 + 39^2 + 25^2}{5} - CM$$

$$= 679.2 - CM = 52.00;$$

SSE = 80.80 − 52.00 = 28.80.

ANOVA

Source	d.f.	SS	MS	F
Areas	3	52.00	17.33	9.63
Error	16	28.80	1.80	
Total	19	80.80		

a. $H_0: \mu_1 = \mu_2 = \mu_3 = \mu_4$; H_a: at least one equality does not hold. $F = 9.63 > F_{.01} = 5.29$. Reject H_0. There is a difference for the four areas.

b. $\bar{T}_1 \pm t_{.025} \sqrt{MSE/n_1}$

$$= 6.2 \pm 2.120 \sqrt{1.80/5}$$

$$= 6.2 \pm 2.12 \, (.6) = 6.2 \pm 1.272.$$

c. $(\bar{T}_1 - \bar{T}_3) \pm t_{.025} \sqrt{MSE\left(\frac{1}{n_1} + \frac{1}{n_3}\right)}$

$$= (6.2 - 7.8) \pm 2.120 \sqrt{1.80(2/5)}$$

$$= -1.6 \pm 2.12 \, (.85) = -1.6 \pm 1.802.$$

Set 13C

1. $CM = (1,453)^2/20 = 105,560.45$; Total SS = 106,399 − CM = 838.55;

$$SST = \frac{362^2 + 343^2 + 366^2 + 382^2}{5} - CM$$

$$= 105,714.6 - CM$$

$$= 154.15;$$

$$SSB = \frac{310^2 + 270^2 + \ldots + 266^2}{4} - CM$$

$$= 106,050.25 - CM = 489.80;$$

SSE = 838.55 − 154.15 − 489.80 = 194.60.

ANOVA

Source	d.f.	SS	MS	F
Litters	4	489.80	122.45	7.55
Additive	3	154.15	51.38	3.17
Error	12	194.60	16.22	
Total	19	838.55		

a. $F = 3.17 < 3.49$. Do not reject H_0. Insufficient evidence to detect a difference due to additives.

b. $F = 7.55 > 3.26$. Reject H_0. There is a difference due to litters. Blocking is desirable.

c. $(\bar{T}_1 - \bar{T}_2) \pm t_{.025} \sqrt{MSE(2/b)}$

$$= (362 - 343)/5 \pm 2.179 \sqrt{6.488}$$

$$= 3.8 \pm 5.6.$$

d. $\bar{T}_3 \pm t_{.025} \sqrt{MSE/b}$

$$= (366/5) \pm 2.179 \sqrt{16.22/5} = 73.2 \pm 3.9.$$

2. a. $CM = (922)^2/12 = 70,840.3333$;

Total SS = 72,432 − CM = 1,591.6667;

$$SST = \frac{453^2 + 469^2}{6} - CM = \frac{425,170}{6} - CM$$

$$= 21.3334;$$

$$SSB = \frac{161^2 + 134^2 + \ldots + 118^2}{2} - CM$$

$$= 72,391 - CM = 1,550.6667;$$

SSE = 19.6666.

ANOVA

Source	d.f.	SS	MS	F
Treatments	1	21.3334	21.3334	5.42
Blocks	5	1,550.6667	310.1333	78.85
Error	5	19.6666	3.9333	
Total	11	1,591.6667		

b. $F = 5.42 = t^2 = (2.328)^2.$

Set 14A

1. Let $p = P$ (paint A shows less wear) and $x =$ number of locations where paint A shows less wear. Since no numerical measure of a response is given, the sign test is appropriate. $H_0: p = 1/2$; $H_a: p < 1/2$. Rejection region: With $n = 25$, $p = 1/2$, reject H_0 if $x \leq 8$ with $\alpha = .054$. (See Table 1e.) Observe $x = 8$; therefore, reject H_0. Paint B is more durable.

2. Rank the times from low to high. Note $n_1 = n_2 = 5$.

Water	Food
2	6
7	8
3	4
1	10
5	9

H_0: no difference in the distributions; H_a: the distributions are different. Rejection region: Use the convention of choosing the smaller of U_1 and U_2, so that we are only concerned with the lower portion of the rejection region. Using Table 11, reject H_0 if the smaller of U_1 and U_2 is less than or equal to 3 with $\alpha/2 = .0278$, so that $\alpha = .0556$. Calculate

$U_1 = 5(5) + (1/2)(5)(6) - 18 = 22$;

$U_2 = 5(5) + (1/2)(5)(6) - 37 = 3$. Reject H_0.

3. $H_0: p = 1/2$, where $p = P$ (variety A exceeds variety B); $H_a: p \neq 1/2$. With $\alpha = .022$ from Table 1, reject H_0 if $x = 0, 1, 9, 10$. Since x = number of plus signs = 8, do not reject H_0. We cannot detect a difference between varieties A and B.

4. Rank the scores from low to high. Let $n_1 = 7$, $n_2 = 8$.

Women (1)	Men (2)
6 (10)	3 (6)
10 (15)	5 (9)
3 (6)	2 (4)
8 (12.5)	0 (1.5)
8 (12.5)	3 (6)
7 (11)	1 (3)
9 (14)	0 (1.5)
	4 (8)

H_0: no difference in the distributions; H_a: scores are higher for women. Rejection region: Women have higher ranks if H_a is true, making $U_1 = n_1 n_2 + (1/2) n_1 (n_1 + 1) - T_1$ small. Using Table 11, reject H_0 if $U_1 \leq 13$ with $\alpha = .0469$. Calculate $U_1 = 7(8) + (1/2)(7)(8) - 81 = 3$. Reject H_0.

Set 14B

1. H_0: no difference in distributions of number of nematodes for varieties A and B; H_a: distribution of number of nematodes differs for varieties A and B. Rank the absolute differences from smallest to largest and calculate T, the smaller of the two (positive and negative) rank sums.

	Location									
	1	2	3	4	5	6	7	8	9	10
d_i	186	138	349	120	-8	92	219	39	-32	52
Rank $|d_i|$	8	7	10	6	1	5	9	3	2	4

Rejection region: With $\alpha = .02$ and a two-sided test, reject H_0 if $T \leq 5$. Since $T = 1 + 2 = 3$, reject H_0. There is a difference between A and B.

2. $H_0: p = 1/2$, where $p = P$ (positive difference) and $n = 24$; $H_a: p > 1/2$. Using $\alpha = .05$ and the normal approximation, H_0 will be rejected if

$$\frac{x - .5n}{.5\sqrt{n}} > 1.645,$$

where x = number of positive differences. Calculate

$$z = \frac{x - 12}{.5\sqrt{24}} = \frac{16 - 12}{2.45} = 1.63. \text{ Do not reject } H_0.$$

3. The ranks of the absolute differences are given along with their corresponding signs:
−12, 13, 6, −4, −10, 3, 14.5, 20, 8, 16, −9, −1, 22, 17, 11, 18, 19, 24, −5, 21, −7, 14.5, −2, 23.
With $\alpha = .05$ and a one-sided test, reject H_0 if $T \leq 92$. Since $T = 50$, reject H_0. Note that the sign test is computationally simple, but the Wilcoxon signed rank test is more efficient since it allows us to reject H_0 while the sign test did not.

Set 14C

1. The data are ranked according to magnitude:

1	2	3
7.1 (13.5)	4.9 (2)	6.7 (10)
6.2 (6.5)	6.6 (9)	6.0 (5)
7.0 (12)	6.8 (11)	7.3 (15)
5.6 (4)	4.6 (1)	6.2 (6.5)
6.4 (8)	5.3 (3)	7.1 (13.5)
$T_1 = 44$	$T_2 = 26$	$T_3 = 50$

Test statistic:

$$H = \frac{12}{15(16)} \left[\frac{44^2 + 26^2 + 50^2}{5} \right] - 3(16)$$

$$= 51.12 - 48 = 3.12$$

Rejection region: With $\alpha = .05$ and $v = k - 1 = 2$, reject H_0 if

$$H \geq \chi^2_{.05} = 5.99.$$

Do not reject H_0. There is insufficient evidence to indicate a difference between rations.

2. The data are ranked according to magnitude:

1	2	3	4
7 (14.5)	1 (1)	7 (14.5)	4 (5)
5 (8.5)	4 (5)	9 (20)	6 (11.5)
5 (8.5)	3 (2.5)	8 (18)	3 (2.5)
6 (11.5)	4 (5)	7 (14.5)	7 (14.5)
8 (18)	5 (8.5)	8 (18)	5 (8.5)
$T_1 = 61$	$T_2 = 22$	$T_3 = 85$	$T_4 = 42$

Test statistic:

$$H = \frac{12}{20(21)} \left[\frac{61^2 + 22^2 + 85^2 + 42^2}{5} \right] - 3(21)$$

$$= 12.39$$

Rejection region: With $\alpha = .01$ and $k - 1 = 3$ degrees of freedom, reject H_0 if $H > \chi^2_{.01} = 11.3449$. Reject H_0. There is evidence to indicate a difference among areas.

Set 14D

1. The responses within litters are ranked from 1 to 4.

	Additive			
Litter	1	2	3	4
1	2.5	1	2.5	4
2	2	1	3.5	3.5
3	3	2	1	4
4	3	1	4	2
5	1	2	3	4
	$T_1 = 11.5$	$T_2 = 7$	$T_3 = 14$	$T_4 = 17.5$

Test statistic:

$$F_r = \frac{12}{5(4)(5)} (11.5^2 + 7^2 + 14^2 + 17.5^2) - 3(5)(5)$$

$$= 82.02 - 75 = 7.02$$

Rejection region: With $\alpha = .01$ and $v = k - 1 = 3$, reject H_0 if

$$F_r \geq \chi^2_{.01} = 11.3449$$

Do not reject H_0. There is insufficient evidence to indicate a difference due to additives.

2. The data are ranked within areas from 1 to 4.

Areas	Brands			
	1	2	3	4
1	1	2	4	3
2	1	4	2	3
3	1	4	2	3
4	1.5	3	1.5	4
5	1	4	3	2
	$T_1 = 5.5$	$T_2 = 17$	$T_3 = 12.5$	$T_4 = 15$

Test statistic:

$$F_r = \frac{12}{5(4)(5)} (5.5^2 + 17^2 + 12.5^2 + 15^2) - 3(5)(5)$$

$$= 84.06 - 75 = 9.06$$

Rejection region: With $\alpha = .05$ and $v = k - 1 = 3$, reject H_0 if

$$F_r \geq 7.81.$$

Reject H_0. There is a difference among the four brands.

Set 14E

1. Rank the examination scores, and note that the interview scores are already in rank order.

Interview Rank (x_i)	Exam Rank (y_i)
4	3
7	2
5	5
3	4
6	1
2	6
1	7

$n = 7$ $\quad \Sigma x_i^2 = 140$
$\Sigma x_i = 28$ $\quad \Sigma y_i^2 = 140$
$\Sigma y_i = 28$ $\quad \Sigma x_i y_i = 88$

$$r_s = \frac{88 - (28)^2/7}{140 - (28)^2/7} = \frac{-24}{28} = -.857.$$

To test $H_0: \rho_s = 0$; $H_a: \rho_s < 0$, the rejection region is $r_s < -.714$. Hence H_0 is rejected with $\alpha = .05$.

2. a. $\Sigma x_i = \Sigma y_i = 45$ $\quad \Sigma x_i^2 = \Sigma y_i^2 = 285$
$\Sigma x_i y_i = 272$

$$r_s = \frac{272 - (45)^2/9}{285 - (45)^2/9} = \frac{47}{60} = .783.$$

b. $H_0: \rho_s = 0$; $H_a: \rho_s > 0$. With $\alpha = .05$, reject H_0 if $r_s > .600$. Reject H_0. They are in basic agreement.

Appendix: Useful Mathematical Notation

Set A

1. $f(4) = 2(4) + 3 = 11$.
2. $g(0) = 0^2 - 2 = -2$; $g(4) = 4^2 - 2 = 14$.
3. $h(0) = 0! = 1$; $h(4) = 4! = (4)(3)(2)(1) = 24$.
4. $g(0) + g(4) = 0! + 4! = 1 + 24 = 25$.

Set B

1. $\sum_{y=1}^{3} h(y) = h(1) + h(2) + h(3)$

$$= 2(1/3)^1 + 2(1/3)^2 + 2(1/3)^3$$
$$= (2/3) + (2/9) + (2/27) = (26/27).$$

2. $\sum_{x=1}^{4} (2x^2 - 5) = 2(1^2) - 5 + 2(2^2) - 5 + 2(3^2) - 5$
$$+ 2(4^2) - 5$$
$$= 2 - 5 + 8 - 5 + 18 - 5 + 32 - 5$$
$$= 40.$$

3. $\sum_{i=1}^{10} x_i = (-1) + 2 + 1 + \ldots + (-2) = 3.$

5. $\sum_{i=1}^{10} x_i/10 = 3/10 = .3.$

Set C

1. $\sum_{x=1}^{10} 3 = 10(3) = 30$, using Theorem A.1.

2. $\sum_{i=1}^{14} 4$ has 14 terms, while $\sum_{i=1}^{6} 4$ has 6 terms.

   $\sum_{i=7}^{14} 4 = \sum_{i=1}^{14} 4 - \sum_{i=1}^{6} 4 = 14(4) - 6(4)$

   $= 8(4) = 32.$

3. a. $\sum_{i=3}^{8} (x_i - 4) = \sum_{i=3}^{8} x_i - 6(4) = 1 + 4 + (-3) + 1 + 6 - 24 = -15.$

   b. $\sum_{i=1}^{10} (x_i - 5)^2 = \sum_{i=1}^{10} x_i^2 - 2(5) \sum_{i=1}^{10} x_i + 10(5)^2$

   $= 97 - 10(3) + 250 = 317.$

   See Example A.13.

   c. $\sum_{i=1}^{5} (x_i^2 - x_i) = \sum_{i=1}^{5} x_i^2 - \sum_{i=1}^{5} x_i = 22 - 6$

   $= 16.$

   d. $\sum_{i=1}^{10} x_i^2 - \left(\sum_{i=1}^{10} x_i \right)^2 \Big/ 10 = 97 - (3^2/10)$

   $= 97 - .9 = 96.1.$

ANSWERS TO EXERCISES

Chapter 2

1. a. Range = 6.8; c. 80; 55; e. median = 19.5;
   f. 10; g. 75 (upper quartile); h. $\bar{x} = 19.03$;
   $s^2 = 2.7937$; $s = 1.67$; i. Yes, since 70%,
   95% and 100% of the measurements lie in the
   intervals $\bar{x} \pm ks$, $k = 1, 2, 3$, respectively;
   j. Yes (see i).
2. Inner fences: 14.925 and 23.125
   Outer fences: 11.85 and 26.20
   There are no outliers.
3. a. Range = 7 b. $m = 2$; $Q_L = .5$; $Q_U = 3$
   c. no d. $\bar{x} = 1.96$; $s^2 = 3.1233$; $s = 1.77$
   f. .44, .32 g. No, not bell-shaped
4. a. 16%. b. 81.5%
5. a. 8. b. 2.45 to 2.95; 5.95 to 6.45.
6. $\mu = 6.6$ ounces.
7. a. s is approximated as 16. b. $\bar{x} = 136.07$;
   $s^2 = 292.4952$; $s = 17.1$. c. $a = 101.82$;
   $b = 170.27$. d. Yes, for approximate calculations. e. No.
9. a. $x = 172$ is a suspected outlier; $z = 2.10$.
   b. Inner fences: 96.25 and 174.25; Outer fences: 67 and 203.5. No outliers detected.
10. a. s is approximated as 5. b. $\bar{x} = 11.67$;
    $s^2 = 13.9523$; $s = 3.74$. c. 14/15.
12. a. Approximately .974. b. Approximately .16.
13. Approximately .025.
14. a. Range = 16 c. $\bar{x} = 82.8$; $s^2 = 20.0$;
    $s = 4.472$ d. .68, .96; yes.

Chapter 3

1. a. $(ABC), (ACB)$. b. $(CAB), (ACB)$.
   c. $(ABC), (ACB), (CAB)$. d. (ACB).
   e. $P(A) = 1/3$; $P(A|B) = 1/2$; A and B are dependent.
2. a. 1/2. b. 1/2. c. 5/6.
3. .1792.
4. .459.
5. 56.
6. 6720.
7. a. 5/32. b. 31/32.

8.

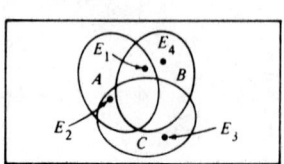

9. a. 1/6. b. 2/3; 1/3. c. 2/3; 2/3.
   d. 5/6; 5/6. e. no; no.
10. a. 56. b. 30. c. 15/28.
11. .41.
12. a. .328. b. .263.
13. .045; yes.
14. a. 14/22; 4/22; 4/22. b. 35/44; 8/44; 1/44.
15. a. 99/991. b. 892/991. c. 3% defective.
16.

x	$p(x)$
0	.1
1	.6
2	.3

17. $p(0) = 3/10$; $p(1) = 6/10$; $p(2) = 1/10$.
18. a.

x	$p(x)$
0	8/27
1	12/27
2	6/27
3	1/27

19. $E(x) = 11,250$; $\sigma = 5,673.4$.
20. $E(R) = 225,000$; $\sigma_R = 113,468$.
21. .65.
22. a. $E(x) = 2$; $\sigma^2 = 1$.
23. a. 1/4. b. 3/16. c. 9/64.
    d. $p(x) = (3/4)^{x-1}(1/4)$. e. Yes.
24. a. $p(0) = 6/15$; $p(1) = 8/15$; $p(2) = 1/15$
    b. $E(x) = 10/15$; $\sigma^2 = .3556$
    c. 1/15.
25. $E(x) = 2.125$; $\sigma = .5995$.
26. $E(x) = 2$; $\sigma^2 = 5$.
27. $6.50.
28. $2.70.
29. No.
30. No; no.

Chapter 4

1. .655360.
2. a. $p(x) = [20!/x!\,(20-x)!]\,(.3)^x\,(.7)^{20-x}$, $x = 0, 1, 2, \ldots, 20$.
   b. .772. c. .780.
3. a. 82. b. 76.
4. a. .122. b. .957.
5. a. n trials are not fixed in advance; not binomial.
   b. Binomial; $p(x) = C_x^5\,(1/15)^x\,(14/15)^{5-x}$
   c. Not binomial; p varies from trial to trial.
   d. Binomial; $p(x) = C_x^5\,(.6)^x\,(.4)^{5-x}$.
6. a. .9801. b. .6400.
7. a. $a = 1$. b. .069.
   c. P (accept lot | p low) is higher and P (reject lot | p high) is lower; more costly.
8. 1202; 1246
9. a. 1.96 to 12.24. b. $p > .071$.
10. a. .794. b. .056. c. 0.34 to 2.66.
11. .083
12. a. .214; .214
    b. .316; .211
13. a. .629 b. .6
14. a. .110803 b. .012
    c. .119
15. a. $44 b. $2200
16. .0758
17. a. 2 to 12 b. unemployment rate is higher than the national rate.
18. a. Geometric with $p = .6$
    b. $\mu = 1.67$ hours
19. a. no b. binomial $n = 4, p = .5$
    c. 2 d. geometric $p = .25$
    e. 4 f. yes

Chapter 5

1. a. .833 b. 1/3
2. a. .135 b. .471 c. 5991.46
3. a. 1; 1 b. .135 c. .693
4. a. .9713. b. .1009. c. .7257. d. .9706.
   e. .8925. f. .5917.
5. a. $z_0 = .70$. b. $z_0 = 2.13$. c. $z_0 = 1.645$.
   d. $z_0 = 1.55$.
6. a. .8413. d. .8944. c. .9876. d. .0401.
7. 87.48
8. $\mu = 10.071$.
9. a. .0139. b. .0668. c. .5764.
   d. .00000269.
10. a. .048. b. .0436.
11. .0017.
12. a. .9838. b. .0000. c. .8686.
13. .3520.

14. a. .1635. b. .0192.
    c. yes, since $P[x \geq 60$ when $p = .2] = .0192$.

Chapter 6

1. .9623
2. .3413
3. .0409
4. a. $65,000 b. $2000
5. .9876
6. .0032
7. .0049
8. .0062
9. a. $110 \pm 2(.99)$ or 108 to 112 b. Approximately 0
10. .0091

Chapter 7

1. The inference; measure of goodness.
2. Unbiasedness; minimum variance.
3. 61.23 ± 1.50.
4. $2.705 \pm .012$.
5. $.030 \pm .033$.
6. $.6 \pm .048$.
7. $2.705 \pm .009$.
8. $21.6 \pm .49$.
9. $.2 \pm .0392$.
10. Approximately 246.
11. Approximately 97.
12. Approximately 97.
13. -8 ± 4.49.
14. 19.3 ± 1.86.
15. $.56 \pm .15;\ -.2475 \pm .22$.

Chapter 8

2. $z = 2.5$; yes.
3. $p = .0062$
4. $z = -5.14$; reject the claim.
5. $p < .002$.
6. $z = 2.8$; yes; $p = .0026$.
7. $z = -3.40$; yes.
8. a. $z = 5.8$; yes. b. $3.0 \pm .85$.
9. $z = 7.30$; no.
10. $p < .001$.
11. $z = -4.59$; yes; $p < .002$.
12. $z = -1.06$; yes
13. a. $z = .80$; do not reject H_0.
    b. $p = .4238$

408 / Answers to Exercises

14. a. 7.2 ± 1.65
    b. $z = 2.19$; Reject $H_0: \mu = 5$
    c. For $\mu = 6$, $\beta = .6906$, $1 - \beta = .3094$; for $\mu = 7$, $\beta = .2578$, $1 - \beta = .7422$.
    d.
μ	8	9	10	12
$1 - \beta$	.9495	.9957	1.0000	1.0000

    e. $\mu > 8$

Chapter 9

1. According to the Central Limit Theorem, these statistics will be approximately normally distributed for large n.
2. i. The parent population has a normal distribution.
   ii. The sample is a random sample.
3. The number of degrees of freedom associated with a . statistic is the denominator of the estimator of σ^2.
4. Do not reject H_0; $t = -.6$.
5. $2.48 < \mu < 4.92$.
6. Do not reject H_0; $t = 1.16$.
7. $-.76 < \mu_1 - \mu_2 < 3.96$.
8. Do not reject H_0; $t = 2.29$.
9. Do not reject H_0; $t = 1.48$.
10. Do not reject H_0; $\chi^2 = 8.19$.
11. $.214 < \sigma^2 < 4.387$.
12. Do not reject H_0; $F = 1.796$.
13. $.565 < \sigma_1^2/\sigma_2^2 < 5.711$.
14. Reject H_0; $F = 2.06$.
15. Do not reject H_0; $t = .95$; p-value $> .10$.
16. a. Reject H_0; $F = 3.88$
    b. Reject H_0; $t^* = 2.13$; $df = 13$.

Chapter 10

1.
	y - intercept	slope
a.	-2	3
b.	0	2
c.	-0.5	-1
d.	2.5	-1.5
e.	2	0

2. a. $\hat{y} = .86 + .71x$.
   c. SSE $= 4/7 = .5714$; $s^2 = .1143$; SSE will be zero only if all of the observed points were to lie on the fitted line.
   d. Reject $H_0: \beta_1 = 0$, since $t = 11.11$.
   e. $.56 < \beta_1 < .86$.
   f. $r = .98$.
   g. Since $r^2 = .96$, the use of the linear model rather than $\bar{y}$ as a predictor for y reduced the sum of squares for error by 96%.
   h. $.82 < y_p < 2.32$.
   i. Reject $H_0: \beta_0 = 0$, since $t = 6.7$.
3. a. $\hat{y} = 8.86 - 1.27x$.
   c. SSE $= 2.34$; $s^2 = .5857$; $s = .76$.
   d. Reject $H_0: \beta_1 = 0$, since $t = -13.9$.
   e. $r^2 = .98$; see problem 2g.
   f. 2.51 ± 1.27; $1.24 < y_p < 3.78$.
   g. $2.51 \pm .48$; $2.03 < E(y|x = 5) < 2.99$.
4. If $r = 1$, the observed points all lie on the fitted line having a positive slope and if $r = -1$, the observed points all lie on the fitted line having a negative slope.
5. a. $\hat{y} = 2 - .875x$.
   c. SSE $= .25$; $s^2 = .0833$; $s = .289$.
   d. Reject $H_0: \beta_1 = 0$, since $t = 12.12$.
   e. $2.525 < E(y|x = -1) < 3.225$.
   f. $r^2 = .98$; see problem 2g.
   g. $-.345 < y_p < 2.595$.
   h. $\bar{x}$.
6. The fitted line may not adequately describe the relationship between x and y outside the experimental region.
7. The error will be a maximum for the values of x at the extremes of the experimental region.
8. a. $\hat{y} = 7.0 + 15.4x$.
   b. SSE $= 50.4$; $s^2 = 8.4$.
   c. Reject $H_0: \beta_1 = 0$, since $t = 16.7$.
   d. $43.0 < E(y|x = 2.5) < 48.0$.
   e. $13.6 < \beta_1 < 17.2$.
   f. $r^2 = .979$; see problem 2g.
9. a. $\hat{y} = 6.96 + 2.31x$.
   c. SSE $= .9751$; $s^2 = .1219$.
   d. Reject $H_0: \beta_1 = 0$, since $t = 19.25$.
   e. $r = .99$.
   f. $r^2 = .979$.
   g. Do not reject H_0; $t = 1.67$.
   h. $9.27 \pm .69$; $8.58 < y_p < 9.96$.
10. a. $\hat{y} = 20.47 - .76x$.
    b. SSE $= 4.658$; $s^2 = .5822$.
    c. Reject $H_0: \beta_1 = 0$, since $t = -22.3$.
    d. $-.86 < \beta_1 < -.66$.
    e. $9.83 \pm .55$; $9.28 < E(y|x = 14) < 10.38$.
    f. $r^2 = .984$; see problem 2g.

Chapter 11

1. a. .998 b. Yes; $F = 1470.84$.
   c. $\hat{y} = -13.01227 + 1.46306x_1$.
   d. $\hat{y} = 23.56$.
2. a. 99.9%. b. Yes; $F = 1676.61$.
   c. Yes, $t = -2.65$. d. Yes, $t = 15.14$.

e. 0.3% of the total variation is explained by the linear term.
3. The contributions of x_1 and x_2 are minimal in the presence of x_3 and x_4.
4. a. .922. b. $F = 23.78$; reject H_0.
   c. $t = -.64$; do not reject H_0.
   d. See part c

Chapter 12

2. a. Yes.
   b. No, since p_i, $i = 1, 2, 3$, changes from trial to trial.
   c. Yes.
3. a. 8.4. b. 18.0. c. 75.6.
4. Reject H_0; $X^2 = 16.535$.
5. Do not reject H_0; $X^2 = 2.300$.
6. Reject H_0; $X^2 = 7.97$.
7. Do not reject H_0: $p_1 = p_2 = p_3 = p_4$: $X^2 = 1.709$.
8. Reject H_0; $X^2 = 9.333$.
9. Do not reject the model; $X^2 = 6.156$.

Chapter 13

1. a. Completely randomized design.
   b.

Source	d.f.	SS	MS	F
Chemicals	2	25.1667	12.5834	2.59
Error	9	43.75	4.8611	
Total	11	68.9167		

   c. No. d. -2.25 ± 3.53
   e. 11 ± 2.02. f. 39.
2. a. Randomized block design.
   b.

Source	d.f.	SS	MS	F
Applications	3	18.9167	6.3056	9.87
Chemicals	2	62.1667	31.0833	48.65
Error	6	3.8333	0.6388	
Total	11	84.9167		

   c. $F = 48.65$; reject H_0; yes. d. 5.25 ± 1.38.
   e. 21. f. Yes.

3. a. Completely randomized design.
   b.

Source	d.f.	SS	MS	F
Treatments	3	1,052.68	350.89	1.76
Error	15	2,997.95	199.86	
Total	18	4,050.63		

   c. $F = 1.76$; no. d. -12.6 ± 19.06.
4. a.

Source	d.f.	SS	MS	F
Programs	2	25,817.49	12,908.74	4.43
Error	13	37,851.51	2,911.65	
Total	15			

   $F = 4.43$; yes.
   b. -63.1 ± 59.9.
5. a.

Source	d.f.	SS	MS	F
Employees	11	477.8889	43.4444	2.9444
Conditions	2	230.7222	115.3611	7.8184
Error	22	324.6111	14.7551	
Total	35			

   $F = 7.8184 > 3.44$. Reject H_0. There is a significant difference between conditions.
   b. $F = 2.944$; yes; yes.
   c. $-9.34 < \mu_2 - \mu_3 < 2.84$.

Chapter 14

1. Sign test: both; Mann-Whitney U test: independent; Wilcoxon signed-rank test: related.
2. $z = -2.54 < -1.645$. Reject H_0.
3. Reject H_0 when $U \leq 28$. $U = 16$. Reject H_0.
4. $r_s = 1$. While $r = 1$ only when the data points all lie on the same straight line, r_s will be 1 whenever y increases steadily with x.
5. a. .738. b. $.738 \geq .643$; reject H_0.
6. With $\alpha = .0703$, reject H_0 if $x = 0, 1, 7, 8$. $x = 6$. Do not reject H_0.
7. Reject H_0 if $T \leq 6$, $\alpha = .10$. $T = 9.5$. Do not reject H_0.
8. Reject H_0 if $U \leq 21$ with $\alpha = .0506$. $U = 15.5$. Reject H_0. Operator A produces a lower percentage defective.
9. a. randomized block design
   b. $F_r = 12.5$, reject H_0; there is a difference in the three production lines.
10. a. completely randomized design b. $H = 4.86$; do not reject H_0.

Appendix

1. u.
2. $3a^3 - 2a$.
3. 3; 3/4.
4. $2/x^2 - 3/x + 5$.
5. $2/v^2 - 3v$.
6. $x_3 + x_4 + x_5 + x_6 - 4a$.
7. $\sum_{i=3}^{6} (x_i - m)^2$.
8. $9x + 9$.
9. 1.
10. 60.
11. 25.
12. 1750.

TABLES

Table 1 Cumulative Binomial Probabilities

Tabulated values are $P(x \leq a) = \sum_{x=0}^{a} p(x)$. (Computations are rounded at the third decimal place.)

$n = 2$

a	0.01	0.05	0.10	0.20	0.30	0.40	0.50	0.60	0.70	0.80	0.90	0.95	0.99	a
0	.980	.902	.810	.640	.490	.360	.250	.160	.090	.040	.010	.002	.000	0
1	1.000	.998	.990	.960	.910	.840	.750	.640	.510	.360	.190	.098	.020	1
2	1.000	1.000	1.000	1.000	1.000	1.000	1.000	1.000	1.000	1.000	1.000	1.000	1.000	2

$n = 3$

a	0.01	0.05	0.10	0.20	0.30	0.40	0.50	0.60	0.70	0.80	0.90	0.95	0.99	a
0	.970	.857	.729	.512	.343	.216	.125	.064	.027	.008	.001	.000	.000	0
1	1.000	.993	.972	.896	.784	.648	.500	.352	.216	.104	.028	.007	.000	1
2	1.000	1.000	.999	.992	.973	.936	.875	.784	.657	.488	.271	.143	.030	2
3	1.000	1.000	1.000	1.000	1.000	1.000	1.000	1.000	1.000	1.000	1.000	1.000	1.000	3

$n = 4$

a	0.01	0.05	0.10	0.20	0.30	0.40	0.50	0.60	0.70	0.80	0.90	0.95	0.99	a
0	.961	.815	.656	.410	.240	.130	.062	.026	.008	.002	.000	.000	.000	0
1	.999	.986	.948	.819	.652	.475	.312	.179	.084	.027	.004	.000	.000	1
2	1.000	1.000	.996	.937	.916	.821	.688	.525	.348	.181	.052	.014	.001	2
3	1.000	1.000	1.000	.998	.992	.974	.938	.870	.760	.590	.344	.185	.039	3
4	1.000	1.000	1.000	1.000	1.000	1.000	1.000	1.000	1.000	1.000	1.000	1.000	1.000	4

(continued)

Table 1 (*Continued*)

$n = 5$

a	0.01	0.05	0.10	0.20	0.30	0.40	0.50	0.60	0.70	0.80	0.90	0.95	0.99	a
0	.951	.774	.590	.328	.168	.078	.031	.010	.002	.000	.000	.000	.000	0
1	.999	.977	.919	.737	.528	.337	.188	.087	.031	.007	.000	.000	.000	1
2	1.000	.999	.991	.942	.837	.683	.500	.317	.163	.058	.009	.001	.000	2
3	1.000	1.000	1.000	.993	.969	.913	.812	.663	.472	.263	.081	.023	.001	3
4	1.000	1.000	1.000	1.000	.998	.990	.969	.922	.832	.672	.410	.226	.049	4
5	1.000	1.000	1.000	1.000	1.000	1.000	1.000	1.000	1.000	1.000	1.000	1.000	1.000	5

$n = 6$

a	0.01	0.05	0.10	0.20	0.30	0.40	0.50	0.60	0.70	0.80	0.90	0.95	0.99	a
0	.941	.735	.531	.262	.118	.047	.016	.004	.001	.000	.000	.000	.000	0
1	.999	.967	.886	.655	.420	.233	.109	.041	.011	.002	.000	.000	.000	1
2	1.000	.998	.984	.901	.744	.544	.344	.179	.070	.017	.001	.000	.000	2
3	1.000	1.000	.999	.983	.930	.821	.656	.456	.256	.099	.016	.002	.000	3
4	1.000	1.000	1.000	.998	.989	.959	.891	.767	.780	.345	.114	.033	.001	4
5	1.000	1.000	1.000	1.000	.999	.996	.984	.953	.882	.738	.469	.265	.059	5
6	1.000	1.000	1.000	1.000	1.000	1.000	1.000	1.000	1.000	1.000	1.000	1.000	1.000	6

$n = 7$

a	0.01	0.05	0.10	0.20	0.30	0.40	0.50	0.60	0.70	0.80	0.90	0.95	0.99	a
0	.932	.698	.478	.210	.082	.028	.008	.002	.000	.000	.000	.000	.000	0
1	.998	.956	.850	.577	.329	.159	.062	.019	.004	.000	.000	.000	.000	1
2	1.000	.996	.974	.852	.647	.420	.227	.096	.029	.005	.000	.000	.000	2
3	1.000	1.000	.997	.967	.874	.710	.500	.290	.126	.033	.003	.000	.000	3
4	1.000	1.000	1.000	.995	.971	.904	.773	.580	.353	.148	.026	.004	.000	4
5	1.000	1.000	1.000	1.000	.996	.981	.938	.841	.671	.423	.150	.044	.002	5
6	1.000	1.000	1.000	1.000	1.000	.998	.992	.972	.918	.790	.522	.302	.068	6
7	1.000	1.000	1.000	1.000	1.000	1.000	1.000	1.000	1.000	1.000	1.000	1.000	1.000	7

Table 1 (*Continued*)

n = 8

a	0.01	0.05	0.10	0.20	0.30	0.40	0.50	0.60	0.70	0.80	0.90	0.95	0.99	a
0	.923	.663	.430	.168	.058	.017	.004	.001	.000	.000	.000	.000	.000	0
1	.997	.943	.813	.503	.255	.106	.035	.009	.001	.000	.000	.000	.000	1
2	1.000	.994	.962	.797	.552	.315	.145	.050	.011	.001	.000	.000	.000	2
3	1.000	1.000	.995	.944	.806	.594	.363	.174	.058	.010	.000	.000	.000	3
4	1.000	1.000	1.000	.990	.942	.826	.637	.406	.194	.056	.005	.000	.000	4
5	1.000	1.000	1.000	.999	.989	.950	.855	.685	.448	.203	.038	.006	.000	5
6	1.000	1.000	1.000	1.000	.999	.991	.965	.894	.745	.497	.187	.057	.003	6
7	1.000	1.000	1.000	1.000	1.000	.999	.996	.983	.942	.832	.570	.337	.077	7
8	1.000	1.000	1.000	1.000	1.000	1.000	1.000	1.000	1.000	1.000	1.000	1.000	1.000	8

n = 9

a	0.01	0.05	0.10	0.20	0.30	0.40	0.50	0.60	0.70	0.80	0.90	0.95	0.99	a
0	.914	.630	.387	.134	.040	.010	.002	.000	.000	.000	.000	.000	.000	0
1	.997	.929	.775	.436	.196	.071	.020	.004	.000	.000	.000	.000	.000	1
2	1.000	.992	.947	.738	.463	.232	.090	.025	.004	.000	.000	.000	.000	2
3	1.000	.999	.992	.914	.730	.483	.254	.099	.025	.003	.000	.000	.000	3
4	1.000	1.000	.999	.980	.901	.733	.500	.267	.099	.020	.001	.000	.000	4
5	1.000	1.000	1.000	.997	.975	.901	.746	.517	.270	.086	.008	.001	.000	5
6	1.000	1.000	1.000	1.000	.996	.975	.910	.768	.537	.262	.053	.008	.000	6
7	1.000	1.000	1.000	1.000	1.000	.996	.980	.929	.804	.564	.225	.071	.003	7
8	1.000	1.000	1.000	1.000	1.000	1.000	.998	.990	.960	.866	.613	.370	.086	8
9	1.000	1.000	1.000	1.000	1.000	1.000	1.000	1.000	1.000	1.000	1.000	1.000	1.000	9

n = 10

a	0.01	0.05	0.10	0.20	0.30	0.40	0.50	0.60	0.70	0.80	0.90	0.95	0.99	a
0	.904	.599	.349	.107	.028	.006	.001	.000	.000	.000	.000	.000	.000	0
1	.996	.914	.736	.376	.149	.046	.011	.002	.000	.000	.000	.000	.000	1
2	1.000	.988	.930	.678	.383	.167	.055	.012	.002	.000	.000	.000	.000	2
3	1.000	.999	.987	.879	.650	.382	.172	.055	.011	.001	.000	.000	.000	3
4	1.000	1.000	.998	.967	.850	.633	.377	.166	.047	.006	.000	.000	.000	4
5	1.000	1.000	1.000	.994	.953	.834	.623	.367	.150	.033	.002	.000	.000	5
6	1.000	1.000	1.000	.999	.989	.945	.828	.618	.350	.121	.013	.001	.000	6
7	1.000	1.000	1.000	1.000	.998	.988	.945	.833	.617	.322	.070	.012	.000	7
8	1.000	1.000	1.000	1.000	1.000	.998	.989	.954	.851	.624	.264	.086	.004	8
9	1.000	1.000	1.000	1.000	1.000	1.000	.999	.994	.972	.893	.651	.401	.096	9
10	1.000	1.000	1.000	1.000	1.000	1.000	1.000	1.000	1.000	1.000	1.000	1.000	1.000	10

(*continued*)

Table 1 (*Continued*)

$n = 11$

a	0.01	0.05	0.10	0.20	0.30	0.40	0.50	0.60	0.70	0.80	0.90	0.95	0.99	a
0	.895	.569	.314	.086	.020	.004	.000	.000	.000	.000	.000	.000	.000	0
1	.995	.898	.697	.322	.113	.030	.006	.001	.000	.000	.000	.000	.000	1
2	1.000	.985	.910	.617	.313	.119	.033	.006	.001	.000	.000	.000	.000	2
3	1.000	.998	.981	.839	.570	.296	.113	.029	.004	.000	.000	.000	.000	3
4	1.000	1.000	.997	.950	.790	.533	.274	.099	.022	.002	.000	.000	.000	4
5	1.000	1.000	1.000	.988	.922	.754	.500	.246	.078	.012	.000	.000	.000	5
6	1.000	1.000	1.000	.998	.978	.901	.726	.467	.210	.050	.003	.000	.000	6
7	1.000	1.000	1.000	1.000	.996	.971	.887	.704	.430	.161	.019	.002	.000	7
8	1.000	1.000	1.000	1.000	.999	.994	.967	.881	.687	.383	.090	.015	.000	8
9	1.000	1.000	1.000	1.000	1.000	.999	.994	.970	.887	.678	.303	.102	.005	9
10	1.000	1.000	1.000	1.000	1.000	1.000	1.000	.996	.980	.914	.686	.431	.105	10
11	1.000	1.000	1.000	1.000	1.000	1.000	1.000	1.000	1.000	1.000	1.000	1.000	1.000	11

$n = 12$

a	0.01	0.05	0.10	0.20	0.30	0.40	0.50	0.60	0.70	0.80	0.90	0.95	0.99	a
0	.886	.540	.282	.069	.014	.002	.000	.000	.000	.000	.000	.000	.000	0
1	.994	.882	.659	.275	.085	.020	.003	.000	.000	.000	.000	.000	.000	1
2	1.000	.980	.889	.558	.253	.083	.019	.003	.000	.000	.000	.000	.000	2
3	1.000	.998	.974	.795	.493	.225	.073	.015	.002	.000	.000	.000	.000	3
4	1.000	1.000	.996	.927	.724	.438	.194	.057	.009	.001	.000	.000	.000	4
5	1.000	1.000	.999	.981	.882	.665	.387	.158	.039	.004	.000	.000	.000	5
6	1.000	1.000	1.000	.996	.961	.842	.613	.335	.118	.019	.001	.000	.000	6
7	1.000	1.000	1.000	.999	.991	.943	.806	.562	.276	.073	.004	.000	.000	7
8	1.000	1.000	1.000	1.000	.998	.985	.927	.775	.507	.205	.026	.002	.000	8
9	1.000	1.000	1.000	1.000	1.000	.997	.981	.917	.747	.442	.111	.020	.000	9
10	1.000	1.000	1.000	1.000	1.000	1.000	.997	.980	.915	.725	.341	.118	.006	10
11	1.000	1.000	1.000	1.000	1.000	1.000	1.000	.998	.986	.931	.718	.460	.114	11
12	1.000	1.000	1.000	1.000	1.000	1.000	1.000	1.000	1.000	1.000	1.000	1.000	1.000	12

Table 1 (*Continued*)

$n = 15$

							p							
a	0.01	0.05	0.10	0.20	0.30	0.40	0.50	0.60	0.70	0.80	0.90	0.95	0.99	a
0	.860	.463	.206	.035	.005	.000	.000	.000	.000	.000	.000	.000	.000	0
1	.990	.829	.549	.167	.035	.005	.000	.000	.000	.000	.000	.000	.000	1
2	1.000	.964	.816	.398	.127	.027	.004	.000	.000	.000	.000	.000	.000	2
3	1.000	.995	.944	.648	.297	.091	.018	.002	.000	.000	.000	.000	.000	3
4	1.000	.999	.987	.836	.515	.217	.059	.009	.001	.000	.000	.000	.000	4
5	1.000	1.000	.998	.939	.722	.403	.151	.034	.004	.000	.000	.000	.000	5
6	1.000	1.000	1.000	.982	.869	.610	.304	.095	.015	.001	.000	.000	.000	6
7	1.000	1.000	1.000	.996	.950	.787	.500	.213	.050	.004	.000	.000	.000	7
8	1.000	1.000	1.000	.999	.985	.905	.696	.390	.131	.018	.000	.000	.000	8
9	1.000	1.000	1.000	1.000	.996	.966	.849	.597	.278	.061	.002	.000	.000	9
10	1.000	1.000	1.000	1.000	.999	.991	.941	.783	.485	.164	.013	.001	.000	10
11	1.000	1.000	1.000	1.000	1.000	.998	.982	.909	.703	.352	.056	.005	.000	11
12	1.000	1.000	1.000	1.000	1.000	1.000	.996	.973	.873	.602	.184	.036	.000	12
13	1.000	1.000	1.000	1.000	1.000	1.000	1.000	.995	.965	.833	.451	.171	.010	13
14	1.000	1.000	1.000	1.000	1.000	1.000	1.000	1.000	.995	.965	.794	.537	.140	14
15	1.000	1.000	1.000	1.000	1.000	1.000	1.000	1.000	1.000	1.000	1.000	1.000	1.000	15

$n = 20$

							p							
a	0.01	0.05	0.10	0.20	0.30	0.40	0.50	0.60	0.70	0.80	0.90	0.95	0.99	a
0	.818	.358	.122	.012	.001	.000	.000	.000	.000	.000	.000	.000	.000	0
1	.983	.736	.392	.069	.008	.001	.000	.000	.000	.000	.000	.000	.000	1
2	.999	.925	.677	.206	.035	.004	.000	.000	.000	.000	.000	.000	.000	2
3	1.000	.984	.867	.411	.107	.016	.001	.000	.000	.000	.000	.000	.000	3
4	1.000	.997	.957	.630	.238	.051	.006	.000	.000	.000	.000	.000	.000	4
5	1.000	1.000	.989	.804	.416	.126	.021	.002	.000	.000	.000	.000	.000	5
6	1.000	1.000	.998	.913	.608	.250	.058	.006	.000	.000	.000	.000	.000	6
7	1.000	1.000	1.000	.968	.772	.416	.132	.021	.001	.000	.000	.000	.000	7
8	1.000	1.000	1.000	.990	.887	.596	.252	.057	.005	.000	.000	.000	.000	8
9	1.000	1.000	1.000	.997	.952	.755	.412	.128	.017	.001	.000	.000	.000	9
10	1.000	1.000	1.000	.999	.983	.872	.588	.245	.048	.003	.000	.000	.000	10
11	1.000	1.000	1.000	1.000	.995	.943	.748	.404	.113	.010	.000	.000	.000	11
12	1.000	1.000	1.000	1.000	.999	.979	.868	.584	.228	.032	.000	.000	.000	12
13	1.000	1.000	1.000	1.000	1.000	.994	.942	.750	.392	.087	.002	.000	.000	13
14	1.000	1.000	1.000	1.000	1.000	.998	.979	.874	.584	.196	.011	.000	.000	14
15	1.000	1.000	1.000	1.000	1.000	1.000	.994	.949	.762	.370	.043	.003	.000	15
16	1.000	1.000	1.000	1.000	1.000	1.000	.999	.984	.893	.589	.133	.016	.000	16
17	1.000	1.000	1.000	1.000	1.000	1.000	1.000	.996	.965	.794	.323	.075	.001	17
18	1.000	1.000	1.000	1.000	1.000	1.000	1.000	.999	.992	.931	.608	.264	.017	18
19	1.000	1.000	1.000	1.000	1.000	1.000	1.000	1.000	.999	.988	.878	.642	.182	19
20	1.000	1.000	1.000	1.000	1.000	1.000	1.000	1.000	1.000	1.000	1.000	1.000	1.000	20

(*continued*)

Table 1 (Continued)

n = 25

a	0.01	0.05	0.10	0.20	0.30	0.40	0.50	0.60	0.70	0.80	0.90	0.95	0.99	a
0	.778	.277	.072	.004	.000	.000	.000	.000	.000	.000	.000	.000	.000	0
1	.974	.642	.271	.027	.002	.000	.000	.000	.000	.000	.000	.000	.000	1
2	.998	.873	.537	.098	.009	.000	.000	.000	.000	.000	.000	.000	.000	2
3	1.000	.966	.764	.234	.033	.002	.000	.000	.000	.000	.000	.000	.000	3
4	1.000	.993	.902	.421	.090	.009	.000	.000	.000	.000	.000	.000	.000	4
5	1.000	.999	.967	.617	.193	.029	.002	.000	.000	.000	.000	.000	.000	5
6	1.000	1.000	.991	.780	.341	.074	.007	.000	.000	.000	.000	.000	.000	6
7	1.000	1.000	.998	.891	.512	.154	.022	.001	.000	.000	.000	.000	.000	7
8	1.000	1.000	1.000	.953	.677	.274	.054	.004	.000	.000	.000	.000	.000	8
9	1.000	1.000	1.000	.983	.811	.425	.115	.013	.000	.000	.000	.000	.000	9
10	1.000	1.000	1.000	.994	.902	.586	.212	.034	.002	.000	.000	.000	.000	10
11	1.000	1.000	1.000	.998	.956	.732	.345	.078	.006	.000	.000	.000	.000	11
12	1.000	1.000	1.000	1.000	.983	.846	.500	.154	.017	.000	.000	.000	.000	12
13	1.000	1.000	1.000	1.000	.994	.922	.655	.268	.044	.002	.000	.000	.000	13
14	1.000	1.000	1.000	1.000	.998	.966	.788	.414	.098	.006	.000	.000	.000	14
15	1.000	1.000	1.000	1.000	1.000	.987	.885	.575	.189	.017	.000	.000	.000	15
16	1.000	1.000	1.000	1.000	1.000	.996	.946	.726	.323	.047	.000	.000	.000	16
17	1.000	1.000	1.000	1.000	1.000	.999	.978	.846	.488	.109	.002	.000	.000	17
18	1.000	1.000	1.000	1.000	1.000	1.000	.993	.926	.659	.220	.009	.000	.000	18
19	1.000	1.000	1.000	1.000	1.000	1.000	.998	.971	.807	.383	.033	.001	.000	19
20	1.000	1.000	1.000	1.000	1.000	1.000	1.000	.991	.910	.579	.098	.007	.000	20
21	1.000	1.000	1.000	1.000	1.000	1.000	1.000	.998	.967	.766	.236	.034	.000	21
22	1.000	1.000	1.000	1.000	1.000	1.000	1.000	1.000	.991	.902	.463	.127	.002	22
23	1.000	1.000	1.000	1.000	1.000	1.000	1.000	1.000	.998	.973	.729	.358	.026	23
24	1.000	1.000	1.000	1.000	1.000	1.000	1.000	1.000	1.000	.996	.928	.723	.222	24
25	1.000	1.000	1.000	1.000	1.000	1.000	1.000	1.000	1.000	1.000	1.000	1.000	1.000	25

Table 2(a) Cumulative Poisson Probabilities

Tabulated values are $P(x \leq a) = \sum_{x=0}^{a} p(x)$. (Computations are rounded at the third decimal place.)

a	0.1	0.2	0.3	0.4	0.5	0.6	0.7	0.8	0.9	1.0	1.5
0	.905	.819	.741	.670	.607	.549	.497	.449	.407	.368	.223
1	.995	.982	.963	.938	.910	.878	.844	.809	.772	.736	.558
2	1.000	.999	.996	.992	.986	.977	.966	.953	.937	.920	.809
3		1.000	1.000	.999	.998	.997	.994	.991	.987	.981	.934
4				1.000	1.000	1.000	.999	.999	.998	.996	.981
5							1.000	1.000	1.000	.999	.996
6										1.000	.999
7											1.000

Table 2(a) (*Continued*)

						μ						
a	2.0	2.5	3.0	3.5	4.0	4.5	5.0	5.5	6.0	6.5	7.0
0	.135	.082	.050	.030	.018	.011	.007	.004	.003	.002	.001
1	.406	.287	.199	.136	.092	.061	.040	.027	.017	.011	.007
2	.677	.544	.423	.321	.238	.174	.125	.088	.062	.043	.030
3	.857	.758	.647	.537	.433	.342	.265	.202	.151	.112	.082
4	.947	.891	.815	.725	.629	.532	.440	.358	.285	.224	.173
5	.983	.958	.916	.858	.785	.703	.616	.529	.446	.369	.301
6	.995	.986	.966	.935	.889	.831	.762	.686	.606	.563	.450
7	.999	.996	.988	.973	.949	.913	.867	.809	.744	.673	.599
8	1.000	.999	.996	.990	.979	.960	.932	.894	.847	.792	.729
9		1.000	.999	.997	.992	.983	.968	.946	.916	.877	.830
10			1.000	.999	.997	.993	.986	.975	.957	.933	.901
11				1.000	.999	.998	.995	.989	.980	.966	.947
12					1.000	.999	.998	.996	.991	.984	.973
13						1.000	.999	.998	.996	.993	.987
14							1.000	.999	.999	.997	.994
15								1.000	.999	.999	.998
16									1.000	1.000	.999
17											1.000

(*continued*)

Table 2(a)
(*Continued*)

a	7.5	8.0	8.5	9.0	9.5	10.0	12.0	15.0	20.0
0	.001	.000	.000	.000	.000	.000	.000	.000	.000
1	.005	.003	.002	.001	.001	.000	.000	.000	.000
2	.020	.014	.009	.006	.004	.003	.001	.000	.000
3	.059	.042	.030	.021	.015	.010	.002	.000	.000
4	.132	.100	.074	.055	.040	.029	.008	.001	.000
5	.241	.191	.150	.116	.089	.067	.020	.003	.000
6	.378	.313	.256	.207	.165	.130	.046	.008	.000
7	.525	.453	.386	.324	.269	.220	.090	.018	.001
8	.662	.593	.523	.456	.392	.333	.155	.037	.002
9	.776	.717	.653	.587	.522	.458	.242	.070	.005
10	.862	.816	.763	.706	.645	.583	.347	.118	.011
11	.921	.888	.849	.803	.752	.697	.462	.185	.021
12	.957	.936	.909	.876	.836	.792	.576	.268	.039
13	.978	.966	.949	.926	.898	.864	.682	.363	.066
14	.990	.983	.973	.959	.940	.917	.772	.466	.105
15	.995	.992	.986	.978	.967	.951	.844	.568	.157
16	.998	.996	.993	.989	.982	.973	.899	.664	.221
17	.999	.998	.997	.995	.991	.986	.937	.749	.297
18	1.000	.999	.999	.998	.996	.993	.963	.819	.381
19		1.000	.999	.999	.998	.997	.979	.875	.470
20			1.000	1.000	.999	.998	.988	.917	.559
21					1.000	.999	.994	.947	.644
22						1.000	.997	.967	.721
23							.999	.981	.787
24							.999	.989	.843
25							1.000	.994	.888
26								.997	.922
27								.998	.948
28								.999	.966
29								1.000	.978
30									.987
31									.992
32									.995
33									.997
34									.999
35									.999
36									1.000

(column header: μ)

Table 2(b) Values of $e^{-\mu}$

μ	$e^{-\mu}$	μ	$e^{-\mu}$	μ	$e^{-\mu}$
0.00	1.000000	2.80	.060810	5.60	.003698
0.05	.951229	2.85	.057844	5.65	.003518
0.10	.904837	2.90	.055023	5.70	.003346
0.15	.860708	2.95	.052340	5.75	.003183
0.20	.818731	3.00	.049787	5.80	.003028
0.25	.778801	3.05	.047359	5.85	.002880
0.30	.740818	3.10	.045049	5.90	.002739
0.35	.704688	3.15	.042852	5.95	.002606
0.40	.670320	3.20	.040762	6.00	.002479
0.45	.637628	3.25	.038774	6.05	.002358
0.50	.606531	3.30	.036883	6.10	.002243
0.55	.576950	3.35	.035084	6.15	.002133
0.60	.548812	3.40	.033373	6.20	.002029
0.65	.522046	3.45	.031746	6.25	.001930
0.70	.496585	3.50	.030197	6.30	.001836
0.75	.472367	3.55	.028725	6.35	.001747
0.80	.449329	3.60	.027324	6.40	.001661
0.85	.427415	3.65	.025991	6.45	.001581
0.90	.406570	3.70	.024724	6.50	.001503
0.95	.386741	3.75	.023518	6.55	.001430
1.00	.367879	3.80	.022371	6.60	.001360
1.05	.349938	3.85	.021280	6.65	.001294
1.10	.332871	3.90	.020242	6.70	.001231
1.15	.316637	3.95	.019255	6.75	.001171
1.20	.301194	4.00	.018316	6.80	.001114
1.25	.286505	4.05	.017422	6.85	.001059
1.30	.272532	4.10	.016573	6.90	.001008
1.35	.259240	4.15	.015764	6.95	.000959
1.40	.246597	4.20	.014996	7.00	.000912
1.45	.234570	4.25	.014264	7.05	.000867
1.50	.223130	4.30	.013569	7.10	.000825
1.55	.212248	4.35	.012907	7.15	.000785
1.60	.201897	4.40	.012277	7.20	.000747
1.65	.192050	4.45	.011679	7.25	.000710
1.70	.182684	4.50	.011109	7.30	.000676
1.75	.173774	4.55	.010567	7.35	.000643
1.80	.165299	4.60	.010052	7.40	.000611
1.85	.157237	4.65	.009562	7.45	.000581
1.90	.149569	4.70	.009095	7.50	.000553
1.95	.142274	4.75	.008652	7.55	.000526
2.00	.135335	4.80	.008230	7.60	.000501
2.05	.128735	4.85	.007828	7.65	.000476
2.10	.122456	4.90	.007447	7.70	.000453
2.15	.116484	4.95	.007083	7.75	.000431
2.20	.110803	5.00	.006738	7.80	.000410
2.25	.105399	5.05	.006409	7.85	.000390
2.30	.100259	5.10	.006097	7.90	.000371
2.35	.095369	5.15	.005799	7.95	.000353
2.40	.090718	5.20	.005517	8.00	.000336
2.45	.086294	5.25	.005248	8.05	.000319
2.50	.082085	5.30	.004992	8.10	.000304
2.55	.078082	5.35	.004748	8.15	.000289
2.60	.074274	5.40	.004517	8.20	.000275
2.65	.070651	5.45	.004296	8.25	.000261
2.70	.067206	5.50	.004087	8.30	.000249
2.75	.063928	5.55	.003887	8.35	.000236

(*continued*)

Table 2(b) *(Continued)*

μ	$e^{-\mu}$	μ	$e^{-\mu}$	μ	$e^{-\mu}$
8.40	.000225	8.95	.000130	9.50	.000075
8.45	.000214	9.00	.000123	9.55	.000071
8.50	.000204	9.05	.000117	9.60	.000068
8.55	.000194	9.10	.000112	9.65	.000064
8.60	.000184	9.15	.000106	9.70	.000061
8.65	.000175	9.20	.000101	9.75	.000058
8.70	.000167	9.25	.000096	9.80	.000056
8.75	.000158	9.30	.000091	9.85	.000053
8.80	.000151	9.35	.000087	9.90	.000050
8.85	.000143	9.40	.000083	9.95	.000048
8.90	.000136	9.45	.000079	10.00	.000045

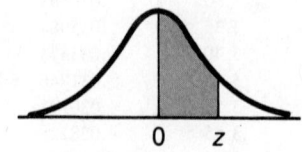

Table 3 Normal Curve Areas

z	.00	.01	.02	.03	.04	.05	.06	.07	.08	.09
0.0	.0000	.0040	.0080	.0120	.0160	.0199	.0239	.0279	.0319	.0359
0.1	.0398	.0438	.0478	.0517	.0557	.0596	.0636	.0675	.0714	.0753
0.2	.0793	.0832	.0871	.0910	.0948	.0987	.1026	.1064	.1103	.1141
0.3	.1179	.1217	.1255	.1293	.1331	.1368	.1406	.1443	.1480	.1517
0.4	.1554	.1591	.1628	.1664	.1700	.1736	.1772	.1808	.1844	.1879
0.5	.1915	.1950	.1985	.2019	.2054	.2088	.2123	.2157	.2190	.2224
0.6	.2257	.2291	.2324	.2357	.2389	.2422	.2454	.2486	.2517	.2549
0.7	.2580	.2611	.2642	.2673	.2704	.2734	.2764	.2794	.2823	.2852
0.8	.2881	.2910	.2939	.2967	.2995	.3023	.3051	.3078	.3106	.3133
0.9	.3159	.3186	.3212	.3238	.3264	.3289	.3315	.3340	.3365	.3389
1.0	.3413	.3438	.3461	.3485	.3508	.3531	.3554	.3577	.3599	.3621
1.1	.3643	.3665	.3686	.3708	.3729	.3749	.3770	.3790	.3810	.3830
1.2	.3849	.3869	.3888	.3907	.3925	.3944	.3962	.3980	.3997	.4015
1.3	.4032	.4049	.4066	.4082	.4099	.4115	.4131	.4147	.4162	.4177
1.4	.4192	.4207	.4222	.4236	.4251	.4265	.4279	.4292	.4306	.4319
1.5	.4332	.4345	.4357	.4370	.4382	.4394	.4406	.4418	.4429	.4441
1.6	.4452	.4463	.4474	.4484	.4495	.4505	.4515	.4525	.4535	.4545
1.7	.4554	.4564	.4573	.4582	.4591	.4599	.4608	.4616	.4625	.4633
1.8	.4641	.4649	.4656	.4664	.4671	.4678	.4686	.4693	.4699	.4706
1.9	.4713	.4719	.4726	.4732	.4738	.4744	.4750	.4756	.4761	.4767
2.0	.4772	.4778	.4783	.4788	.4793	.4798	.4803	.4808	.4812	.4817
2.1	.4821	.4826	.4830	.4834	.4838	.4842	.4846	.4850	.4854	.4857
2.2	.4861	.4864	.4868	.4871	.4875	.4878	.4881	.4884	.4887	.4890
2.3	.4893	.4896	.4898	.4901	.4904	.4906	.4909	.4911	.4913	.4916
2.4	.4918	.4920	.4922	.4925	.4927	.4929	.4931	.4932	.4934	.4936
2.5	.4938	.4940	.4941	.4943	.4945	.4946	.4948	.4949	.4951	.4952
2.6	.4953	.4955	.4956	.4957	.4959	.4960	.4961	.4962	.4963	.4964
2.7	.4965	.4966	.4967	.4968	.4969	.4970	.4971	.4972	.4973	.4974
2.8	.4974	.4975	.4976	.4977	.4977	.4978	.4979	.4979	.4980	.4981
2.9	.4981	.4982	.4982	.4983	.4984	.4984	.4985	.4985	.4986	.4986
3.0	.4987	.4987	.4987	.4988	.4988	.4989	.4989	.4989	.4990	.4990

Source: This table is abridged from Table 1 of *Statistical Tables and Formulas,* by A. Hald (New York: Wiley, 1952). Reproduced by permission of A. Hald and the publisher, John Wiley & Sons, Inc.

**Table 4
Critical Values of t**

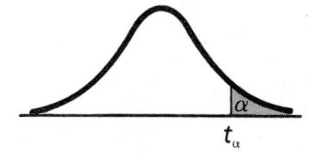

d.f.	$t_{.100}$	$t_{.050}$	$t_{.025}$	$t_{.010}$	$t_{.005}$	d.f.
1	3.078	6.314	12.706	31.821	63.657	1
2	1.886	2.920	4.303	6.965	9.925	2
3	1.638	2.353	3.182	4.541	5.841	3
4	1.533	2.132	2.776	3.747	4.604	4
5	1.476	2.015	2.571	3.365	4.032	5
6	1.440	1.943	2.447	3.143	3.707	6
7	1.415	1.895	2.365	2.998	3.499	7
8	1.397	1.860	2.306	2.896	3.355	8
9	1.383	1.833	2.262	2.821	3.250	9
10	1.372	1.812	2.228	2.764	3.169	10
11	1.363	1.796	2.201	2.718	3.106	11
12	1.356	1.782	2.179	2.681	3.055	12
13	1.350	1.771	2.160	2.650	3.012	13
14	1.345	1.761	2.145	2.624	2.977	14
15	1.341	1.753	2.131	2.602	2.947	15
16	1.337	1.746	2.120	2.583	2.921	16
17	1.333	1.740	2.110	2.567	2.898	17
18	1.330	1.734	2.101	2.552	2.878	18
19	1.328	1.729	2.093	2.539	2.861	19
20	1.325	1.725	2.086	2.528	2.845	20
21	1.323	1.721	2.080	2.518	2.831	21
22	1.321	1.717	2.074	2.508	2.819	22
23	1.319	1.714	2.069	2.500	2.807	23
24	1.318	1.711	2.064	2.492	2.797	24
25	1.316	1.708	2.060	2.485	2.787	25
26	1.315	1.706	2.056	2.479	2.779	26
27	1.314	1.703	2.052	2.473	2.771	27
28	1.313	1.701	2.048	2.467	2.763	28
29	1.311	1.699	2.045	2.462	2.756	29
inf.	1.282	1.645	1.960	2.326	2.576	inf.

Source: From "Table of Percentage Points of the t-Distribution," *Biometrika* 32 (1941) 300. Reproduced by permission of the *Biometrika* Trustees.

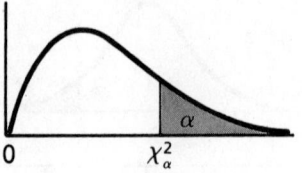

Table 5
Critical Values of Chi-square

d.f.	$\chi^2_{0.995}$	$\chi^2_{0.990}$	$\chi^2_{0.975}$	$\chi^2_{0.950}$	$\chi^2_{0.900}$
1	0.0000393	0.0001571	0.0009821	0.0039321	0.0157908
2	0.0100251	0.0201007	0.0506356	0.102587	0.210720
3	0.0717212	0.114832	0.215795	0.351846	0.584375
4	0.206990	0.297110	0.484419	0.710721	1.063623
5	0.411740	0.554300	0.831211	1.145476	1.61031
6	0.675727	0.872085	1.237347	1.63539	2.20413
7	0.989265	1.239043	1.68987	2.16735	2.83311
8	1.344419	1.646482	2.17973	2.73264	3.48954
9	1.734926	2.087912	2.70039	3.32511	4.16816
10	2.15585	2.55821	3.24697	3.94030	4.86518
11	2.60321	3.05347	3.81575	4.57481	5.57779
12	3.07382	3.57056	4.40379	5.22603	6.30380
13	3.56503	4.10691	5.00874	5.89186	7.04150
14	4.07468	4.66043	5.62872	6.57063	7.78953
15	4.60094	5.22935	6.26214	7.26094	8.54675
16	5.14224	5.81221	6.90766	7.96164	9.31223
17	5.69724	6.40776	7.56418	8.67176	10.0852
18	6.26481	7.01491	8.23075	9.39046	10.8649
19	6.84398	7.63273	8.90655	10.1170	11.6509
20	7.43386	8.26040	9.59083	10.8508	12.4426
21	8.03366	8.89720	10.28293	11.5913	13.2396
22	8.64272	9.54249	10.9823	12.3380	14.0415
23	9.26042	10.19567	11.6885	13.0905	14.8479
24	9.88623	10.8564	12.4011	13.8484	15.6587
25	10.5197	11.5240	13.1197	14.6114	16.4734
26	11.1603	12.1981	13.8439	15.3791	17.2919
27	11.8076	12.8786	14.5733	16.1513	18.1138
28	12.4613	13.5648	15.3079	16.9279	18.9392
29	13.1211	14.2565	16.0471	17.7083	19.7677
30	13.7867	14.9535	16.7908	18.4926	20.5992
40	20.7065	22.1643	24.4331	26.5093	29.0505
50	27.9907	29.7067	32.3574	34.7642	37.6886
60	35.5346	37.4848	40.4817	43.1879	46.4589
70	43.2752	45.4418	48.7576	51.7393	55.3290
80	51.1720	53.5400	57.1532	60.3915	64.2778
90	59.1963	61.7541	65.6466	69.1260	73.2912
100	67.3276	70.0648	74.2219	77.9295	82.3581

Source: From "Tables of the Percentage Points of the χ^2-Distribution," *Biometrika Tables for Statisticians* 1, 3rd ed. (1966). Reproduced by permission of the *Biometrika* Trustees.

Table 5
(*Continued*)

$\chi^2_{0.100}$	$\chi^2_{0.050}$	$\chi^2_{0.025}$	$\chi^2_{0.010}$	$\chi^2_{0.005}$	d.f.
2.70554	3.84146	5.02389	6.63490	7.87944	1
4.60517	5.99147	7.37776	9.21034	10.5966	2
6.25139	7.81473	9.34840	11.3449	12.8381	3
7.77944	9.48773	11.1433	13.2767	14.8602	4
9.23635	11.0705	12.8325	15.0863	16.7496	5
10.6446	12.5916	14.4494	16.8119	18.5476	6
12.0170	14.0671	16.0128	18.4753	20.2777	7
13.3616	15.5073	17.5346	20.0902	21.9550	8
14.6837	16.9190	19.0228	21.6660	23.5893	9
15.9871	18.3070	20.4831	23.2093	25.1882	10
17.2750	19.6751	21.9200	24.7250	26.7569	11
18.5494	21.0261	23.3367	26.2170	28.2995	12
19.8119	23.3621	24.7356	27.6883	29.8194	13
21.0642	23.6848	26.1190	29.1413	31.3193	14
22.3072	24.9958	27.4884	30.5779	32.8013	15
23.5418	26.2962	28.8454	31.9999	34.2672	16
24.7690	27.5871	30.1910	33.4087	35.7185	17
25.9894	28.8693	31.5264	34.8053	37.1564	18
27.2036	30.1435	32.8523	36.1908	38.5822	19
28.4120	31.4104	34.1696	37.5662	39.9968	20
29.6151	32.6705	35.4789	38.9321	41.4010	21
30.8133	33.9244	36.7807	40.2894	42.7956	22
32.0069	35.1725	38.0757	41.6384	44.1813	23
33.1963	36.4151	39.3641	42.9798	45.5585	24
34.3816	37.6525	40.6465	44.3141	46.9278	25
35.5631	38.8852	41.9232	45.6417	48.2899	26
36.7412	40.1133	43.1944	46.9630	49.6449	27
37.9159	41.3372	44.4607	48.2782	50.9933	28
39.0875	42.5569	45.7222	49.5879	52.3356	29
40.2560	43.7729	46.9792	50.8922	53.6720	30
51.8050	55.7585	59.3417	63.6907	66.7659	40
63.1671	67.5048	71.4202	76.1539	79.4900	50
74.3970	79.0819	83.2976	88.3794	91.9517	60
85.5271	90.5312	95.0231	100.425	104.215	70
96.5782	101.879	106.629	112.329	116.321	80
107.565	113.145	118.136	124.116	128.299	90
118.498	124.342	129.561	135.807	140.169	100

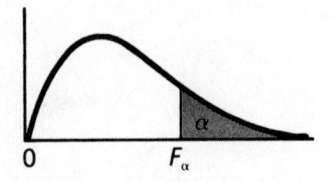

Table 6
Percentage Points of the F Distribution: $\alpha = .10$

v_2 (d.f.)	1	2	3	4	5	6	7	8	9
1	39.86	49.50	53.59	55.83	57.24	58.20	58.91	59.44	59.86
2	8.53	9.00	9.16	9.24	9.29	9.33	9.35	9.37	9.38
3	5.54	5.46	5.39	5.34	5.31	5.28	5.27	5.25	5.24
4	4.54	4.32	4.19	4.11	4.05	4.01	3.98	3.95	3.94
5	4.06	3.78	3.62	3.52	3.45	3.40	3.37	3.34	3.32
6	3.78	3.46	3.29	3.18	3.11	3.05	3.01	2.98	2.96
7	3.59	3.26	3.07	2.96	2.88	2.83	2.78	2.75	2.72
8	3.46	3.11	2.92	2.81	2.73	2.67	2.62	2.59	2.56
9	3.36	3.01	2.81	2.69	2.61	2.55	2.51	2.47	2.44
10	3.29	2.92	2.73	2.61	2.52	2.46	2.41	2.38	2.35
11	3.23	2.86	2.66	2.54	2.45	2.39	2.34	2.30	2.27
12	3.18	2.81	2.61	2.48	2.39	2.33	2.28	2.24	2.21
13	3.14	2.76	2.56	2.43	2.35	2.28	2.23	2.20	2.16
14	3.10	2.73	2.52	2.39	2.31	2.24	2.19	2.15	2.12
15	3.07	2.70	2.49	2.36	2.27	2.21	2.16	2.12	2.09
16	3.05	2.67	2.46	2.33	2.24	2.18	2.13	2.09	2.06
17	3.03	2.64	2.44	2.31	2.22	2.15	2.10	2.06	2.03
18	3.01	2.62	2.42	2.29	2.20	2.13	2.08	2.04	2.00
19	2.99	2.61	2.40	2.27	2.18	2.11	2.06	2.02	1.98
20	2.97	2.59	2.38	2.25	2.16	2.09	2.04	2.00	1.96
21	2.96	2.57	2.36	2.23	2.14	2.08	2.02	1.98	1.95
22	2.95	2.56	2.35	2.22	2.13	2.06	2.01	1.97	1.93
23	2.94	2.55	2.34	2.21	2.11	2.05	1.99	1.95	1.92
24	2.93	2.54	2.33	2.19	2.10	2.04	1.98	1.94	1.91
25	2.92	2.53	2.32	2.18	2.09	2.02	1.97	1.93	1.89
26	2.91	2.52	2.31	2.17	2.08	2.01	1.96	1.92	1.88
27	2.90	2.51	2.30	2.17	2.07	2.00	1.95	1.91	1.87
28	2.89	2.50	2.29	2.16	2.06	2.00	1.94	1.90	1.87
29	2.89	2.50	2.28	2.15	2.06	1.99	1.93	1.89	1.86
30	2.88	2.49	2.28	2.14	2.05	1.98	1.93	1.88	1.85
40	2.84	2.44	2.23	2.09	2.00	1.93	1.87	1.83	1.79
60	2.79	2.39	2.18	2.04	1.95	1.87	1.82	1.77	1.74
120	2.75	2.35	2.13	1.99	1.90	1.82	1.77	1.72	1.68
∞	2.71	2.30	2.08	1.94	1.85	1.77	1.72	1.67	1.63

Source: From "Tables of Percentage Points of the Inverted Beta (F)-Distribution," *Biometrika* 33 (1943) 73–88. by Maxine Merrington and Catherine M. Thompson. Reproduced by permission of the *Biometrika* Trustees.

Table 6
(*Continued*)

10	12	15	20	24	30	40	60	120	∞	v_2 (d.f.)
60.19	60.71	61.22	61.74	62.00	62.26	62.53	62.79	63.06	63.33	1
9.39	9.41	9.42	9.44	9.45	9.46	9.47	9.47	9.48	9.49	2
5.23	5.22	5.20	5.18	5.18	5.17	5.16	5.15	5.14	5.13	3
3.92	3.90	3.87	3.84	3.83	3.82	3.80	3.79	3.78	3.76	4
3.30	3.27	3.24	3.21	3.19	3.17	3.16	3.14	3.12	3.10	5
2.94	2.90	2.87	2.84	2.82	2.80	2.78	2.76	2.74	2.72	6
2.70	2.67	2.63	2.59	2.58	2.56	2.54	2.51	2.49	2.47	7
2.54	2.50	2.46	2.42	2.40	2.38	2.36	2.34	2.32	2.29	8
2.42	2.38	2.34	2.30	2.28	2.25	2.23	2.21	2.18	2.16	9
2.32	2.28	2.24	2.20	2.18	2.16	2.13	2.11	2.08	2.06	10
2.25	2.21	2.17	2.12	2.10	2.08	2.05	2.03	2.00	1.97	11
2.19	2.15	2.10	2.06	2.04	2.01	1.99	1.96	1.93	1.90	12
2.14	2.10	2.05	2.01	1.98	1.96	1.93	1.90	1.88	1.85	13
2.10	2.05	2.01	1.96	1.94	1.91	1.89	1.86	1.83	1.80	14
2.06	2.02	1.97	1.92	1.90	1.87	1.85	1.82	1.79	1.76	15
2.03	1.99	1.94	1.89	1.87	1.84	1.81	1.78	1.75	1.72	16
2.00	1.96	1.91	1.86	1.84	1.81	1.78	1.75	1.72	1.69	17
1.98	1.93	1.89	1.84	1.81	1.78	1.75	1.72	1.69	1.66	18
1.96	1.91	1.86	1.81	1.79	1.76	1.73	1.70	1.67	1.63	19
1.94	1.89	1.84	1.79	1.77	1.74	1.71	1.68	1.64	1.61	20
1.92	1.87	1.83	1.78	1.75	1.72	1.69	1.66	1.62	1.59	21
1.90	1.86	1.81	1.76	1.73	1.70	1.67	1.64	1.60	1.57	22
1.89	1.84	1.80	1.74	1.72	1.69	1.66	1.62	1.59	1.55	23
1.88	1.83	1.78	1.73	1.70	1.67	1.64	1.61	1.57	1.53	24
1.87	1.82	1.77	1.72	1.69	1.66	1.63	1.59	1.56	1.52	25
1.86	1.81	1.76	1.71	1.68	1.65	1.61	1.58	1.54	1.50	26
1.85	1.80	1.75	1.70	1.67	1.64	1.60	1.57	1.53	1.49	27
1.84	1.79	1.74	1.69	1.66	1.63	1.59	1.56	1.52	1.48	28
1.83	1.78	1.73	1.68	1.65	1.62	1.58	1.55	1.51	1.47	29
1.82	1.77	1.72	1.67	1.64	1.61	1.57	1.54	1.50	1.46	30
1.76	1.71	1.66	1.61	1.57	1.54	1.51	1.47	1.42	1.38	40
1.71	1.66	1.60	1.54	1.51	1.48	1.44	1.40	1.35	1.29	60
1.65	1.60	1.55	1.48	1.45	1.41	1.37	1.32	1.26	1.19	120
1.60	1.55	1.49	1.42	1.38	1.34	1.30	1.24	1.17	1.00	∞

v_1 (d.f.)

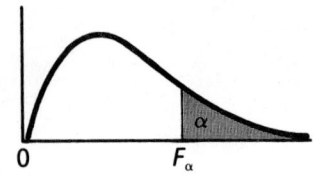

Table 7
Percentage Points of the F Distribution: $\alpha = .05$

v_2 (d.f.)	1	2	3	4	5	6	7	8	9
1	161.4	199.5	215.7	224.6	230.2	234.0	236.8	238.9	240.5
2	18.51	19.00	19.16	19.25	19.30	19.33	19.35	19.37	19.38
3	10.13	9.55	9.28	9.12	9.01	8.94	8.89	8.85	8.81
4	7.71	6.94	6.59	6.39	6.26	6.16	6.09	6.04	6.00
5	6.61	5.79	5.41	5.19	5.05	4.95	4.88	4.82	4.77
6	5.99	5.14	4.76	4.53	4.39	4.28	4.21	4.15	4.10
7	5.59	4.74	4.35	4.12	3.97	3.87	3.79	3.73	3.68
8	5.32	4.46	4.07	3.84	3.69	3.58	3.50	3.44	3.39
9	5.12	4.26	3.86	3.63	3.48	3.37	3.29	3.23	3.18
10	4.96	4.10	3.71	3.48	3.33	3.22	3.14	3.07	3.02
11	4.84	3.98	3.59	3.36	3.20	3.09	3.01	2.95	2.90
12	4.75	3.89	3.49	3.26	3.11	3.00	2.91	2.85	2.80
13	4.67	3.81	3.41	3.18	3.03	2.92	2.83	2.77	2.71
14	4.60	3.74	3.34	3.11	2.96	2.85	2.76	2.70	2.65
15	4.54	3.68	3.29	3.06	2.90	2.79	2.71	2.64	2.59
16	4.49	3.63	3.24	3.01	2.85	2.74	2.66	2.59	2.54
17	4.45	3.59	3.20	2.96	2.81	2.70	2.61	2.55	2.49
18	4.41	3.55	3.16	2.93	2.77	2.66	2.58	2.51	2.46
19	4.38	3.52	3.13	2.90	2.74	2.63	2.54	2.48	2.42
20	4.35	3.49	3.10	2.87	2.71	2.60	2.51	2.45	2.39
21	4.32	3.47	3.07	2.84	2.68	2.57	2.49	2.42	2.37
22	4.30	3.44	3.05	2.82	2.66	2.55	2.46	2.40	2.34
23	4.28	3.42	3.03	2.80	2.64	2.53	2.44	2.37	2.32
24	4.26	3.40	3.01	2.78	2.62	2.51	2.42	2.36	2.30
25	4.24	3.39	2.99	2.76	2.60	2.49	2.40	2.34	2.28
26	4.23	3.37	2.98	2.74	2.59	2.47	2.39	2.32	2.27
27	4.21	3.35	2.96	2.73	2.57	2.46	2.37	2.31	2.25
28	4.20	3.34	2.95	2.71	2.56	2.45	2.36	2.29	2.24
29	4.18	3.33	2.93	2.70	2.55	2.43	2.35	2.28	2.22
30	4.17	3.32	2.92	2.69	2.53	2.42	2.33	2.27	2.21
40	4.08	3.23	2.84	2.61	2.45	2.34	2.25	2.18	2.12
60	4.00	3.15	2.76	2.53	2.37	2.25	2.17	2.10	2.04
120	3.92	3.07	2.68	2.45	2.29	2.17	2.09	2.02	1.96
∞	3.84	3.00	2.60	2.37	2.21	2.10	2.01	1.94	1.88

Source: From "Tables of Percentage Points of the Inverted Beta (F)-Distribution," *Biometrika* 33 (1943) 73–88, by Maxine Merrington and Catherine M. Thompson. Reproduced by permission of the *Biometrika* Trustees.

Table 7
(*Continued*)

\	\	\	v_1 (d.f.)	\	\	\	\	\	\	\
10	12	15	20	24	30	40	60	120	∞	v_2 (d.f.)
241.9	243.9	245.9	248.0	249.1	250.1	251.1	252.2	253.3	254.3	1
19.40	19.41	19.43	19.45	19.45	19.46	19.47	19.48	19.49	19.50	2
8.79	8.74	8.70	8.66	8.64	8.62	8.59	8.57	8.55	8.53	3
5.96	5.91	5.86	5.80	5.77	5.75	5.72	5.69	5.66	5.63	4
4.74	4.68	4.62	4.56	4.53	4.50	4.46	4.43	4.40	4.36	5
4.06	4.00	3.94	3.87	3.84	3.81	3.77	3.74	3.70	3.67	6
3.64	3.57	3.51	3.44	3.41	3.38	3.34	3.30	3.27	3.23	7
3.35	3.28	3.22	3.15	3.12	3.08	3.04	3.01	2.97	2.93	8
3.14	3.07	3.01	2.94	2.90	2.86	2.83	2.79	2.75	2.71	9
2.98	2.91	2.85	2.77	2.74	2.70	2.66	2.62	2.58	2.54	10
2.85	2.79	2.72	2.65	2.61	2.57	2.53	2.49	2.45	2.40	11
2.75	2.69	2.62	2.54	2.51	2.47	2.43	2.38	2.34	2.30	12
2.67	2.60	2.53	2.46	2.42	2.38	2.34	2.30	2.25	2.21	13
2.60	2.53	2.46	2.39	2.35	2.31	2.27	2.22	2.18	2.13	14
2.54	2.48	2.40	2.33	2.29	2.25	2.20	2.16	2.11	2.07	15
2.49	2.42	2.35	2.28	2.24	2.19	2.15	2.11	2.06	2.01	16
2.45	2.38	2.31	2.23	2.19	2.15	2.10	2.06	2.01	1.96	17
2.41	2.34	2.27	2.19	2.15	2.11	2.06	2.02	1.97	1.92	18
2.38	2.31	2.23	2.16	2.11	2.07	2.03	1.98	1.93	1.88	19
2.35	2.28	2.20	2.12	2.08	2.04	1.99	1.95	1.90	1.84	20
2.32	2.25	2.18	2.10	2.05	2.01	1.96	1.92	1.87	1.81	21
2.30	2.23	2.15	2.07	2.03	1.98	1.94	1.89	1.84	1.78	22
2.27	2.20	2.13	2.05	2.01	1.96	1.91	1.86	1.81	1.76	23
2.25	2.18	2.11	2.03	1.98	1.94	1.89	1.84	1.79	1.73	24
2.24	2.16	2.09	2.01	1.96	1.92	1.87	1.82	1.77	1.71	25
2.22	2.15	2.07	1.99	1.95	1.90	1.85	1.80	1.75	1.69	26
2.20	2.13	2.06	1.97	1.93	1.88	1.84	1.79	1.73	1.67	27
2.19	2.12	2.04	1.96	1.91	1.87	1.82	1.77	1.71	1.65	28
2.18	2.10	2.03	1.94	1.90	1.85	1.81	1.75	1.70	1.64	29
2.16	2.09	2.01	1.93	1.89	1.84	1.79	1.74	1.68	1.62	30
2.08	2.00	1.92	1.84	1.79	1.74	1.69	1.64	1.58	1.51	40
1.99	1.92	1.84	1.75	1.70	1.65	1.59	1.53	1.47	1.39	60
1.91	1.83	1.75	1.66	1.61	1.55	1.50	1.43	1.35	1.25	120
1.83	1.75	1.67	1.57	1.52	1.46	1.39	1.32	1.22	1.00	∞

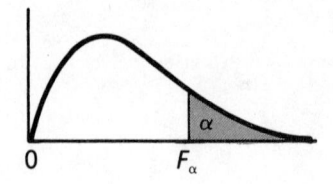

Table 8
Percentage Points of the F Distribution: $\alpha = .025$

v_2 (d.f.)	\multicolumn{9}{c}{v_1 (d.f.)}
	1	2	3	4	5	6	7	8	9
1	647.8	799.5	864.2	899.6	921.8	937.1	948.2	956.7	963.3
2	38.51	39.00	39.17	39.25	39.30	39.33	39.36	39.37	39.39
3	17.44	16.04	15.44	15.10	14.88	14.73	14.62	14.54	14.47
4	12.22	10.65	9.98	9.60	9.36	9.20	9.07	8.98	8.90
5	10.01	8.43	7.76	7.39	7.15	6.98	6.85	6.76	6.68
6	8.81	7.26	6.60	6.23	5.99	5.82	5.70	5.60	5.52
7	8.07	6.54	5.89	5.52	5.29	5.12	4.99	4.90	4.82
8	7.57	6.06	5.42	5.05	4.82	4.65	4.53	4.43	4.36
9	7.21	5.71	5.08	4.72	4.48	4.32	4.20	4.10	4.03
10	6.94	5.46	4.83	4.47	4.24	4.07	3.95	3.85	3.78
11	6.72	5.26	4.63	4.28	4.04	3.88	3.76	3.66	3.59
12	6.55	5.10	4.47	4.12	3.89	3.73	3.61	3.51	3.44
13	6.41	4.97	4.35	4.00	3.77	3.60	3.48	3.39	3.31
14	6.30	4.86	4.24	3.89	3.66	3.50	3.38	3.29	3.21
15	6.20	4.77	4.15	3.80	3.58	3.41	3.29	3.20	3.12
16	6.12	4.69	4.08	3.73	3.50	3.34	3.22	3.12	3.05
17	6.04	4.62	4.01	3.66	3.44	3.28	3.16	3.06	2.98
18	5.98	4.56	3.95	3.61	3.38	3.22	3.10	3.01	2.93
19	5.92	4.51	3.90	3.56	3.33	3.17	3.05	2.96	2.88
20	5.87	4.46	3.86	3.51	3.29	3.13	3.01	2.91	2.84
21	5.83	4.42	3.82	3.48	3.25	3.09	2.97	2.87	2.80
22	5.79	4.38	3.78	3.44	3.22	3.05	2.93	2.84	2.76
23	5.75	4.35	3.75	3.41	3.18	3.02	2.90	2.81	2.73
24	5.72	4.32	3.72	3.38	3.15	2.99	2.87	2.78	2.70
25	5.69	4.29	3.69	3.35	3.13	2.97	2.85	2.75	2.68
26	5.66	4.27	3.67	3.33	3.10	2.94	2.82	2.73	2.65
27	5.63	4.24	3.65	3.31	3.08	2.92	2.80	2.71	2.63
28	5.61	4.22	3.63	3.29	3.06	2.90	2.78	2.69	2.61
29	5.59	4.20	3.61	3.27	3.04	2.88	2.76	2.67	2.59
30	5.57	4.18	3.59	3.25	3.03	2.87	2.75	2.65	2.57
40	5.42	4.05	3.46	3.13	2.90	2.74	2.62	2.53	2.45
60	5.29	3.93	3.34	3.01	2.79	2.63	2.51	2.41	2.33
120	5.15	3.80	3.23	2.89	2.67	2.52	2.39	2.30	2.22
∞	5.02	3.69	3.12	2.79	2.57	2.41	2.29	2.19	2.11

Source: From "Tables of Percentage Points of the Inverted Beta (F)-Distribution," *Biometrika* 33 (1943) 73–88, by Maxine Merrington and Catherine M. Thompson. Reproduced by permission of the *Biometrika* Trustees.

Table 8 (*Continued*)

				v_1 (d.f.)							
10	12	15	20	24	30	40	60	120	∞	v_2 (d.f.)
968.6	976.7	984.9	993.1	997.2	1001	1006	1010	1014	1018	1
39.40	39.41	39.43	39.45	39.46	39.46	39.47	39.48	39.49	39.50	2
14.42	14.34	14.25	14.17	14.12	14.08	14.04	13.99	13.95	13.90	3
8.84	8.75	8.66	8.56	8.51	8.46	8.41	8.36	8.31	8.26	4
6.62	6.52	6.43	6.33	6.28	6.23	6.18	6.12	6.07	6.02	5
5.46	5.37	5.27	5.17	5.12	5.07	5.01	4.96	4.90	4.85	6
4.76	4.67	4.57	4.47	4.42	4.36	4.31	4.25	4.20	4.14	7
4.30	4.20	4.10	4.00	3.95	3.89	3.84	3.78	3.73	3.67	8
3.96	3.87	3.77	3.67	3.61	3.56	3.51	3.45	3.39	3.33	9
3.72	3.62	3.52	3.42	3.37	3.31	3.26	3.20	3.14	3.08	10
3.53	3.43	3.33	3.23	3.17	3.12	3.06	3.00	2.94	2.88	11
3.37	3.28	3.18	3.07	3.02	2.96	2.91	2.85	2.79	2.72	12
3.25	3.15	3.05	2.95	2.89	2.84	2.78	2.72	2.66	2.60	13
3.15	3.05	2.95	2.84	2.79	2.73	2.67	2.61	2.55	2.49	14
3.06	2.96	2.86	2.76	2.70	2.64	2.59	2.52	2.46	2.40	15
2.99	2.89	2.79	2.68	2.63	2.57	2.51	2.45	2.38	2.32	16
2.92	2.82	2.72	2.62	2.56	2.50	2.44	2.38	2.32	2.25	17
2.87	2.77	2.67	2.56	2.50	2.44	2.38	2.32	2.26	2.19	18
2.82	2.72	2.62	2.51	2.45	2.39	2.33	2.27	2.20	2.13	19
2.77	2.68	2.57	2.46	2.41	2.35	2.29	2.22	2.16	2.09	20
2.73	2.64	2.53	2.42	2.37	2.31	2.25	2.18	2.11	2.04	21
2.70	2.60	2.50	2.39	2.33	2.27	2.21	2.14	2.08	2.00	22
2.67	2.57	2.47	2.36	2.30	2.24	2.18	2.11	2.04	1.97	23
2.64	2.54	2.44	2.33	2.27	2.21	2.15	2.08	2.01	1.94	24
2.61	2.51	2.41	2.30	2.24	2.18	2.12	2.05	1.98	1.91	25
2.59	2.49	2.39	2.28	2.22	2.16	2.09	2.03	1.95	1.88	26
2.57	2.47	2.36	2.25	2.19	2.13	2.07	2.00	1.93	1.85	27
2.55	2.45	2.34	2.23	2.17	2.11	2.05	1.98	1.91	1.83	28
2.53	2.43	2.32	2.21	2.15	2.09	2.03	1.96	1.89	1.81	29
2.51	2.41	2.31	2.20	2.14	2.07	2.01	1.94	1.87	1.79	30
2.39	2.29	2.18	2.07	2.01	1.94	1.88	1.80	1.72	1.64	40
2.27	2.17	2.06	1.94	1.88	1.82	1.74	1.67	1.58	1.48	60
2.16	2.05	1.94	1.82	1.76	1.69	1.61	1.53	1.43	1.31	120
2.05	1.94	1.83	1.71	1.64	1.57	1.48	1.39	1.27	1.00	∞

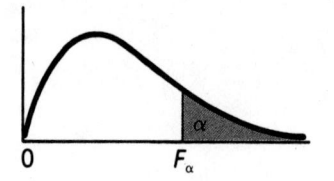

Table 9 Percentage Points of the F Distribution: $\alpha = .01$

v_2 (d.f.)	1	2	3	4	5	6	7	8	9
				v_1 (d.f.)					
1	4052	4999.5	5403	5625	5764	5859	5928	5982	6022
2	98.50	99.00	99.17	99.25	99.30	99.33	99.36	99.37	99.39
3	34.12	30.82	29.46	28.71	28.24	27.91	27.67	27.49	27.35
4	21.20	18.00	16.69	15.98	15.52	15.21	14.98	14.80	14.66
5	16.26	13.27	12.06	11.39	10.97	10.67	10.46	10.29	10.16
6	13.75	10.92	9.78	9.15	8.75	8.47	8.26	8.10	7.98
7	12.25	9.55	8.45	7.85	7.46	7.19	6.99	6.84	6.72
8	11.26	8.65	7.59	7.01	6.63	6.37	6.18	6.03	5.91
9	10.56	8.02	6.99	6.42	6.06	5.80	5.61	5.47	5.35
10	10.04	7.56	6.55	5.99	5.64	5.39	5.20	5.06	4.94
11	9.65	7.21	6.22	5.67	5.32	5.07	4.89	4.74	4.63
12	9.33	6.93	5.95	5.41	5.06	4.82	4.64	4.50	4.39
13	9.07	6.70	5.74	5.21	4.86	4.62	4.44	4.30	4.19
14	8.86	6.51	5.56	5.04	4.69	4.46	4.28	4.14	4.03
15	8.68	6.36	5.42	4.89	4.56	4.32	4.14	4.00	3.89
16	8.53	6.23	5.29	4.77	4.44	4.20	4.03	3.89	3.78
17	8.40	6.11	5.18	4.67	4.34	4.10	3.93	3.79	3.68
18	8.29	6.01	5.09	4.58	4.25	4.01	3.84	3.71	3.60
19	8.18	5.93	5.01	4.50	4.17	3.94	3.77	3.63	3.52
20	8.10	5.85	4.94	4.43	4.10	3.87	3.70	3.56	3.46
21	8.02	5.78	4.87	4.37	4.04	3.81	3.64	3.51	3.40
22	7.95	5.72	4.82	4.31	3.99	3.76	3.59	3.45	3.35
23	7.88	5.66	4.76	4.26	3.94	3.71	3.54	3.41	3.30
24	7.82	5.61	4.72	4.22	3.90	3.67	3.50	3.36	3.26
25	7.77	5.57	4.68	4.18	3.85	3.63	3.46	3.32	3.22
26	7.72	5.53	4.64	4.14	3.82	3.59	3.42	3.29	3.18
27	7.68	5.49	4.60	4.11	3.78	3.56	3.39	3.26	3.15
28	7.64	5.45	4.57	4.07	3.75	3.53	3.36	3.23	3.12
29	7.60	5.42	4.54	4.04	3.73	3.50	3.33	3.20	3.09
30	7.56	5.39	4.51	4.02	3.70	3.47	3.30	3.17	3.07
40	7.31	5.18	4.31	3.83	3.51	3.29	3.12	2.99	2.89
60	7.08	4.98	4.13	3.65	3.34	3.12	2.95	2.82	2.72
120	6.85	4.79	3.95	3.48	3.17	2.96	2.79	2.66	2.56
∞	6.63	4.61	3.78	3.32	3.02	2.80	2.64	2.51	2.41

Source: From "Tables of Percentage Points of the Inverted Beta (F)-Distribution," *Biometrika* 33 (1943) 73–88, by Maxine Merrington and Catherine M. Thompson. Reproduced by permission of the *Biometrika* Trustees.

Table 9 (*Continued*)

ν_1 (d.f.)										
10	*12*	*15*	*20*	*24*	*30*	*40*	*60*	*120*	∞	ν_2 (d.f.)
6056	6106	6157	6209	6235	6261	6287	6313	6339	6366	1
99.40	99.42	99.43	99.45	99.46	99.47	99.47	99.48	99.49	99.50	2
27.23	27.05	26.87	26.69	26.60	26.50	26.41	26.32	26.22	26.13	3
14.55	14.37	14.20	14.02	13.93	13.84	13.75	13.65	13.56	13.46	4
10.05	9.89	9.72	9.55	9.47	9.38	9.29	9.20	9.11	9.02	5
7.87	7.72	7.56	7.40	7.31	7.23	7.14	7.06	6.97	6.88	6
6.62	6.47	6.31	6.16	6.07	5.99	5.91	5.82	5.74	5.65	7
5.81	5.67	5.52	5.36	5.28	5.20	5.12	5.03	4.95	4.86	8
5.26	5.11	4.96	4.81	4.73	4.65	4.57	4.48	4.40	4.31	9
4.85	4.71	4.56	4.41	4.33	4.25	4.17	4.08	4.00	3.91	10
4.54	4.40	4.25	4.10	4.02	3.94	3.86	3.78	3.69	3.60	11
4.30	4.16	4.01	3.86	3.78	3.70	3.62	3.54	3.45	3.36	12
4.10	3.96	3.82	3.66	3.59	3.51	3.43	3.34	3.25	3.17	13
3.94	3.80	3.66	3.51	3.43	3.35	3.27	3.18	3.09	3.00	14
3.80	3.67	3.52	3.37	3.29	3.21	3.13	3.05	2.96	2.87	15
3.69	3.55	3.41	3.26	3.18	3.10	3.02	2.93	2.84	2.75	16
3.59	3.46	3.31	3.16	3.08	3.00	2.92	2.83	2.75	2.65	17
3.51	3.37	3.23	3.08	3.00	2.92	2.84	2.75	2.66	2.57	18
3.43	3.30	3.15	3.00	2.92	2.84	2.76	2.67	2.58	2.49	19
3.37	3.23	3.09	2.94	2.86	2.78	2.69	2.61	2.52	2.42	20
3.31	3.17	3.03	2.88	2.80	2.72	2.64	2.55	2.46	2.36	21
3.26	3.12	2.98	2.83	2.75	2.67	2.58	2.50	2.40	2.31	22
3.21	3.07	2.93	2.78	2.70	2.62	2.54	2.45	2.35	2.26	23
3.17	3.03	2.89	2.74	2.66	2.58	2.49	2.40	2.31	2.21	24
3.13	2.99	2.85	2.70	2.62	2.54	2.45	2.36	2.27	2.17	25
3.09	2.96	2.81	2.66	2.58	2.50	2.42	2.33	2.23	2.13	26
3.06	2.93	2.78	2.63	2.55	2.47	2.38	2.29	2.20	2.10	27
3.03	2.90	2.75	2.60	2.52	2.44	2.35	2.26	2.17	2.06	28
3.00	2.87	2.73	2.57	2.49	2.41	2.33	2.23	2.14	2.03	29
2.98	2.84	2.70	2.55	2.47	2.39	2.30	2.21	2.11	2.01	30
2.80	2.66	2.52	2.37	2.29	2.20	2.11	2.02	1.92	1.80	40
2.63	2.50	2.35	2.20	2.12	2.03	1.94	1.84	1.73	1.60	60
2.47	2.34	2.19	2.03	1.95	1.86	1.76	1.66	1.53	1.38	120
2.32	2.18	2.04	1.88	1.79	1.70	1.59	1.47	1.32	1.00	∞

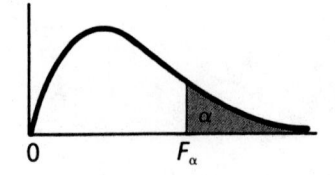

Table 10 Percentage Points of the F Distribution: $\alpha = .005$

v_2 (d.f.)	1	2	3	4	5	6	7	8	9
1	16211	20000	21615	22500	23056	23437	23715	23925	24091
2	198.5	199.0	199.2	199.2	199.3	199.3	199.4	199.4	199.4
3	55.55	49.80	47.47	46.19	45.39	44.84	44.43	44.13	43.88
4	31.33	26.28	24.26	23.15	22.46	21.97	21.62	21.35	21.14
5	22.78	18.31	16.53	15.56	14.94	14.51	14.20	13.96	13.77
6	18.63	14.54	12.92	12.03	11.46	11.07	10.79	10.57	10.39
7	16.24	12.40	10.88	10.05	9.52	9.16	8.89	8.68	8.51
8	14.69	11.04	9.60	8.81	8.30	7.95	7.69	7.50	7.34
9	13.61	10.11	8.72	7.96	7.47	7.13	6.88	6.69	6.54
10	12.83	9.43	8.08	7.34	6.87	6.54	6.30	6.12	5.97
11	12.23	8.91	7.60	6.88	6.42	6.10	5.86	5.68	5.54
12	11.75	8.51	7.23	6.52	6.07	5.76	5.52	5.35	5.20
13	11.37	8.19	6.93	6.23	5.79	5.48	5.25	5.08	4.94
14	11.06	7.92	6.68	6.00	5.56	5.26	5.03	4.86	4.72
15	10.80	7.70	6.48	5.80	5.37	5.07	4.85	4.67	4.54
16	10.58	7.51	6.30	5.64	5.21	4.91	4.69	4.52	4.38
17	10.38	7.35	6.16	5.50	5.07	4.78	4.56	4.39	4.25
18	10.22	7.21	6.03	5.37	4.96	4.66	4.44	4.28	4.14
19	10.17	7.09	5.92	5.27	4.85	4.56	4.34	4.18	4.04
20	9.94	6.99	5.82	5.17	4.76	4.47	4.26	4.09	3.96
21	9.83	6.89	5.73	5.09	4.68	4.39	4.18	4.01	3.88
22	9.73	6.81	5.65	5.02	4.61	4.32	4.11	3.94	3.81
23	9.63	6.73	5.58	4.95	4.54	4.26	4.05	3.88	3.75
24	9.55	6.66	5.52	4.89	4.49	4.20	3.99	3.83	3.69
25	9.48	6.60	5.46	4.84	4.43	4.15	3.94	3.78	3.64
26	9.41	6.54	5.41	4.79	4.38	4.10	3.89	3.73	3.60
27	9.34	6.49	5.36	4.74	4.34	4.06	3.85	3.69	3.56
28	9.28	6.44	5.32	4.70	4.30	4.02	3.81	3.65	3.52
29	9.23	6.40	5.28	4.66	4.26	3.98	3.77	3.61	3.48
30	9.18	6.35	5.24	4.62	4.23	3.95	3.74	3.58	3.45
40	8.83	6.07	4.98	4.37	3.99	3.71	3.51	3.35	3.22
60	8.49	5.79	4.73	4.14	3.76	3.49	3.29	3.13	3.01
120	8.18	5.54	4.50	3.92	3.55	3.28	3.09	2.93	2.81
∞	7.88	5.30	4.28	3.72	3.35	3.09	2.90	2.74	2.62

Source: From "Tables of Percentage Points of the Inverted Beta (F)-Distribution," *Biometrika* 33 (1943) 73–88, by Maxine Merrington and Catherine M. Thompson. Reproduced by permission of the *Biometrika* Trustees.

Table 10 (*Continued*)

10	12	15	20	24	30	40	60	120	∞	v_2 (d.f.)
\multicolumn{11}{c}{v_1 (d.f.)}
24224	24426	24630	24836	24940	25044	25148	25253	25359	25465	1
199.4	199.4	199.4	199.4	199.5	199.5	199.5	199.5	199.5	199.5	2
43.69	43.39	43.08	42.78	42.62	42.47	42.31	42.15	41.99	41.83	3
20.97	20.70	20.44	20.17	20.03	19.89	19.75	19.61	19.47	19.32	4
13.62	13.38	13.15	12.90	12.78	12.66	12.53	12.40	12.27	12.14	5
10.25	10.03	9.81	9.59	9.47	9.36	9.24	9.12	9.00	8.88	6
8.38	8.18	7.97	7.75	7.65	7.53	7.42	7.31	7.19	7.08	7
7.21	7.01	6.81	6.61	6.50	6.40	6.29	6.18	6.06	5.95	8
6.42	6.23	6.03	5.83	5.73	5.62	5.52	5.41	5.30	5.19	9
5.85	5.66	5.47	5.27	5.17	5.07	4.97	4.86	4.75	4.64	10
5.42	5.24	5.05	4.86	4.76	4.65	4.55	4.44	4.34	4.23	11
5.09	4.91	4.72	4.53	4.43	4.33	4.23	4.12	4.01	3.90	12
4.82	4.64	4.46	4.27	4.17	4.07	3.97	3.87	3.76	3.65	13
4.60	4.43	4.25	4.06	3.96	3.86	3.76	3.66	3.55	3.44	14
4.42	4.25	4.07	3.88	3.79	3.69	3.58	3.48	3.37	3.26	15
4.27	4.10	3.92	3.73	3.64	3.54	3.44	3.33	3.22	3.11	16
4.14	3.97	3.79	3.61	3.51	3.41	3.31	3.21	3.10	2.98	17
4.03	3.86	3.68	3.50	3.40	3.30	3.20	3.10	2.99	2.87	18
3.93	3.76	3.59	3.40	3.31	3.21	3.11	3.00	2.89	2.78	19
3.85	3.68	3.50	3.32	3.22	3.12	3.02	2.92	2.81	2.69	20
3.77	3.60	3.43	3.24	3.15	3.05	2.95	2.84	2.73	2.61	21
3.70	3.54	3.36	3.18	3.08	2.98	2.88	2.77	2.66	2.55	22
3.64	3.47	3.30	3.12	3.02	2.92	2.82	2.71	2.60	2.48	23
3.59	3.42	3.25	3.06	2.97	2.87	2.77	2.66	2.55	2.43	24
3.54	3.37	3.20	3.01	2.92	2.82	2.72	2.61	2.50	2.38	25
3.49	3.33	3.15	2.97	2.87	2.77	2.67	2.56	2.45	2.33	26
3.45	3.28	3.11	2.93	2.83	2.73	2.63	2.52	2.41	2.29	27
3.41	3.25	3.07	2.89	2.79	2.69	2.59	2.48	2.37	2.25	28
3.38	3.21	3.04	2.86	2.76	2.66	2.56	2.45	2.33	2.21	29
3.34	3.18	3.01	2.82	2.73	2.63	2.52	2.42	2.30	2.18	30
3.12	2.95	2.78	2.60	2.50	2.40	2.30	2.18	2.06	1.93	40
2.90	2.74	2.57	2.39	2.29	2.19	2.08	1.96	1.83	1.69	60
2.71	2.54	2.37	2.19	2.09	1.98	1.87	1.75	1.61	1.43	120
2.52	2.36	2.19	2.00	1.90	1.79	1.67	1.53	1.36	1.00	∞

Table 11
Distribution Function of U,
$P(U \leq U_0)$;
U_0 is the argument;
$n_1 \leq n_2$; $3 \leq n_2 \leq 10$.

$n_2 = 3$

U_0	\multicolumn{3}{c}{n_1}		
	1	2	3
0	.25	.10	.05
1	.50	.20	.10
2		.40	.20
3		.60	.35
4			.50

$n_2 = 4$

U_0	\multicolumn{4}{c}{n_1}			
	1	2	3	4
0	.2000	.0667	.0286	.0143
1	.4000	.1333	.0571	.0286
2	.6000	.2667	.1143	.0571
3		.4000	.2000	.1000
4		.6000	.3143	.1714
5			.4286	.2429
6			.5714	.3429
7				.4429
8				.5571

Note: Computed by M. Pagano, Department of Biostatistics, Harvard School of Public Health.

$n_2 = 5$

U_0	\multicolumn{5}{c}{n_1}				
	1	2	3	4	5
0	.1667	.0476	.0179	.0079	.0040
1	.3333	.0952	.0357	.0159	.0079
2	.5000	.1905	.0714	.0317	.0159
3		.2857	.1250	.0556	.0278
4		.4286	.1964	.0952	.0476
5		.5714	.2857	.1429	.0754
6			.3929	.2063	.1111
7			.5000	.2778	.1548
8				.3651	.2103
9				.4524	.2738
10				.5476	.3452
11					.4206
12					.5000

Table 11 (Continued)

$n_2 = 6$

U_0	n_1=1	2	3	4	5	6
0	.1429	.0357	.0119	.0048	.0022	.0011
1	.2857	.0714	.0238	.0095	.0043	.0022
2	.4286	.1429	.0476	.0190	.0087	.0043
3	.5714	.2143	.0833	.0333	.0152	.0076
4		.3214	.1310	.0571	.0260	.0130
5		.4286	.1905	.0857	.0411	.0206
6		.5714	.2738	.1286	.0628	.0325
7			.3571	.1762	.0887	.0465
8			.4524	.2381	.1234	.0660
9			.5476	.3048	.1645	.0898
10				.3810	.2143	.1201
11				.4571	.2684	.1548
12				.5429	.3312	.1970
13					.3961	.2424
14					.4654	.2944
15					.5346	.3496
16						.4091
17						.4686
18						.5314

$n_2 = 7$

U_0	n_1=1	2	3	4	5	6	7
0	.1250	.0278	.0083	.0030	.0013	.0006	.0003
1	.2500	.0556	.0167	.0061	.0025	.0012	.0006
2	.3750	.1111	.0333	.0121	.0051	.0023	.0012
3	.5000	.1667	.0583	.0212	.0088	.0041	.0020
4		.2500	.0917	.0364	.0152	.0070	.0035
5		.3333	.1333	.0545	.0240	.0111	.0055
6		.4444	.1917	.0818	.0366	.0175	.0087
7		.5556	.2583	.1152	.0530	.0256	.0131
8			.3333	.1576	.0745	.0367	.0189
9			.4167	.2061	.1010	.0507	.0265
10			.5000	.2636	.1338	.0688	.0364
11				.3242	.1717	.0903	.0487
12				.3939	.2159	.1171	.0641
13				.4636	.2652	.1474	.0825
14				.5364	.3194	.1830	.1043
15					.3775	.2226	.1297
16					.4381	.2669	.1588
17					.5000	.3141	.1914
18						.3654	.2279
19						.4178	.2675
20						.4726	.3100
21						.5274	.3552
22							.4024
23							.4508
24							.5000

(continued)

Table 11
(Continued)

$n_2 = 8$

U_0	n_1=1	2	3	4	5	6	7	8
0	.1111	.0222	.0061	.0020	.0008	.0003	.0002	.0001
1	.2222	.0444	.0121	.0040	.0016	.0007	.0003	.0002
2	.3333	.0889	.0242	.0081	.0031	.0013	.0006	.0003
3	.4444	.1333	.0424	.0141	.0054	.0023	.0011	.0005
4	.5556	.2000	.0667	.0242	.0093	.0040	.0019	.0009
5		.2667	.0970	.0364	.0148	.0063	.0030	.0015
6		.3556	.1394	.0545	.0225	.0100	.0047	.0023
7		.4444	.1879	.0768	.0326	.0147	.0070	.0035
8		.5556	.2485	.1071	.0466	.0213	.0103	.0052
9			.3152	.1414	.0637	.0296	.0145	.0074
10			.3879	.1838	.0855	.0406	.0200	.0103
11			.4606	.2303	.1111	.0539	.0270	.0141
12			.5394	.2848	.1422	.0709	.0361	.0190
13				.3414	.1772	.0906	.0469	.0249
14				.4040	.2176	.1142	.0603	.0325
15				.4667	.2618	.1412	.0760	.0415
16				.5333	.3108	.1725	.0946	.0524
17					.3621	.2068	.1159	.0652
18					.4165	.2454	.1405	.0803
19					.4716	.2864	.1678	.0974
20					.5284	.3310	.1984	.1172
21						.3773	.2317	.1393
22						.4259	.2679	.1641
23						.4749	.3063	.1911
24						.5251	.3472	.2209
25							.3894	.2527
26							.4333	.2869
27							.4775	.3227
28							.5225	.3605
29								.3992
30								.4392
31								.4796
32								.5204

Table 11
(*Continued*)

$n_2 = 9$

U_0	n_1=1	2	3	4	5	6	7	8	9
0	.1000	.0182	.0045	.0014	.0005	.0002	.0001	.0000	.0000
1	.2000	.0364	.0091	.0028	.0010	.0004	.0002	.0001	.0000
2	.3000	.0727	.0182	.0056	.0020	.0008	.0003	.0002	.0001
3	.4000	.1091	.0318	.0098	.0035	.0014	.0006	.0003	.0001
4	.5000	.1636	.0500	.0168	.0060	.0024	.0010	.0005	.0002
5		.2182	.0727	.0252	.0095	.0038	.0017	.0008	.0004
6		.2909	.1045	.0378	.0145	.0060	.0026	.0012	.0006
7		.3636	.1409	.0531	.0210	.0088	.0039	.0019	.0009
8		.4545	.1864	.0741	.0300	.0128	.0058	.0028	.0014
9		.5455	.2409	.0993	.0415	.0180	.0082	.0039	.0020
10			.3000	.1301	.0559	.0248	.0115	.0056	.0028
11			.3636	.1650	.0734	.0332	.0156	.0076	.0039
12			.4318	.2070	.0949	.0440	.0209	.0103	.0053
13			.5000	.2517	.1199	.0567	.0274	.0137	.0071
14				.3021	.1489	.0723	.0356	.0180	.0094
15				.3552	.1818	.0905	.0454	.0232	.0122
16				.4126	.2188	.1119	.0571	.0296	.0157
17				.4699	.2592	.1361	.0708	.0372	.0200
18				.5301	.3032	.1638	.0869	.0464	.0252
19					.3497	.1942	.1052	.0570	.0313
20					.3986	.2280	.1261	.0694	.0385
21					.4491	.2643	.1496	.0836	.0470
22					.5000	.3035	.1755	.0998	.0567
23						.3445	.2039	.1179	.0680
24						.3878	.2349	.1383	.0807
25						.4320	.2680	.1606	.0951
26						.4773	.3032	.1852	.1112
27						.5227	.3403	.2117	.1290
28							.3788	.2404	.1487
29							.4185	.2707	.1701
30							.4591	.3029	.1933
31							.5000	.3365	.2181
32								.3715	.2447
33								.4074	.2729
34								.4442	.3024
35								.4813	.3332
36								.5187	.3652
37									.3981
38									.4317
39									.4657
40									.5000

(*continued*)

Table 11
(*Continued*)

$n_2 = 10$

U_0	n_1=1	2	3	4	5	6	7	8	9	10
0	.0909	.0152	.0035	.0010	.0003	.0001	.0001	.0000	.0000	.0000
1	.1818	.0303	.0070	.0020	.0007	.0002	.0001	.0000	.0000	.0000
2	.2727	.0606	.0140	.0040	.0013	.0005	.0002	.0001	.0000	.0000
3	.3636	.0909	.0245	.0070	.0023	.0009	.0004	.0002	.0001	.0000
4	.4545	.1364	.0385	.0120	.0040	.0015	.0006	.0003	.0001	.0001
5	.5455	.1818	.0559	.0180	.0063	.0024	.0010	.0004	.0002	.0001
6		.2424	.0804	.0270	.0097	.0037	.0015	.0007	.0003	.0002
7		.3030	.1084	.0380	.0140	.0055	.0023	.0010	.0005	.0002
8		.3788	.1434	.0529	.0200	.0080	.0034	.0015	.0007	.0004
9		.4545	.1853	.0709	.0276	.0112	.0048	.0022	.0011	.0005
10		.5455	.2343	.0939	.0376	.0156	.0068	.0031	.0015	.0008
11			.2867	.1199	.0496	.0210	.0093	.0043	.0021	.0010
12			.3462	.1518	.0646	.0280	.0125	.0058	.0028	.0014
13			.4056	.1868	.0823	.0363	.0165	.0078	.0038	.0019
14			.4685	.2268	.1032	.0467	.0215	.0103	.0051	.0026
15			.5315	.2697	.1272	.0589	.0277	.0133	.0066	.0034
16				.3177	.1548	.0736	.0351	.0171	.0086	.0045
17				.3666	.1855	.0903	.0439	.0217	.0110	.0057
18				.4196	.2198	.1099	.0544	.0273	.0140	.0073
19				.4725	.2567	.1317	.0665	.0338	.0175	.0093
20				.5275	.2970	.1566	.0806	.0416	.0217	.0116
21					.3393	.1838	.0966	.0506	.0267	.0144
22					.3839	.2139	.1148	.0610	.0326	.0177
23					.4296	.2461	.1349	.0729	.0394	.0216
24					.4765	.2811	.1574	.0864	.0474	.0262
25					.5235	.3177	.1819	.1015	.0564	.0315
26						.3564	.2087	.1185	.0667	.0376
27						.3962	.2374	.1371	.0782	.0446
28						.4374	.2681	.1577	.0912	.0526
29						.4789	.3004	.1800	.1055	.0615
30						.5211	.3345	.2041	.1214	.0716
31							.3698	.2299	.1388	.0827
32							.4063	.2574	.1577	.0952
33							.4434	.2863	.1781	.1088
34							.4811	.3167	.2001	.1237
35							.5189	.3482	.2235	.1399
36								.3809	.2483	.1575
37								.4143	.2745	.1763
38								.4484	.3019	.1965
39								.4827	.3304	.2179
40								.5173	.3598	.2406
41									.3901	.2644
42									.4211	.2894
43									.4524	.3153
44									.4841	.3421
45									.5159	.3697
46										.3980
47										.4267
48										.4559
49										.4853
50										.5147

Table 12
Critical Values of T in the Wilcoxon Signed-rank Test; $n = 5(1)50$

One-sided	Two-sided	$n = 5$	$n = 6$	$n = 7$	$n = 8$	$n = 9$	$n = 10$
$P = .05$	$P = .10$	1	2	4	6	8	11
$P = .025$	$P = .05$		1	2	4	6	8
$P = .01$	$P = .02$			0	2	3	5
$P = .005$	$P = .01$				0	2	3

One-sided	Two-sided	$n = 11$	$n = 12$	$n = 13$	$n = 14$	$n = 15$	$n = 16$
$P = .05$	$P = .10$	14	17	21	26	30	36
$P = .025$	$P = .05$	11	14	17	21	25	30
$P = .01$	$P = .02$	7	10	13	16	20	24
$P = .005$	$P = .01$	5	7	10	13	16	19

One-sided	Two-sided	$n = 17$	$n = 18$	$n = 19$	$n = 20$	$n = 21$	$n = 22$
$P = .05$	$P = .10$	41	47	54	60	68	75
$P = .025$	$P = .05$	35	40	46	52	59	66
$P = .01$	$P = .02$	28	33	38	43	49	56
$P = .005$	$P = .01$	23	28	32	37	43	49

One-sided	Two-sided	$n = 23$	$n = 24$	$n = 25$	$n = 26$	$n = 27$	$n = 28$
$P = .05$	$P = .10$	83	92	101	110	120	130
$P = .025$	$P = .05$	73	81	90	98	107	117
$P = .01$	$P = .02$	62	69	77	85	93	102
$P = .005$	$P = .01$	55	68	68	76	84	92

One-sided	Two-sided	$n = 29$	$n = 30$	$n = 31$	$n = 32$	$n = 33$	$n = 34$
$P = .05$	$P = .10$	141	152	163	175	188	201
$P = .025$	$P = .05$	127	137	148	159	171	183
$P = .01$	$P = .02$	111	120	130	141	151	162
$P = .005$	$P = .01$	100	109	118	128	138	149

One-sided	Two-sided	$n = 35$	$n = 36$	$n = 37$	$n = 38$	$n = 39$	$n = 40$
$P = .05$	$P = .10$	214	228	242	256	271	287
$P = .025$	$P = .05$	195	208	222	235	250	264
$P = .01$	$P = .02$	174	186	198	211	224	238
$P = .005$	$P = .01$	160	171	183	195	208	221

One-sided	Two-sided	$n = 41$	$n = 42$	$n = 43$	$n = 44$	$n = 45$	$n = 46$
$P = .05$	$P = .10$	303	319	336	353	371	389
$P = .025$	$P = .05$	279	295	311	327	344	361
$P = .01$	$P = .02$	252	267	281	297	313	329
$P = .005$	$P = .01$	234	248	262	277	292	307

One-sided	Two-sided	$n = 47$	$n = 48$	$n = 49$	$n = 50$
$P = .05$	$P = .10$	408	427	446	466
$P = .025$	$P = .05$	379	397	415	434
$P = .01$	$P = .02$	345	362	380	398
$P = .005$	$P = .01$	323	339	356	373

Source: From "Some Rapid Approximate Statistical Procedures" (1964) 28, by F. Wilcoxon and R. A. Wilcox. Reproduced with the kind permission of Lederle Laboratories, a division of American Cyanamid Company.

Table 13
Critical Values of Spearman's Rank Correlation Coefficient for a One-tailed Test

n	$\alpha = .05$	$\alpha = .025$	$\alpha = .01$	$\alpha = .005$
5	0.900	—	—	—
6	0.829	0.886	0.943	—
7	0.714	0.786	0.893	—
8	0.643	0.738	0.833	0.881
9	0.600	0.683	0.783	0.833
10	0.564	0.648	0.745	0.794
11	0.523	0.623	0.736	0.818
12	0.497	0.591	0.703	0.780
13	0.475	0.566	0.673	0.745
14	0.457	0.545	0.646	0.716
15	0.441	0.525	0.623	0.689
16	0.425	0.507	0.601	0.666
17	0.412	0.490	0.582	0.645
18	0.399	0.476	0.564	0.625
19	0.388	0.462	0.549	0.608
20	0.377	0.450	0.534	0.591
21	0.368	0.438	0.521	0.576
22	0.359	0.428	0.508	0.562
23	0.351	0.418	0.496	0.549
24	0.343	0.409	0.485	0.537
25	0.336	0.400	0.475	0.526
26	0.329	0.392	0.465	0.515
27	0.323	0.385	0.456	0.505
28	0.317	0.377	0.448	0.496
29	0.311	0.370	0.440	0.487
30	0.305	0.364	0.432	0.478

Source: From "Distribution of Sums of Squares of Rank Differences for Small Samples," by E. G. Olds, *Annals of Mathematical Statistics* 9 (1938). Reproduced with the kind permission of the Editor, *Annals of Mathematical Statistics*.